DISTANCE SAMPLING

* Current address:
National Marine Mammal Laboratory, Alaska Fisheries Science Center,
NMFS, 7600 Sand Point Way N.E., Seattle, WA 98115, USA

DISTANCE SAMPLING

Estimating abundance of biological populations

S. T. Buckland

Environmental Modelling Unit
Scottish Agricultural Statistics Service
Macaulay Land Use Research Institute, Craigiebuckler, Aberdeen
Scotland AB9 2QJ

D. R. Anderson

K. P. Burnham

Colorado Cooperative Fish and Wildlife Research Unit
U.S. Fish and Wildlife Service, Fort Collins,
Colorado 80523, USA

and

J. L. Laake*

Colorado Cooperative Fish and Wildlife Research Unit
Colorado State University, Fort Collins,
Colorado 80523, USA

CHAPMAN & HALL

London · Glasgow · New York · Tokyo · Melbourne · Madras

Published by Chapman & Hall, 2–6 Boundary Row, London SE1 8HN

Chapman & Hall, 2–6 Boundary Row, London SE1 8HN, UK

Blackie Academic & Professional, Wester Cleddens Road,
Bishopbriggs, Glasgow G64 2NZ, UK

Chapman & Hall Inc., 29 West 35th Street, New York NY10001, USA

Chapman & Hall Japan, Thomson Publishing Japan, Hirakawacho
Nemoto Building, 6F, 1–7–11 Hirakawa-cho, Chiyoda-ku, Tokyo 102,
Japan

Chapman & Hall Australia, Thomas Nelson Australia, 102 Dodds
Street, South Melbourne, Victoria 3205, Australia

Chapman & Hall India, R. Seshadri, 32 Second Main Road, CIT East,
Madras 600 035, India

First edition 1993

© 1993 S. T. Buckland, D. R. Anderson, K. P. Burham and
J. L. Laake

Typeset in 10/12pt Times New Roman by Pure Tech Corporation,
Pondicherry, India
Printed in Great Britain by St Edmundsbury Press, Bury St Edmunds,
Suffolk

ISBN 0 412 42660 9(HB)   0 412 42670 6(PB)

A catalogue record for this book is available from the British Library

Library of Congress Cataloging-in-Publication data
Distance sampling : estimating abundance of biological populations /
S. T. Buckland ... [et al.]. – 1st ed.
      p.   cm.
   Includes bibliographical references and index.
   ISBN 0–412–42660–9. – ISBN 0–412–42670–6 (pbk.)
   1. Animal populations–Statistical methods. 2. Sampling
(Statistics) I. Buckland, S. T. (Stephen T.)
QL752.D57 1993   92–39560
591.5′248–dc20   CIP

Contents

CONTENTS

Foreword

Our environment and natural food resources are continually coming under threat so that the monitoring of population trends is essential today. Whaling is a good example. Here politics and conservation often clash, and over the years more and more restrictions have been applied through the efforts of the International Whaling Commission in an endeavour to save some of our whale species from extinction. Localized fisheries also need to be monitored and quotas set each year. In some countries, sports fishing and hunting are popular so that information is needed about the populations being exploited in order to determine such things as the duration of hunting season and bag limits. Methods of estimating animal abundance have been developing steadily since the 1940s but over the last 20 years activity in this area has intensified and the subject has begun to blossom. At the centre of this growth were two of the authors of this book, David Anderson and Kenneth Burnham, who have widely published in this field. The need for computers in this area was soon recognized and David and Ken were joined by Jeffrey Laake who, with his computing expertise, helped to develop suitable software packages for implementing some of the new techniques. In the 1980s Stephen Buckland entered the arena and began to make his presence felt. Among other contributions, he firmly established the role of Monte Carlo and bootstrapping techniques in population estimation where the unique role of the computer could be fully exploited. He also turned his attention to the difficult problem of monitoring marine mammals such as dolphins and whales. Many of the early methods of estimating animal abundance involved the tagging of animals. However, it has since been found that for such methods to be effective, large numbers of animals have to be tagged and high proportions of the population need to be caught on each sampling occasion. One area where such methods have been particularly successful is bird banding. However, for animals like the whale, these so-called capture–recapture methods are woefully inadequate and there has been a need for the development of alternative methods. 'Distance' methods, the subject of this book, based on animal distances from points or lines, provide such alternatives. In essence, one proceeds down a randomly chosen path called a line transect and measures or estimates the perpendicular distances from the line to the animals actually detected. Alternatively, one can choose a point instead and measure the radial distances of the animals detected. It is very appropriate that the leading exponents in this field have come together to produce an authoritative description on 'how to do it'. They bring with them many years of experience in this research area. This book is a must for all those involved in estimating animal abundance as the methods can be used for such a wide variety of animal species including birds and marine mammals. The methods also apply to clusters of animals such as schools of dolphins and to animal signs. The beauty of such methods lies in the fact that not every animal has to be seen when a population is investigated. At the heart of the methodology is a 'detectability' function which is estimated in some robust fashion from the distances to the animals actually seen. Many species are not always visible and may be detected by the sounds they make or by being flushed out into the open. Clearly animals can have widely different behaviour patterns so that different models will be needed for different situations. This book provides a tool box of such methods with a computer package which helps the researcher to select the right tool for each occasion. The authors have a reputation for being very thorough and, typically, they endeavour to cover every conceivable situation that might be encountered in the field. They bring to the book a practical as well as a head knowledge of the subject matter so that their book is well laced with real examples. One strength of their work is their chapter on experimental design, which looks at each aspect of setting up a 'distance' experiment. Sadly, aspects of design are often omitted from books on statistical ecology, usually because of the inherent difficulty of designing experiments. Such a chapter is refreshing. There are eight chapters in all, covering the basic concepts, background, and statistical theory, together with separate chapters on line and point transects, study design and field methods. A whole chapter is devoted to illustrative examples, which is most welcome, and there is a chapter looking at extensions and related work. This latter chapter, of perhaps less relevance to the practitioner, is important in that it highlights the fact that the subject is still developing. We welcome these additional insights from those who have spent so much time working in this topic. In conclusion I would like to congratulate the authors for all their hard work in bringing to the scientific community such a detailed and helpful book.

G. A. F. Seber
April 1992

Preface

This book is about the use of distance sampling to estimate the density or abundance of biological populations. Line and point transect sampling are the primary distance methods. Here, lines or points are surveyed in the field and the observer records a distance to those objects of interest that are detected. The sample data are the set of **distances** of detected objects and any relevant covariates; however, many objects may remain undetected during the course of the survey. Distance sampling provides a way to obtain reliable estimates of density of objects under fairly mild assumptions. Distance sampling is an extension of plot sampling methods where it is assumed that **all** objects within sample plots are counted.

The objects of interest are typically various vertebrate species, including those that exist in coveys or schools, or inanimate objects such as bird nests, mammal burrows or dead animals. The range of application is quite broad, includes a variety of surveys of terrestrial and aquatic species, and several innovative approaches are reviewed. Distance sampling often provides a practical, cost-effective class of methods for estimating population density. For objects distributed sparsely across large geographic areas, there are often no competing methods.

Line and point transect sampling is well named because the important focus must be on accurate distance measurements of **all** objects near the line or point. It is the area near the line or point that is critical in nearly all aspects. Within this framework, many extensions and special cases are developed and illustrated.

The objective of this book is to provide a comprehensive treatment of distance sampling theory and application. Much work has been done on this subject since 1976. Development of fundamental new theory has diminished recently and it is timely to provide a state-of-the-art treatment of the information. Currently, there is no other book or monograph that provides a comprehensive synthesis of this material. A comprehensive computer software package, called DISTANCE, is also introduced.

This book covers the theory and application of distance sampling with emphasis on line and point transects. Specialized applications are noted briefly, such as trapping webs and cue counts. General considerations are given to the design of distance sampling surveys. Many examples are provided to illustrate the application of the theory. The book is written for both statisticians and biologists and this objective imposed a few obvious compromises.

The book contains eight chapters. Chapters 1 and 2 are introductory. Chapter 3 presents the general theory for both line and point transect sampling, including modelling, estimation, testing and inference. Chapters 4 and 5 provide insight into the application of the theory for line and point transects, respectively. These chapters are meant to stand alone, thus there is some duplication of the material. Extensions to the theory are given in Chapter 6, along with some new research directions. Chapter 7 provides material on the design of studies employing distance sampling. The emphasis here is on ways to assure that the key assumptions are met. Chapter 8 provides several comprehensive examples. Over 300 references to the published literature are listed.

The main concepts in this book are not complex; however, some of the statistical theory may be difficult for non-statisticians. We hope biologists will not be deterred by the quantitative theory chapter and hope that statisticians will understand that we are presenting methods intended to be useful and usable given all the practicalities a biologist faces in field sampling. We assume that the reader has some familiarity with basic statistical methods, including point and variance estimation. Knowledge of sampling theory would be useful, as would some acquaintance with numerical methods. Some experience with likelihood inference would be useful. The following guidelines are provided for a first reading of the book.

Everyone should read Chapters 1 and 2. While statisticians will want to study Chapters 3 and 6, Chapters 4 (line transects) and 5 (point transects) will be of more interest to biologists. Biologists should study Chapter 7 (design) in detail. Everyone might benefit from the illustrative examples and case studies in Chapter 8, where readers will find guidance on advanced applications involving several data sets.

Our interest in these subjects dates back to 1966 (DRA), 1974 (KPB), 1977 (JLL) and 1980 (STB). We have all contributed to the theory, been involved with field sampling, and had substantial interaction with the analysis of real sample data. Jointly, we have published around 50 papers in the literature on distance sampling. Computer software packages TRANSECT (now superseded) and DISTANCE have been the domain of JLL.

The contribution of Steve Buckland to this book was partially supported with funds from the Scottish Office Agriculture and Fisheries Department, through the Scottish Agricultural Statistics Service. David Anderson and Ken Burnham are grateful to the U.S. Fish and Wildlife Service for support and freedom in their research. Jeff Laake and the development of DISTANCE were funded by the Colorado Division of Wildlife and the U.S. National Marine Fisheries Service. Thomas Drummer, Eric Rexstad and Tore Schweder provided reviews of an early draft of this material and their help and support are gratefully acknowledged. David Bowden, Robert Parmenter and George Seber also provided review comments. Several biologists generously allowed us to use their research data as examples and in this regard we appreciate the contributions of Roger Bergstedt, Colin Bibby, Eric Bollinger, Graeme Coulson, Fritz Knopf and Robert Parmenter. We also gratefully acknowledge the following organizations for funding research to address the practical problems of distance sampling and for allowing us to use their data: the Inter-American Tropical Tuna Commission; the International Whaling Commission; the Marine Research Institute of Iceland; and the U.S. National Marine Fisheries Service. We have all benefited from the use of Les Robinette's data sets. David Carlile provided the photo of the DELTA II submersible, Fred Lindzey provided photos of aircraft used to survey pronghorn in Wyoming, and John Reinhardt allowed the use of a photo of the survey aircraft shown in Fig. 7.9. Tom Drummer and Charles Gates helped us with their software, SIZETRAN and LINETRAN, respectively. David Gilbert provided help with the Monte Vista duck nest data. Karen Cattanach carried out some of the analyses in the marine mammal examples and generated the corresponding figures. Finally, Barb Knopf's assistance in manuscript preparation and Eric Rexstad's help with many of the figures is appreciated.

We plan to continue our work and interest in distance sampling issues. We welcome comments and suggestions from those readers who share our interests.

S. T. Buckland
D. R. Anderson
K. P. Burnham
J. L. Laake
April 1992

1

Introductory concepts

1.1 Introduction

Ecology is the study of the distribution and abundance of plants and animals and their interactions with their environment. Many studies of biological populations require estimates of population density (D) or size (N), or rate of population change $\lambda_t = D_{t+1}/D_t = N_{t+1}/N_t$. These parameters vary in time and over space as well as by species, sex and age. Further, population dynamics and hence these parameters often depend on environmental factors.

This book is a synthesis of the state-of-the-art theory and application of distance sampling and analysis. The fundamental parameter of interest is density (D = number per unit area). Density and population size are related as $N = D \cdot A$ where A is area. Thus, attention can be focused on D.

Consider a population of N objects distributed according to some spatial stochastic process, not necessarily Poisson, in a field of size A. A traditional approach has been to establish a number of plots or quadrats at random (e.g. circular, square or long rectangular) and **census** the population within these plots. Conceptually, if n objects are counted within plots of total area a, then an estimator of density, termed $\hat{D}$, is

$$\hat{D} = n/a$$

Under certain reasonable assumptions, $\hat{D}$ is an estimator of the parameter $D = N/A$. This is the finite population sampling approach (Cochran 1977) and was fully developed for most situations many years ago. This approach asks the following question:

Given a fixed area (i.e. the total area of the sample plots), how many objects are in it (Fig. 1.1)?

Distance sampling theory extends the finite population sampling approach. Again, consider a population of N objects distributed according

1

to some stochastic process in a field of size A. In distance sampling theory, a set of randomly placed lines (Fig. 1.2) or points (Fig. 1.3) is established and distances are measured to those objects detected by travelling the line or surveying the points. The theory allows for the fact that some, perhaps many, of the objects will go undetected. In addition, there is a marked tendency for detectability to decrease with increasing distance from the transect line or point. The distance sampling approach asks the following question:

> Given the detection of n objects, how many objects are estimated to be within the sampled area?

Two differences can be noted in comparing distance sampling theory with classical finite population sampling theory: (1) the size of the sample area is sometimes unknown, and (2) many objects may not be detected for whatever reason. One of the major advantages of distance sampling is that objects can remain undetected (i.e. it can be used when a census is not possible). As a particular object is detected, its distance to the randomly chosen line or point is measured. Thus, distances are sampled. Upon completion of a simple survey, n objects have been

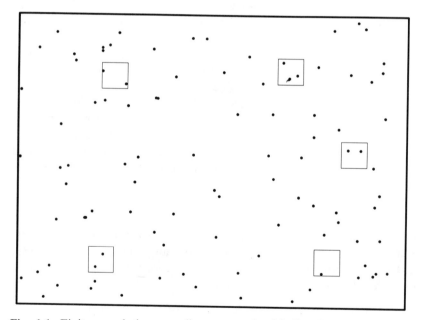

Fig. 1.1. Finite population sampling approach with five 1 m square quadrats placed at random in a population containing 100 objects of interest. Σa_i = 5, $\Sigma n_i = 10$, and $\hat{D} = 2$ objects/m^2. In this illustration, the population is confined within a well-defined area.

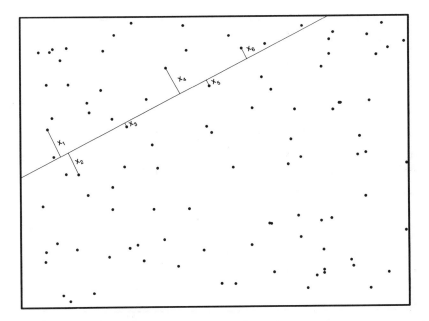

Fig. 1.2. Line transect sampling approach with a single, randomly placed, line of length L. Six objects ($n = 6$) were detected at distances $x_1, x_2, \ldots, x_6$. Those objects detected are denoted by a line showing the perpendicular distance measured. In practical applications, several lines would be used to sample the population.

detected and their associated distances $y_1, y_2, \ldots, y_n$ recorded. The variable y will be used as a general symbol for a distance measurement, while x will denote a perpendicular distance and r will denote a radial distance. Unbiased estimates of density can be made from these distance data **if certain assumptions are met**.

Distance sampling theory includes two main approaches to the estimation of density: line transects and point transects. Traditional sampling theory may be considered a special case of distance sampling theory. An application of point transect theory is the sampling method called a trapping web, which is potentially useful in animal trapping studies. Cue counting is another application of point transect theory and was developed for marine mammal surveys. Nearest neighbour and point-to-object methods are similar in character to point transects, but are generally less useful for estimating object density.

1.1.1 Strip transects

Strip transects are long, narrow plots or quadrats and are typically used in conjunction with finite population sampling theory. Viewed differently,

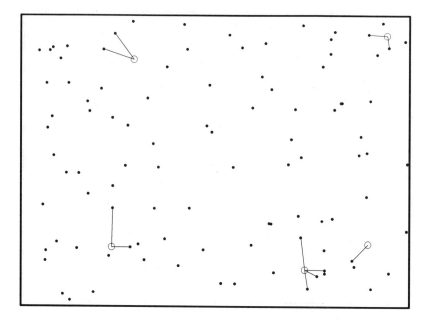

Fig. 1.3. Point transect sampling approach with five randomly placed points ($k = 5$), denoted by the open circles. Eleven objects were detected and the 11 sighting distances $r_1, r_2, \ldots, r_{11}$ are shown.

they represent a very special case of distance sampling theory. Consider a strip of length L and of width $2w$ (the width of the area censused). Then, it is **assumed** that all objects are detected out to distance w either side of the centreline, a complete census of the strip. No distances are measured; instead, the strong assumption is made that all objects in the strip are detected. Detections beyond w are ignored. Line and point transect surveys allow a relaxation of the strong assumptions required for strip (i.e. plot or quadrat) sampling (Burnham and Anderson 1984). Note the distinction here between a **census**, in which all objects in an area are counted, and a **survey**, where only some proportion of the objects in the sampled area is detected and recorded.

1.1.2 Line transects

Line transects are a generalization of strip transects. In strip transect sampling one assumes that the entire strip is censused, whereas in line transect sampling, one must only assume a narrow strip around the centreline is censused; that is, except near the centreline, there is no assumption that all objects are detected. A straight line transect is

4

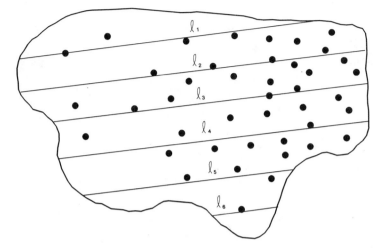

Fig. 1.4. A population of objects with a gradient in density is sampled with lines parallel to the direction of the gradient. In this case, there are $k = 6$ lines of length $l_1, l_2, \ldots, l_6$, and $\Sigma \, l_i = L$.

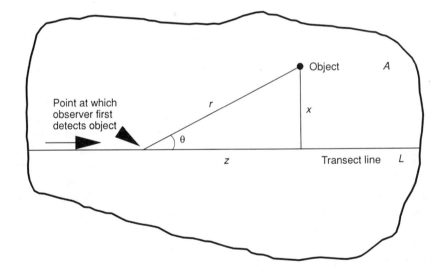

Fig. 1.5. Basic measurements that can be taken in line transect surveys. Here an area of size A is sampled by a single line of length L. If sighting distances r are to be taken in the field, one should also measure the sighting angles θ, to allow analysis of perpendicular distances x, calculated as $x = r \cdot \sin(\theta)$. The distance of the object from the observer parallel to the transect at the moment of detection is $z = r \cdot \cos(\theta)$.

5

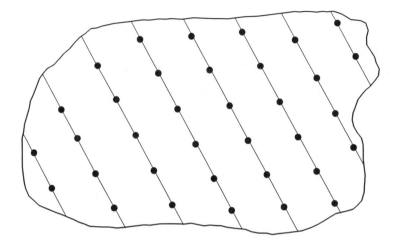

Fig. 1.6. Point transect surveys are often based on points laid out systematically along parallel lines. Alternatively, the points could be placed completely at random or in a stratified design. Variance estimation is dependent upon the point placement design.

traversed by an observer and perpendicular distances are measured from the line to each detected object (Fig. 1.2). The line is to be placed at random and is of known length, L. In practice, a number of lines of lengths $l_1, l_2, \ldots, l_k$ are used and their total length is denoted as L (Fig. 1.4). Objects away from the line may go undetected and, if distances are recorded accurately, reliable estimates of density can be computed.

It is often convenient to measure the sighting distance r_i and sighting angle θ_i, rather than the perpendicular distance x_i, for each of the n objects detected (Fig. 1.5). The x_i are then found by simple trigonometry: $x_i = r_i \cdot \sin(\theta_i)$. Methods exist to allow estimation of density based directly on r_i and θ_i. They are reviewed by Hayes and Buckland (1983), who show that they perform poorly relative to methods based on perpendicular distances, because they require more restrictive, and generally implausible, assumptions. In addition, observations made behind the observer (i.e. $\theta_i > 90°$) are problematic for models based on sighting distances and angles.

1.1.3 Point transects

The term point transect was coined because it may be considered as a line transect of zero length (i.e. a point). This analogy is only of limited conceptual use because there are several differences between line and point transect theory. Point transects are often termed variable circular plots in the ornithological literature, where the points are often placed at intervals along straight line transects (Fig. 1.6). We consider a series of k points positioned randomly. An observer measures the sighting (radial) distance r_i from the random point to each of the objects detected. Upon completion of the survey of the k points, one has distance measurements to the detected objects. Point transects are a generalization of traditional circular plot surveys. In circular plot sampling, an area of πw^2 is censused, whereas in point transect sampling, only the area close to the random point must be fully censused; a proportion of objects away from the random point but within the survey area remains undetected.

The area searched in strip and line transect sampling is $2wL$, whereas the area searched in circular plot and point transect sampling is $k\pi w^2$ (assuming, for the moment, that w is finite). In strip and traditional circular plot sampling, it is assumed that these areas are censused, i.e. all objects of interest are detected. In line and point transect sampling, only a relatively small percentage of the objects might be detected within the searched area (of width $2w$ for line transects or radius w for point transects), possibly as few as 10–30%. Because objects can remain undetected, distance sampling methods provide biologists with a powerful yet practical methodology for estimating density of populations.

1.1.4 Special applications

Distance sampling theory has been extended in two ways that deserve mention here: trapping webs and cue counts. These important applications are useful in restricted contexts and are direct applications of existing distance sampling theory. Two spatial modelling methods sometimes termed 'distance sampling' are more familiar to many botanists, but have limited use for estimating object density. These methods are point-to-object and nearest neighbour methods; they have some similarities to distance sampling as defined in this book, but differ in that there is no analogy to the detection function $g(y)$.

(a) *Trapping webs* Trapping webs (Anderson *et al.* 1983; Wilson and Anderson 1985b) represent a particular application of distance sampling theory and provide a new approach to density estimation for animal trapping studies. Traps are placed along lines radiating from randomly chosen points (Fig. 1.7); the traditionally used rectangular trapping grid cannot be used as a trapping web. Here 'detection' by an observer is replaced by animals being caught in traps at a known distance from the centre of a trapping web. Trapping continues for t occasions and only the data from the initial capture of each animal are analysed. Trapping webs provide an alternative to traditional capture–recapture sampling where density is of primary interest.

(b) *Cue counting* Cue counting (Hiby 1985) was developed as an alternative to line transect sampling for estimating whale abundance from sighting surveys. Observers on a ship or aircraft record all sighting cues within a sector ahead of the platform and their distance from the platform. The cue used depends on species, but might be the blow of a

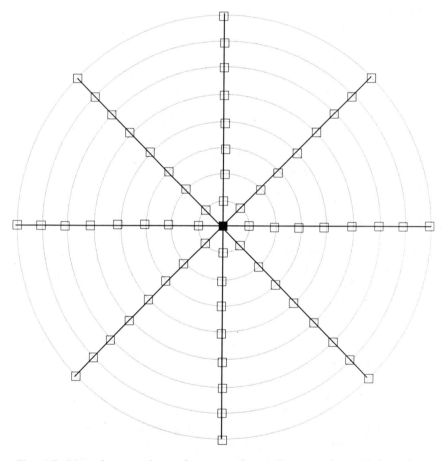

Fig. 1.7. Use of a trapping web to sample small mammal populations is an extension of point transect theory. Traps (e.g. live traps, snap traps or pitfall traps), represented as □, are placed at the intersections of the radial lines with the concentric circles.

whale at the surface. The sighting distances are converted into the estimated number of cues per unit time per unit area using point transect models. The cue rate (usually corresponding to blow rate) is estimated from separate experiments, in which individual animals or pods are monitored over a period of time.

(c) *Point-to-object methods* In point transect sampling, the distance of each detected object from the point is recorded. In point-to-object methods, the distance of the **nearest** object from the point is recorded (Clark and Evans 1954; Eberhardt 1967). The method may be extended, so that the distances of the n nearest objects to the point are recorded (Holgate 1964; Diggle 1983). Thus the number of detected objects from a point is predetermined, and the area around the point must be searched exhaustively to ensure that no objects are missed closer to the point than the farthest of the n identified objects. Generally the method is inefficient for estimating density, and estimators are prone to bias.

(d) *Nearest neighbour methods* Nearest neighbour methods are closely similar to point-to-object methods, but distances are measured from a random object, not a random point (Diggle 1983). If objects are randomly distributed, the methods are equivalent, whereas if objects are aggregated, distances under this method will be smaller on average. Diggle (1983) summarizes *ad hoc* estimators that improve robustness by combining data from both methods; if the assumption that objects are randomly distributed is violated, biases in the point-to-object and nearest neighbour density estimates tend to be in opposite directions.

1.1.5 The detection function

Central to the concept of distance sampling is the detection function $g(y)$:

$g(y)$ = the probability of detecting an object, given that it is at distance y from the random line or point
= prob {detection|distance y}.

The distance y refers to either the perpendicular distance x for line transects or the sighting (radial) distance r for point transects. Generally, the detection function decreases with increasing distance, but $0 \leq g(y) \leq 1$ always. In the development to follow we usually assume that $g(0) = 1$, i.e. objects on the line or point are seen with certainty (i.e. probability 1). Typical graphs of $g(y)$ are shown in Fig. 1.8. Often, only a small

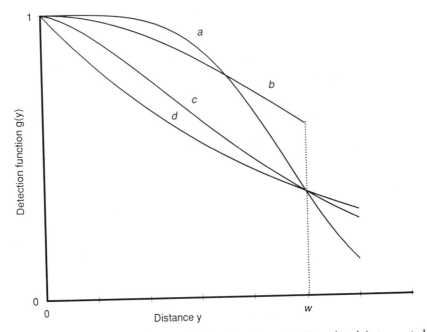

Fig. 1.8. Some examples of the detection function $g(y)$. Function b is truncated at w and thus takes the value zero for all $y > w$. Functions with shapes similar to a, b and c are common in distance sampling. Function d usually results from poor survey design and conduct, and is problematic.

percentage of the objects of interest are detected in field surveys. However, a proper analysis of the associated distances allows reliable estimates of true density to be made. The detection function $g(y)$ could be written as $g(y|v)$, where v is the collection of variables other than distance affecting detection, such as object size. We will not use this explicit notation, but it is understood.

1.1.6 Summary

Distance sampling is a class of methods that allow the estimation of density (D = number per unit area) of biological populations. The critical data collected are distances y_i from a randomly placed line or point to objects of interest. A large proportion of the objects may go undetected, but the theory allows accurate estimates of density to be made under mild assumptions. Underlying the theory is the concept of a detection function $g(y)$ = prob {detection|distance y}. Detectability usually decreases with increasing distance from the random line or point.

10

1.2 Range of applications

1.2.1 Objects of interest

Studies of birds represent a major use of both point and line transect studies. Birds are often conspicuous by their bright coloration or distinctive song or call, thus making detection possible even in dense habitats. Surveys in open habitats often use line transects, whereas surveys in more closed habitats with high canopies often use point transects. Distance sampling methods have seen use in studying populations of many species of gamebirds, raptors, passerines and shorebirds.

Many terrestrial mammals have been successfully surveyed using distance sampling methods (e.g. pronghorn, feral pigs, fruit bats, mice, and several species of deer, rabbits, hares, primates and African ungulates). Marine mammals (several species of dolphin, porpoise, seal and whale) have been the subject of many surveys reported in the literature. Reptiles, amphibians, beetles and wolf spiders have all been the subject of distance sampling surveys, and fish (in coral reefs) and red crab densities have been estimated from underwater survey data.

Many inanimate objects have been surveyed using distance sampling, including birds' nests, mammal burrows, and dead deer and pigs. Plant populations and even plant diseases are candidates for density estimation using distance sampling theory. One military application is estimation of the number of mines anchored to the seabed in mine fields.

1.2.2 Method of transect coverage

Distance sampling methods have found use in many situations. Specific applications are still being developed from the existing theory. The versatility of the method is partially due to the variety of ways in which the transect line can be traversed. Historically, line transects were traversed on foot by a trained observer. In recent years, terrestrial studies have used trail bikes, all terrain vehicles, or horses. Transect surveys have been conducted using fixed wing aircraft and helicopters; 'ultralight' aircraft are also appropriate in some instances.

Transect surveys in aquatic environments can be conducted by divers with snorkels or scuba gear, or from surface vessels ranging in size from small boats to large ships, or various aircraft, or by sleds with mounted video units pulled underwater by vessels on the surface. Small submarines may have utility in line or point transect surveys if proper visibility can be achieved. Remote sensing may find extensive use as the technology develops (e.g. acoustic instruments, radar, remotely controlled cameras, multispectral scanners).

11

In general, the observer can traverse a line transect at a variable speed, travelling more slowly to search heavy cover. The observer may leave the line and walk an irregular path, keeping within w on each side of the line. However, the investigator must ensure that all objects on the line are detected, and that the recorded effort L is the length of the line, not the total distance the observer travels. Point transects are usually surveyed for a fixed time (e.g. 10 minutes per sample point).

1.2.3 Clustered populations

Distance sampling is best explained in terms of 'objects of interest', rather than a particular species of bird or mammal. Objects of interest might be dead deer, birds' nests, jackrabbits, etc. Often, however, interest lies in populations whose members are naturally aggregated into clusters. Here we will take clusters as a generic term to indicate herds of mammals, flocks of birds, coveys of quail, pods of whales, prides of lions, schools of fish, etc. A cluster is a relatively tight aggregation of objects of interest, as opposed to a loosely clumped spatial distribution of objects. More commonly, 'group' is used, but we prefer 'cluster' to avoid confusion with the term 'grouped data', defined below.

Surveying clustered populations differs in a subtle but important way between strip transect sampling and line or point transect sampling. In strip transect sampling, all individuals inside the strip are censused; essentially one ignores the fact that the objects occur in clusters. In contrast, in distance sampling with a fixed w, one records all clusters detected if the centre of the cluster is inside the strip (i.e. 0 to w). If the centre of the cluster is inside the strip, then the count of the size of the cluster must include all individuals in the cluster, even if some individuals are beyond w. On the other hand, if the centre of the cluster is outside the strip, then no observation is recorded, even though some individuals in the cluster are inside the strip.

In distance sampling theory, the clusters must be considered to be the object of interest and distances should be measured from the line or point to the geometric centre of the cluster. Then, estimation of the density of clusters is straightforward. The sample size n is the number of clusters detected during the survey. If a count is also made of the number of individuals (s) in each observed cluster, one can estimate the average cluster size, $E(s)$. The density of individuals D can be computed as a product of the density of clusters D_s times the average cluster size:

$$D = D_s \cdot E(s)$$

12

A complication arises if detection is a function of cluster size. This relationship is evident if most of the clusters detected at a substantial distance from the line or point are relatively large in size. Typically, the estimator of D_s is still unbiased, but using the mean cluster size $\bar{s}$ to estimate $E(s)$ results in a positive bias in the estimator (because the smaller clusters tend to go undetected toward w).

A well-developed general theory exists for the analysis of distance data from clustered populations. Here the detection probability is dependent on both distance from the line or point and cluster size (this phenomenon is called size-biased sampling). Several approaches are possible: (1) stratify by cluster size and apply the usual methods within each stratum, then sum the estimated densities of individuals; (2) treat cluster size as a covariate and use parametric models for the bivariate distance–cluster size data (Drummer and McDonald 1987); (3) truncate the distance data to reduce the correlation between detection distance and cluster size and then apply robust semiparametric line transect analysis methods; (4) first estimate cluster density, then regress cluster size on $\hat{g}(y)$ to estimate mean cluster size where detection is certain ($\hat{g}(y) = 1$); (5) attempt an analysis by individual object rather than cluster, and use robust inference methods to allow for failure of the assumption of independent detections. Strategy (3) is straightforward and generally quite robust; appropriate data truncation after data collection can greatly reduce the dependence of detection probability on cluster size, and more severe truncation can be used for mean cluster size estimation than for fitting the line transect model, thus reducing the bias in $\bar{s}$ further. We have also found strategy (4) to be effective.

1.3 Types of data

Distance data can be recorded accurately or grouped. Rounding errors in measurements often cause the data to be grouped to some degree, but they must then be analysed as if they had been recorded accurately, or grouped further, in an attempt to reduce the effects of rounding on bias. Distances are often assigned to predetermined distance intervals, and must then be analysed using methods developed for the analysis of frequency data.

1.3.1 Ungrouped data

Two types of ungrouped data can be taken in line transect surveys: perpendicular distances x_i or sighting distances r_i and angles θ_i. If sighting distances and angles are taken, they should be transformed to perpendicular distances for analysis. Only sighting distances r_i are used in the estimation of density in point transects. Trapping webs use the same type of measurement r_i, which is then the distance from the centre of the web to the trap containing animal i. The cue counting method also requires sighting distances r_i, although only those within a sector ahead of the observer are recorded. Angles (0 to 360° from some arbitrary baseline) are potentially useful in testing assumptions in point transects and trapping webs, but have not usually been taken. In cue counting also, angles (sighting angles θ_i) are not usually recorded, except to ensure that they fall between $\pm \phi$, where 2ϕ is the sector angle. In all cases we will assume that n distances $\{y_1, y_2, \ldots y_n\}$ are measured corresponding to the n detected objects. Of course, n itself is usually a random variable, although one could design a survey in which searching continues until a pre-specified number of objects n is detected; L is then random and the theory is modified slightly (Rao 1984).

Sample size n should generally be at least 60–80, although for some purposes, as few as 40 might be adequate. Formulae are available to determine the sample size that one expects to achieve with a given level of precision (measured, for example, by the coefficient of variation). A pilot survey is valuable in predicting sample sizes required, and will usually show that a sample as small as 40 for an entire study is unlikely to achieve the desired precision.

1.3.2 Grouped data

Data grouping arises in distance sampling in two ways. First, ungrouped data y_i, $i = 1, \ldots, n$, may be taken in the field, but analysed after deliberate grouping into frequency counts n_i, $i = 1, \ldots, u$, where u is the number of groups. Such grouping into distance intervals is often done to achieve robustness in the analysis of data showing systematic errors such as heaping (i.e. rounding errors). Grouping the r_i and θ_i data by intervals in the field or for analysis in line transect surveys is not recommended because it complicates calculation of perpendicular distances, although techniques (e.g. 'smearing') exist to handle such grouped data.

Second, the data might be taken in the field only by distance categories or intervals. For example, in aerial surveys it may only be practical to count the number of objects detected in the following distance intervals: 0–20, 20–50, 50–100, 100–200, and 200–500 m. Thus, the exact distance of an object detected anywhere from 0 to 20 m from the line or point would not be recorded, but only that the object was in the first distance category. The resulting data are a set of frequency counts n_i by specified distance categories rather than the set of exact distances, and total sample size is equal to $n = \Sigma \, n_i$.

Distance categories are defined by boundaries called cutpoints c_i. For u such boundaries, one has cutpoints $0 < c_1 < c_2 < \ldots < c_u$. By convention, let $c_0 = 0$ and $c_u = w$, where w can be finite or infinite (i.e. unbounded). Typically in line transect sampling the intervals defined by the cutpoints will be wider near w and narrower near the centreline. However, in point transect sampling, the first interval may be quite wide because the area corresponding to it is relatively small. The sum of the counts in each distance category equals the total number of detections n, which is the sample size. In the example above, $u = 5$, and the cutpoints are $0 < c_1 = 20 < c_2 = 50 < c_3 = 100 < c_4 = 200 < c_5 = w = 500$. Suppose the frequency counts n_i are 80, 72, 60, 45 and 25, respectively. Then $n = \Sigma\ n_i = 282$ detections.

1.3.3 Data truncation

In designing a line transect survey, one can establish a distance $y = w$ whereby objects at distances greater than w are ignored. In this case, the width of the transect to be searched is $2w$, and the area searched is of size $2wL$. In point transects, a radius w can similarly be established, giving the total area searched as $k\pi w^2$. In the general theory, w may be assumed to be infinite so that objects may be detected at quite large distances. In such cases, the width of the transect or radius around the point is unbounded.

Distance data can be truncated (i.e. discarded) prior to analysis. Data can be truncated beyond some distance w to delete outliers that make modelling of the detection function $g(y)$ difficult (Fig. 1.9). For example, w might be chosen such that $\hat{g}(w) = 0.15$. Such a rule might eliminate many detections in some point transect surveys, but only relatively few detections in line transect surveys. A simpler rule might be to truncate 5–10% of the objects detected at the largest distances. If data are truncated in the field, further truncation may be carried out at the analysis stage if this seems useful.

General methodology is available for 'left-truncation' (Alldredge and Gates 1985). This theory is potentially useful in aerial surveys if visibility directly below the aircraft is limited and, thus, $g(0) < 1$. Quang and Lanctot (1991) provide an alternative solution to this problem. Selection of a model for the distance data is critical under left-truncation because estimation may be very model dependent. Other alternatives exist at the survey design stage and we hesitate to recommend left-truncation except in special circumstances, such as the case where there is evidence of a wide shoulder in the detection function.

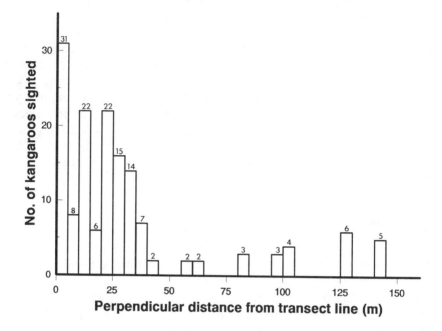

Fig. 1.9. Histogram of the number of eastern grey kangaroos detected as a function of distance from a line transect survey on Rotamah Island, Australia (redrawn from Coulson and Raines 1985). These data illustrate some heaping in the first, third and fifth distance classes, and the need to truncate observations beyond about 50 m.

1.3.4 Units of measurement

The derivation of the theory assumes that the units of y_i, L and D are all on the same measurement scale. Thus, if the distances y_i are measured in metres, then L should be in metres and density will be in numbers per square metre. In practice it is a simple but important matter to convert the y_i, l_i or D from any unit of measure into any other; in fact, computer software facilitates such conversions (e.g. feet to metres or acres to square kilometres or numbers/m^2 to numbers/km^2).

1.3.5 Ancillary data

In some cases, there is interest in age or sex ratios of animals detected, in which case these ancillary data must be recorded. Cluster size is a type of ancillary data. Size of the animal, its reproductive state (e.g.

kangaroos carrying young in the pouch), or presence of a marker or radio transmitter are other examples of ancillary data collected during a survey. Such ancillary information can be incorporated in a variety of ways. If probability of detection is a function of the ancillary variable, then it might be used to stratify the data, or it might enter the analysis as a covariate, to improve estimation.

1.4 Known constants and parameters

1.4.1 Known constants

Several known constants are used in this book and their notation is given below:

A = area occupied by the population of interest;

k = number of lines or points surveyed;

l_i = length of the ith transect line, $i = 1, \ldots, k$;

L = total line length = Σl_i;

and $\quad w$ = the width of the area searched on each side of the line transect, or the radius searched around a point transect, or the truncation point beyond which data are not used in the analysis.

1.4.2 Parameters

In line and point transect surveys there are only a few unknown parameters of interest. These are defined below:

D = density (number per unit area);

N = population size in the study area;

$E(s)$ = mean cluster size in the population (not the same as, but often estimated by, the sample mean $\bar{s}$ of detected objects);

$f(0)$ = the probability density function of distances from the line, evaluated at zero distance;

$h(0)$ = the slope of the probability density function of distances from the point, evaluated at zero distance;

17

and $g(0)$ = probability of detection on the line or point, usually assumed to be 1. For some applications (e.g. species of whale which spend substantial periods underwater and thus avoid detection, even on the line or point), this parameter must be estimated from other types of information.

Density D may be used in preference to population size N in cases where the size of the area is not well defined. Often an encounter rate n/L is computed as an index for sample size considerations or even as a crude relative density index.

1.5 Assumptions

Statistical inference in distance sampling rests on the validity of several assumptions. First, the survey must be competently designed and conducted. No analysis or inference theory can make up for fundamental flaws in survey procedure. Second, the physical setting is idealized:

1. Objects are spatially distributed in the area to be sampled according to some stochastic process with rate parameter D (= number per unit area).
2. Randomly placed lines or points are surveyed and a sample of n objects is detected, measured and recorded.

It is not necessary that the objects be randomly (i.e. Poisson) distributed. Rather, it is critical that the line or point be placed randomly with respect to the distribution of objects. Random line or point placement ensures a representative sample of the relevant distances and hence a valid density estimate. The use of transects along trails or roads does not constitute a random sample and represents poor survey practice. In practice, a systematic grid of lines or points, randomly placed in the study area, suffices.

Three assumptions are essential for reliable estimation of density from line or point transect sampling. These assumptions are given in order from most to least critical:

1. Objects directly on the line or point are always detected (i.e. they are detected with probability 1, or $g(0) = 1$).
2. Objects are detected at their initial location, prior to any movement in response to the observer.
3. Distances (and angles where relevant) are measured accurately (ungrouped data) or objects are correctly counted in the proper distance category (grouped data).

Some investigators include the assumption that one must be able to identify the object of interest correctly. In rich communities of song-birds, this problem is often substantial. Marine mammals often occur in mixed schools, so it is necessary both to identify all species present and to count the number of each species separately. In rigorous theoretical developments, assumption (2) is taken to be that objects are immobile. However, slow movement relative to the speed of the observer causes few problems in line transects. In contrast, responsive movement of animals to the approaching observer can create serious problems. In point transects, undetected movement of animals is always problematic because the observer is stationary.

The effects of partial failure of these assumptions will be covered at length in later sections, including the condition $g(0) < 1$; estimation in this circumstance is one of the main areas of current methodological development. All of these assumptions can be relaxed under certain circumstances. These extensions are covered in the following chapters. We note that no assumption is made regarding symmetry of $g(y)$ on the two sides of the line or around the point, although extreme asymmetry would be problematic. Generally, we believe that asymmetry near the line or point will seldom be large, although topography may sometimes cause difficulty. If data are pooled to give a reasonable sample size, such problems can probably be ignored.

1.6 Fundamental concept

It may seem counterintuitive that a survey be conducted, fail to detect perhaps 60–90% of the objects of interest in the survey plots (strips of dimension L by $2w$ or circular plots of size πw^2), and still obtain accurate estimates of population density. The following two sections provide insights into how distances are the key to the estimation of density when some of the objects remain undetected. We will illustrate the intuitive ideas for the case of line transect sampling; those for point transects are similar.

Consider an arbitrary area of size A with objects of interest distributed according to some random process. Assume a randomly placed line and grouped data taken in each of eight 1-foot distance intervals from the line on either side, so that $w = 8$. If all objects were detected, we would expect, on average, a histogram of the observations to be uniform as in Fig. 1.10a. In other words, on average, one would not expect many more or fewer observations to fall, say, within the seventh interval than the first interval, or any other interval.

In contrast, distance data from a survey of duck (*Anas* and *Aythya* spp.) nests at the Monte Vista National Wildlife Refuge in Colorado,

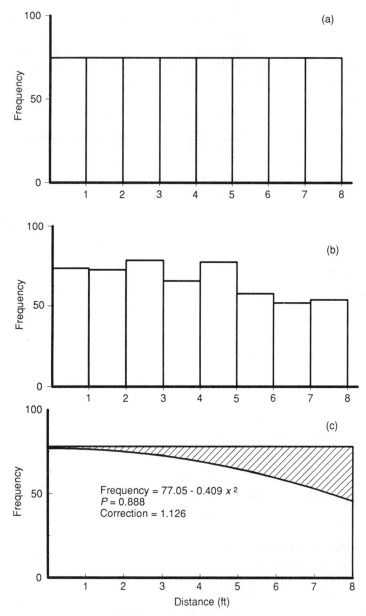

Fig. 1.10. Conceptual basis for line transect sampling: (a) the expected number of objects detected in eight distance classes if no objects were left undetected; (b) real data where a tendency to detect fewer objects at greater distances from the line can be noticed; (c) simple methods can be used to estimate the proportion of the objects left undetected (shaded area). The proportion detected, P, can be estimated from the distance data.

20

USA (Anderson and Pospahala 1970) are shown in Fig. 1.10b as a histogram. Approximately 10 000 acres of the refuge were surveyed using $L = 1600$ miles of transect, and an area $w = 8$ feet on each side of the transect was searched. A total of 534 nests was found during 1967 and 1968 and the distance data were grouped for analysis into 1-foot intervals. Clearly there is evidence from this large survey that some nests went undetected in the outer three feet of the area searched. Visual inspection might suggest that about 10% of the nests were missed during the survey. Note that the intuitive evidence that nests were missed is contained in the distances, here plotted as a histogram.

Examination of such a histogram suggests that a 'correction factor', based on the distance data, is needed to correct for undetected objects. Note that such a correction factor would be impossible if the distances (or some other ancillary information) were not recorded. Anderson and Pospahala (1970) fitted a simple quadratic equation to the midpoints of each histogram class to obtain an objective estimate of the number of nests not detected (Fig. 1.10c). Their equation, fitted by least squares, was

$$\text{frequency} = 77.05 - 0.4039x^2$$

The proportion (P) of nests detected was computed as the unshaded area in Fig. 1.10c divided by the total area (shaded + unshaded). (The areas were computed using calculus, but several simpler approximations could be used.) The estimated proportion of nests detected from 0 to 8 feet can be computed to be 0.888, suggesting a correction factor of 1.126 ($= 1/0.888$) be applied to the total count of $n = 534$. Thus, the estimated number of nests within eight feet of the sample transects was $n/P = 601$, and because the transects sampled 5.5% of the refuge, the estimate of the total number of nests on the refuge during the 2-year period was $601/0.055 = 10\,927$. This procedure provides the intuition that distances are important in reliable density estimates even if most of the objects are not detected. The Anderson–Pospahala method is no longer recommended since superior analysis methods are now available, but it illustrates the principle underlying the theory. The next two chapters will put this intuitive argument on a more formal basis.

1.7 Detection

When a survey has been conducted, n objects will have been detected. Considerable confusion regarding the meaning of n exists in the literature. Here an attempt is made to factor n into its fundamental components.

Burnham *et al.* (1981), Dawson (1981) and Morgan (1986: 9–25) give a discussion of these issues. Ramsey (1981) provides an informative example.

The number detected, *n*, is a confounding of true density and probability of detection. The latter is a function of many factors, including cue production by, or characteristics of, the object of interest, observer effectiveness, and the environment. Of these factors, one could hope that only the first, density, influences the count. While this might be ideal, it is rarely true.

1.7.1 Cue production

The object of interest often provides cues that lead to its detection by the survey observer. Obvious cues may be a loud or distinctive song or call. A splash made by a marine mammal or flock of sea birds above a school of dolphins are other examples of cues. Large size, bright or contrasting colouring, movement or other behaviour may be causes for detection. These cues are frequently species-specific and may vary by age or sex of the animal, time of day, or season of the year. Thus, the total count *n* can vary for reasons unrelated to density (Mayfield 1981; Richards 1981; Bollinger *et al.* 1988). Most often, the probability of detection of objects based on some cue diminishes as distance from the observer increases.

1.7.2 Observer effectiveness

Observer variability is well known in the literature on biological surveys. Interest in the survey, training and experience are among the dominant reasons why observers vary widely in their ability to detect objects of interest. However, both vision and hearing acuity may be major variables which are often age-specific (Ramsey and Scott 1981a; Scott, *et al.* 1981). Fatigue is a factor on long or difficult surveys. Even differing heights of observers may be important for surveys carried out on foot, with tall observers detecting objects at a higher rate. Generally, the detection of objects decreases with increasing distance due to observer effectiveness.

1.7.3 Environment

Environmental variables often influence the number of objects detected (Best 1981; Ralph 1981; Verner 1985). The habitat type and its phenology are clearly important (Bibby and Buckland 1987). Physical conditions often inhibit detection: wind, precipitation, darkness, sun angle, etc. Cue production varies by time of day, which can have a tenfold effect in the detectability of some avian species (Robbins 1981; Skirvin 1981). Often, these variables interact to cause further variability in detection and the count n.

Distance sampling provides a general and comprehensive approach to the estimation of population density. The distances y_i allow reliable estimates of density in the face of variability in detection due to factors such as cue production, observer effectiveness and environmental differences. The specific reasons why an object was not detected are unimportant. Furthermore, it seems unnecessary to research the influence of these environmental variables or to standardize survey protocol for them, if distances are taken properly and appropriate analysis carried out. Distance sampling methods fully allow for the fact that many objects will remain undetected, as long as they are not on the line or point. For example, in Laake's stake surveys (Burnham et al. 1980) only 27–67% of the stakes present were detected and recorded by various surveyors traversing the line. Still, accurate estimates of stake density were made using distance sampling theory.

1.8 History of methods

1.8.1 Line transects

In the 1930s, R.T. King recognized that not all animals were seen on strip transect surveys and presumably tried to estimate an **effective** width of the transect. He recognized that distances were useful and used the average sighting distance $\bar{r}$ as the effective width surveyed (Leopold 1933; Gates 1979). The early literature tried to conceptualize the idea of effective area sampled. Finally, Gates (1979) provided a formal definition for the effective strip width (μ): the distance for which unseen animals located closer to the line than μ equals the number of animals seen at distances greater than μ. Then, $D = n/A'$, where $A' = 2\mu L$ and is the estimated area 'effectively' sampled. Note that μ is actually one-half the effective strip width, i.e. only one side of the line.

Kelker (1945) took an alternative approach that is still sometimes used. Instead of trying to retain the total sample of n distances and estimate the 'area' effectively sampled, Kelker determined a strip width Δ on each side of the transect centreline, within which all animals were probably seen. The value of Δ was judged subjectively from an inspection of the histogram of the perpendicular distance data. Once Δ was chosen, density was estimated as a strip transect with $W = \Delta$ and n the number of objects detected from 0 to Δ on each side of the line transect. Distance data exceeding Δ were not used further.

No attempts were made to formulate a firm conceptual and mathematical foundation for line transects until Hayne's paper in 1949. All estimators then in use were *ad hoc* and generally based on either the concept of the effective strip width or the related idea of determining a strip width narrow enough such that no animals were undetected in that strip. Variations of these approaches are still being used and sometimes 'rediscovered' today, even though better methods have existed for many years.

Hayne (1949) provided the first estimator that has a rigorous justification in statistical theory. While Hayne's method rests on only the use of sighting distances r_i, the critical assumption made can only be tested using the sighting angles θ_i. Hayne's (1949) method is poor if $\bar{\theta}$ is not approximately 32.7° and may not perform well even if $\bar{\theta}$ falls close to this value, i.e. not a robust method.

After Hayne's (1949) paper, almost no significant theoretical advances appeared until 1968. During that 20 year period, line transect sampling was used frequently, and on a variety of species. The assumptions behind the method were sharpened in the wildlife literature and some evaluations of the method were presented (e.g. Robinette *et al.* 1956).

In 1968, two important papers were published in which some of the basic ideas and conceptual approaches to line transect sampling finally appeared (Eberhardt 1968; Gates *et al.* 1968). Gates *et al.* (1968) published the first truly rigorous statistical development of a line transect estimator, applicable only to untruncated and ungrouped perpendicular distance data. They proposed that $f(x)$ be a negative exponential form, $f(x) = a \cdot \exp(-ax)$, where a is an unknown parameter to be estimated. Under that model, $f(0) = a$. Gates *et al.* (1968) developed the optimal estimator of a based on a sample of perpendicular distances and provided an estimator of the sampling variance. For the first time, rigorous consideration was given to questions such as optimal estimation under the model, construction of confidence intervals, and tests of assumptions. The one weakness was that because the assumed detection function was very restrictive and might easily be inappropriate, the resulting estimate of density could be severely biased.

In contrast, Eberhardt (1968) conceptualized a fairly general model in which the probabilities of detection decreased with increasing perpendicular distance. He reflected on the shape of the detection function $g(x)$, and suggested both that there was a lack of information about the appropriate shape and that the shape might change from survey to survey. Consequently, he suggested that the appropriate approach would be to adopt a family of curves to model $g(x)$. He suggested two such families, a power series and a modified logistic, both of which are fairly flexible parametric functions. His statistical development of these models was limited, but important considerations had been advanced.

Since 1968, line transect sampling has been developed along rigorous statistical inference principles. Parametric approaches to modelling $g(x)$ were predominant, with the notable exception of Anderson and Pospahala (1970), who rather inadvertently introduced some of the basic ideas that underlie a non-parametric or semiparametric approach to the analysis of line transect data. Emlen (1971) proposed an *ad hoc* method that found use in avian studies.

A general model structure for line transect sampling based on perpendicular distances was presented by Seber (1973: 28–30). For an arbitrary detection function, Seber gave the probability distribution of the distances $x_1, \ldots, x_n$ and the general form of the estimator of animal density D. This development was left at the conceptual stage and not pursued to the final step of a workable general approach for deriving line transect estimators, and the approach was still based on the concept of an effective strip width.

More work on sighting distance estimators appeared (Gates 1969; Overton and Davis 1969). There was a tendency to think of approaches based on perpendicular distances as appropriate for inanimate or non-responsive objects, whereas methods for flushing animals were to be based on sighting distances and angles (Eberhardt 1968, 1978a). This artificial distinction tended to prevent the development of a unified theory for line transect sampling. By the mid-1970s, line transect sampling remained a relatively unexplored methodology for the estimation of animal density. Robinette *et al.* (1974) reported on a series of field evaluations of various line transect methods. Their field results were influential in the development of the general theory.

Burnham and Anderson (1976) pursued the general formulation of line transect sampling and gave a basis for the general construction of line transect estimators. They developed the general result $\hat{D} = n \cdot \hat{f}(0)/2L$, wherein the parameter $f(0)$ is a well-defined function of the distance data. The key problem of line transect data analysis was seen to be the modelling of $g(x)$ or $f(x)$ and the subsequent estimation of $f(0)$. The nature of the specific data (grouped or ungrouped, truncated or untruncated) is irrelevant to the basic estimation problem. Consequently, their formulation is applicable for the development of any parametric or semiparametric line transect estimator. Further, the general theory is applicable to point transect sampling with some modification (Buckland 1987a).

Burnham and Anderson's (1976) paper heralded a period of new statistical theory. Major contributions published during the 1976–80 period include Schweder (1977), Crain *et al.* (1978, 1979), Pollock (1978), Patil *et al.* (1979b), Quinn (1979), Ramsey (1979), Seber (1979) and Quinn and Gallucci (1980). Other papers developing methodology during this short period include Anderson *et al.* (1978, 1979, 1980), Eberhardt (1978a, b, 1979), Sen *et al.* (1978), Burnham *et al.* (1979), Patil *et al.* (1979a) and Smith (1979). Anderson *et al.* (1979) provided guidelines for field sampling, including practical considerations. Burdick (1979) produced an advanced method to estimate spatial patterns of abundance from line transect sampling where there are major gradients in population density. Laake *et al.* (1979) and Gates (1980) produced comprehensive computer software packages, TRANSECT and LINETRAN respectively, for the analysis of line transect data.

Gates (1979) provided a readable summary of line transect sampling theory and Ramsey's (1979) paper presents a more mathematical treatment of parametric approaches. Hayes (1977) gave an excellent summary of methodology and provided many useful insights at that time.

Burnham *et al.* (1980) published a major monograph on line transect sampling theory and application. This work provided a review of previous methods, gave guidelines for field use, and identified a small class of estimators that seemed generally useful. Usefulness was based on four criteria: **model robustness, pooling robustness**, a **shape criterion**, and **estimator efficiency**. Theoretical and Monte Carlo studies led them to suggest the use of estimators based on the Fourier series (Crain *et al.* 1978, 1979), the exponential power series (Pollock 1978), and the exponential quadratic model.

Since 1980, more theory has been developed on a wide variety of issues. Seber (1986) and Ramsey *et al.* (1988) give brief reviews. Major contributions during the 1980s include Butterworth (1982a, b), Patil *et al.* (1982), Hayes and Buckland (1983), Buckland (1985), Burnham *et al.* (1985), Johnson and Routledge (1985), Quinn (1985), Drummer and McDonald (1987), Ramsey *et al.* (1987), Thompson and Ramsey (1987) and Zahl (1989). Other papers during the decade include Buckland (1982), Stoyan (1982), Burnham and Anderson (1984), Anderson *et al.* (1985a, b) and Gates *et al.* (1985). Several interesting field evaluations where density was known have appeared since 1980, including Burnham *et al.* (1981), Hone (1986, 1988), White *et al.* (1989), Bergstedt and Anderson (1990) and Otto and Pollock (1990). In addition, other field evaluations where the true density was not known have been published, but these results are difficult to interpret.

A great deal of statistical theory has been developed since 1976, but new theory may have started to decrease by the late 1980s. Field studies using line transect sampling have increased and new applications have appeared in the literature. No attempt to discuss all of the recent developments will be given in this chapter. At the present time, there are several good models for fitting $g(x)$. There now exist sound approaches for analysing grouped or ungrouped data with truncated or untruncated transect widths, under various extensions (e.g. clustered populations). Estimation based on sighting distances and angles has been shown to be problematic and we recommend transforming such data to perpendicular distances prior to analysis. Current areas of development include estimation when $g(0) < 1$, when responsive movement to the observer occurs, when the objects occur in clusters, leading to size-biased sampling, and when there is covariate information on factors such as sighting conditions or habitat.

1.8.2 Point transects

Point transect sampling has had a much shorter history. The method can be traced to the paper by Wiens and Nussbaum (1975) and their application of what they called a variable circular plot census. They drew heavily on the paper on line transects by Emlen (1971). Ramsey and Scott (1979) provided a statistical formalism for the general method and noted several close relationships to line transect sampling. Following the 'effective area' thinking, they noted 'The methods are similar in spirit to line transect methods, in that the total number of detections divided by an estimate of the area surveyed is the estimate of the population density.' Ramsey and Scott (1979) provided a summary of the assumptions and derived a general theory for density estimation, including sampling variances. This represented a landmark paper at the time.

Reynolds et al. (1980) presented additional information on the variable circular plot method. Burnham et al. (1980) and Buckland (1987a) also noted the close links between line transects and point transects (i.e. variable circular plots). Buckland (1987a) developed other models, evaluated the Fourier series, Hermite polynomial and hazard-rate estimators, and provided an evaluation of the efficiency of binomial models (where objects of interest are grouped into two categories, within or beyond a specified distance c_1). The general theory for line and point transects is somewhat similar because they both involve sampling distances. Thus, the term point transect will be used rather than the variable circular plot 'census'.

1.9 Program DISTANCE

The computation for most estimators is arduous and prone to errors if done by hand. Estimators of sampling variances and covariances are similarly tedious. Data should be plotted and estimates of $f(y)$ should usually be graphed for visual comparison with the observed distance data.

Program DISTANCE (Laake *et al.* 1993) was developed to allow comprehensive analyses of the type of distance data we discuss here. The program is written in FORTRAN and runs on any IBM PC compatible microcomputer with 640 K of RAM. A math coprocessor is desirable, but not required. Program DISTANCE allows researchers to focus on the biology of the population, its habitat and the survey operation; one can concentrate on the results and interpretation, rather than on computational details. Almost all the examples presented in this book were analysed using program DISTANCE; the distance data and associated program commands for some of the examples are available as an aid to data analysts. The program is useful both for data analysis and as a research tool. Only occasional references to DISTANCE are made throughout this book because a comprehensive manual on the program is available (Laake *et al.* 1993).

2

Assumptions and modelling philosophy

2.1 Assumptions

This section provides material for a deeper understanding of the assumptions required for the successful application of distance sampling theory. The validity of the assumptions allows the investigator assurance that valid inference can be made concerning the density of the population sampled. The existing theory covers a very broad application area and makes it difficult to present a simple list of all the assumptions that are generally true for all applications. Three primary assumptions are emphasized, but first two initial conditions are mentioned.

First, it is assumed that a population comprises objects of interest that are distributed in the area to be sampled according to some stochastic process with rate parameter D (= expected number per unit area). In particular, it is not necessary (in any practically significant way) that the objects be randomly (i.e. Poisson) distributed, although this is mistakenly given in several places in the literature. Rather, it is critical that the lines or points be placed randomly **with respect to the distribution of objects**. Random line or point placement justifies the extrapolation of the sample statistics to the population of interest. The area to be sampled must be defined, but its size need not be measured if only object density (rather than abundance) is to be estimated. Further, the observer must be able to recognize and correctly identify the objects of interest. This requirement seems almost trite, but in rich avian communities, the problem can be substantial. The distances from the line or point to the identified objects must be measured without bias.

Second, the design and conduct of the survey must pay due regard to good survey practice, as outlined in Chapter 7. If the survey is poorly designed or executed, the estimates may be of little value. Sound theory and analysis procedures cannot change this.

29

Three assumptions are critical to achieving reliable estimates of density from line or point transect sampling. These assumptions are given roughly in order of importance from most to least critical. The effects of partial failure of these assumptions and corresponding theoretical extensions are covered at length in later sections. All three assumptions can be relaxed under certain circumstances.

2.1.1 Assumption 1: objects on the line or point are detected with certainty

It is assumed that all objects at zero distance are detected, that is $g(0) = 1$. In practice, detection on or near the line or point should be nearly certain. Design of surveys must fully consider ways to assure that this assumption is met; its importance cannot be overemphasized.

It is sometimes possible to obtain an independent estimate of the probability of detection on the centreline in a line transect survey, for example by assigning two (or more) independent observers to each leg of search effort. Chapter 6 summarizes methods which have been developed for estimating $g(0)$, and Chapter 3 shows how the estimate can be incorporated in the estimation of density. It is important to note that $g(0)$ cannot be estimated from the distances y_i alone, and attempts to estimate $g(0)$ with low bias or adequate precision when it is known to be less than unity have seldom been successful. This issue should be addressed during the design of surveys, so that observation protocol will assure that $g(0) = 1$ or that a procedure for estimating $g(0)$ is incorporated into the design.

In fact, the theory can be generalized such that density can be computed if the value of $g(y)$ is known for **some** value of y. However, this result is of little practical significance in biological sampling unless an assumption that $g(y) = 1$ for some $y > 0$ is made (Quang and Lanctot 1991).

If objects on or near the line or point are missed, the estimate will be biased low (i.e. $E(\hat{D}) < D$). The bias is a simple function of $g(0)$: $E(\hat{D}) - D = - [1 - g(0)] \cdot D$, which is zero (unbiased) when $g(0) = 1$. Many things can be done in the field to help ensure that $g(0) = 1$. For example, video cameras have been used in aerial and underwater surveys to allow a check of objects on or very near the line; the video can be monitored after completion of the field survey. Trained dogs have been used in ground surveys to aid in detection of grouse close to the line.

Although we stress that every effort should be made to ensure $g(0) = 1$, the practice of 'guarding the centreline' during shipboard or aerial line transect surveys can be counterproductive. For example, suppose that most search effort is carried out using $20 \times$ or $25 \times$ tripod-

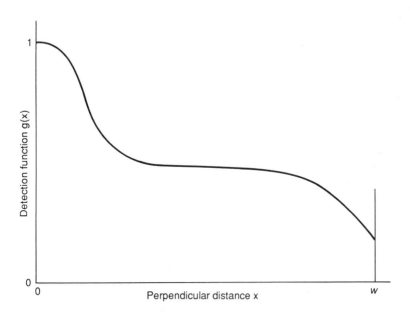

Fig. 2.1. Hypothetical detection function illustrating the danger of assigning an observer to 'guard the centreline'. This problem is most common in shipboard and aircraft surveys involving more than one observer.

mounted binoculars on a ship, but an observer is assigned to search with the naked eye, to ensure animals very close to the ship are not missed. If $g(0)$ in the absence of this observer is appreciably below 1, then the detection function may be as illustrated in Fig. 2.1. This function violates the **shape criterion** described later, and no line transect model can reliably estimate density in this case. The problem may be exacerbated if animals are attracted to the ship; the observer guarding the centreline may only detect animals as they move in toward the bow. Polacheck and Smith (unpublished) argued that if effort is concentrated close to the centreline, large bias can arise. Thus, field procedures should ensure both that $g(0) = 1$ and that the detection function does not fall steeply with distance from the line or point.

2.1.2 *Assumption 2: objects are detected at their initial location*

In studies of mobile animals, it is possible that an animal moves from its original location for some distance prior to being detected. The

31

measured distance is then from the random line or point to the location of the detection, not the animal's original location. If such undetected movements prior to detection were random (see Yapp 1956), no serious problem would result, provided that the animal's movement is slow relative to the speed of the observer. If movement is not slow, its effect must be modelled (Schweder 1977), or field procedures must be modified (Section 7.6). However, movement may be in response to the observer. If the movement is away from the transect line being traversed by the observer, the density estimator is biased low, whereas if the movement is toward the observer (e.g. some songbirds and marine mammals), the estimator of density will be biased high. Substantial movement away from the observer can often be detected in a histogram of the distance data (Fig. 2.2). However, if some animals move a considerable perpendicular distance and others remain in their original location, then the effect may not be detectable from the data. Ideally, the observer on a line transect survey would try to minimize such movement by looking well ahead as the area is searched. Field procedures should try to ensure that most detections occur beyond the likely range of the effect of the observer on the animals. In point transect surveys, one must be careful not to disturb animals as the sample point is approached, or perhaps wait a while upon reaching the point.

The theory of distance sampling and analysis is idealized in terms of dimensionless points or 'objects of interest'. Surveys of dead deer, plants or duck nests are easily handled in this framework. More generally, movement independent of the observer causes no problems, unless the object is counted more than once on the same unit of transect sampling effort (usually the line or point) or if it is moving at roughly half the speed of the observer or faster. Animals such as jackrabbits or pheasants will flush suddenly as an observer approaches. The measurement must be taken to the animal's original location. In these cases, the flush is often the cue that leads to detection. Animal movement after detection is not a problem, as long as the original location can be established accurately and the appropriate distance measured. Similarly, it is of no concern if an animal is detected more than once on different occasions of sampling the same transect. Animals that move to the vicinity of the next transect in response to disturbance by the observer are problematic. If the observer unknowingly records the same animal several times while traversing a transect, due to undetected movement ahead of him, bias can be large.

The assumption of no movement before detection is not met when animals take evasive movement prior to detection. A jackrabbit might hop several metres away from the observer into heavy cover and wait. As the observer moves closer, the rabbit might eventually flush. If the

32

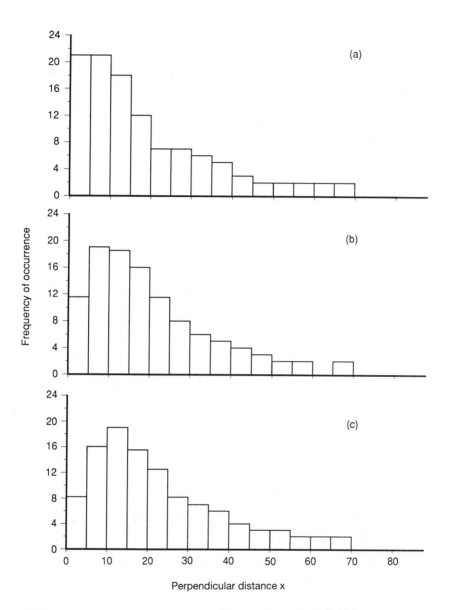

Fig. 2.2. Three histograms of perpendicular distance data, for equally spaced cutpoints, illustrating the effect of evasive movement prior to detection. Expected values are shown for the case where relatively little movement away from the observer was experienced prior to detection (a), while (b) and (c) illustrate cases where movement prior to detection was more pronounced. Data taken from Laake (1978).

new location is thought to be the original location and this distance is measured and recorded, then the assumption is violated. If this condition happens in only, say, 5% of the observations, then the bias is likely to be trivial. If a substantial portion of the population moves further from the line prior to detection, this movement will often be apparent from examination of the histogram of the distance data. If evasive movement occurs prior to detection, the estimator will be biased low ($E(\hat{D}) < D$) (Fig. 2.2b or c). Less frequently, members of a species will be attracted to the observer (Bollinger *et al.* 1988; Buckland and Turnock in press). If animals move toward the observer prior to being detected, a positive bias in estimated density can be expected ($E(\hat{D}) > D$). However, in this case, the movement is unlikely to be detected in the histogram, even if it is severe. It seems unlikely that methods will be developed for the reliable estimation of density for cases where a high proportion of the objects moves in response to the observer prior to detection without making some very critical and untestable assumptions (e.g. Smith 1979), unless relevant and reliable ancillary data can be gathered (Turnock and Quinn 1991; Buckland and Turnock in press).

2.1.3 Assumption 3: measurements are exact

Ideally, recorded distances (and angles, where relevant) are exact, without measurement errors, recording errors or heaping. For grouped data, detected objects are assumed to be correctly assigned to distance categories. Reliable estimates of density may be possible even if the assumption is violated. Although the effect of inaccurate measurements of distances or angles can often be reduced by careful analysis (e.g. grouping), it is better to gather good data in the field, rather than to rely on analytical methods. It is important that measurements near the line or point are made accurately. Rounding errors in measuring angles near zero are problematic, especially in the analysis of ungrouped data, and for shipboard surveys. If errors in distance measurements are random and not too large, then reliable density estimates are still likely, especially if the sample size is large (Gates *et al.* 1985). Biased measurements pose a larger problem (e.g. a strong tendency to overestimate the distances using ocular judgements), and field methods should be considered to minimize this bias.

For duck nests and other stationary objects, distances can be measured with a steel tape or similar device, but distances are often merely paced or estimated, taken with a rangefinder or estimated using binocular reticles. These approximate methods compromise the quality of the data, but are often forced by practical considerations. A useful alternative is to take grouped data in, say, 5–7 distance categories, such that the width of the categories increases toward w (i.e. $[c_1 - 0] \leqslant [c_2 - c_1] \leqslant [c_3 - c_2] \leqslant \ldots$). Thus, careful measurement is required only near the cutpoints c_i.

(a) *Heaping* Often, when distances are estimated (e.g. ocular estimates, 'eyeballing'), the observer may 'round' to convenient values (e.g. 5, 10, 50 or 100) when recording the result. Thus, a review of the n distance values will frequently result in many 'heaped' values and relatively few numbers such as 3, 4, 7, 8 or 11. Heaping is common in sighting angles, which are often strongly heaped at 0, 15, 30, 45, 60 and 90 degrees. A histogram of the data will often reveal evidence of heaping. Often some judicious grouping of the data will allow better estimates of density, i.e. the analysis can often be improved by proper grouping of the distance data. Cutpoints for grouping distances from the line or point should be selected so that large 'heaps' fall approximately at the midpoints of the groups. For line transects, sighting distances and angles should not be grouped prior to conversion into perpendicular distances. Heaping can be avoided in the field by measuring distances, rather than merely estimating them. The effects of heaping can be reduced during the analysis by smearing (Butterworth 1982b). Heaping at perpendicular distance zero can result in serious overestimation of density. This problem is sometimes reduced if a model is used that always satisfies the **shape criterion** (Section 2.3.2), although accurate measurement is the most effective solution.

(b) *Systematic bias* When distances are estimated, it is possible that the errors are systematic rather than random. For example, there is sometimes a strong tendency to underestimate distances at sea. Each distance may tend to be over- or underestimated. In surveys where only grouped data are taken, the counts may be in error because the cutpoints c_i are in effect $c_i + \delta_i$ where δ_i is some systematic increment. Thus, n_1 is not the count of objects detected between perpendicular distances 0 and c_1, it is the count of objects detected between 0 and $c_1 + \delta_1$. Little can be done to reduce the effect of these biased measurements in the analysis of the data unless experiments are carried out to estimate the bias; a calibration equation then allows the biased measurements to be corrected. Again, careful measurements are preferable to rough estimates of distances.

(c) *Outliers* If data are collected with no fixed width w, it is possible that a few extreme outliers will be recorded. A histogram of the data will reveal outliers. These data values contain little information about the density and will frequently be difficult to fit (Fig. 1.9). Generally, such extreme values will not be useful in the final analysis of density, and should be truncated. It is often recommended that the 5–10% of the largest observations be routinely truncated prior to analysis.

2.1.4 Other assumptions

Other aspects of the theory can be considered as assumptions. The assumption that detections are (statistically) independent events is often mentioned. If detections are somewhat dependent (e.g. 'string' flushes of quail), then the theoretical variances will be underestimated. However, we recommend that empirically based estimates of sampling variances be made, thus alleviating the need for this assumption. That is, if var(n) is estimated from independent replicate lines or points, then the assumption of within line or (point) independence is not problematic, provided the dependence is over short distances relative to the distance between replicate lines or points. Independence of detection of individual animals is clearly violated in clustered populations. This is handled by defining the cluster as the object of interest and measuring the ancillary variable, cluster size. This solution can be unsatisfactory for objects that occur in loose, poorly defined clusters, so that the location and size of the cluster may be difficult to determine or estimate without bias. The assumption of independence is a minor one in a properly designed survey, unless the clusters are poorly defined.

Statistical inference methods used here (e.g. maximum likelihood estimators of parameters, theoretical sampling variance estimators, and goodness of fit tests) assume independence among detections. Failure of the assumption of independence has little effect on the point estimators, but causes a bias (underestimation) in theoretical variance estimates (Cox and Snell 1989). The assumption of independence can fail because objects do not have a random (Poisson) distribution in space and this pattern could result in a dependency in the detections. Non-random distribution, by itself, is not necessarily a cause of lack of independence. If the transects are placed at random and a robust estimator of the sampling variance is used, then the assumption of independence can be ignored. At least in practice, it is not at all important that the objects be randomly distributed on the study area. Similarly, it is of little concern if detection on either side of the line or around the point is not symmetric, provided that the asymmetry is not extreme, such that modelling $g(y)$ is difficult.

A more practically important consideration relates to the shape of the detection function near zero distance. This shape can often be judged by examining histograms of the distance data using different groupings. Distance sampling theory performs well when a 'shoulder' in detectability exists near the line or around the point. That is, detectability is certain near the line or point and stays certain or nearly certain for some distance. This will be defined as the 'shape criterion' in Section 2.3. If detectability falls sharply just off the line or point, then estimation tends to be poor, even if the true model for the data is known. Thus if data are to be analysed reliably, the detection function from which they come should possess a shoulder; to this extent, the shape criterion is an assumption.

Some papers imply that an object should not be counted on more than one line or point. This, by itself, is not true as no such assumption is required. In surveys with $w = \infty$, an object of interest (e.g. a dead elk) can be detected from two different lines without violating any assumptions. As noted above, if in line transect sampling an animal moves ahead of the observer and is counted repeatedly, abundance will be overestimated. This is undetected movement in response to the observer; double counting, by itself, is not a cause of bias if such counts correspond to different units of counting effort. Bias is likely to be small unless repeated counting is common during a survey. Detections made behind the observer in line transect sampling may be utilized, unless the object is located before the start of a transect leg, in which case it is outside the rectangular strip being surveyed.

These assumptions, their importance, models robust to partial violations of assumptions, and field methods to meet assumptions adequately will be addressed in the material that follows.

2.2 Fundamental models

This section provides a glimpse of the theory underlying line and point transect sampling. This material is an extension of Section 1.6.

2.2.1 Line transects

In strip transect sampling, if strips of width $2w$ and total length L are surveyed, an area of size $a = 2wL$ is censused. All n objects within the strips are enumerated, and estimated density is the expected number of objects per unit area:

$$\hat{D} = n/2wL$$

In line transect sampling, only a proportion of the objects in the area a surveyed is detected. Let this unknown proportion be P_a. If P_a can be estimated from the distance data, the estimate of density could be written as

$$\hat{D} = n/2wL\hat{P}_a \tag{2.1}$$

Now, some formalism is needed for the estimation of P_a from the distances. The unconditional probability of detecting an object in the strip (of area $a = 2wL$) is

$$P_a = \frac{\int_0^w g(x)dx}{w} \tag{2.2}$$

In the duck nest example of Chapter 1, $g(x)$ was found by dividing the estimated quadratic equation by the intercept (77.05), to give

$$\hat{g}(x) = 1 - 0.0052x^2$$

Note that $\hat{g}(8) = 0.66$, indicating that approximately one-third of the nests near the edges of the transect were never detected. Then

$$\hat{P}_a = \frac{\int_0^8 (1 - 0.0052x^2)dx}{8}$$

$$= 0.888$$

Substituting the estimator of P_a from Equation 2.2 into $\hat{D}$ from Equation 2.1 gives

$$\hat{D} = \frac{n}{2L\int_0^w \hat{g}(x)dx} \tag{2.3}$$

because the w and $1/w$ cancel out. Then the integral $\int_0^w g(x)dx$ becomes the critical quantity and is denoted as μ for simplicity. Thus,

$$\hat{D} = n/2L\hat{\mu}$$

There is a very convenient way to estimate the quantity $1/\mu$. The derivation begins by noting that the probability density function (pdf) of the perpendicular distance data, conditional on the object being detected, is merely

$$f(x) = \frac{g(x)}{\int_0^w g(x)\,dx} \tag{2.4}$$

This result follows because the expected number of objects (including those that are not detected) at distance x from the line is independent of x. This implies that the density function is identical in shape to the detection function; it can thus be obtained by rescaling, so that the function integrates to unity.

By assumption, $g(0) = 1$, so that the pdf, evaluated at zero distance, is

$$f(0) = \frac{1}{\int_0^w g(x)\,dx}$$

$$= 1/\mu$$

The parameter $\mu = \int_0^w g(x)\,dx$ is a function of the measured distances. Therefore, we will often write the general estimator of density for line transect sampling simply as

$$\hat{D} = \frac{n \cdot \hat{f}(0)}{2L} \tag{2.5}$$

$$= \frac{n}{2L\hat{\mu}}$$

This estimator can be further generalized, but the conceptual approach remains the same. $\hat{D}$ is valid whether w is bounded or unbounded (infinite) and when the data are grouped or ungrouped. Note that either form of Equation 2.5 is equivalent to $\hat{D} = n/2wL\hat{P}_a$ (Equation 2.1).

For the example, an estimate of Gates' (1979) effective strip width is $\hat{\mu} = w\hat{P}_a = 8(0.888) = 7.10$ ft, and $\hat{D} = 534/(2 \times 1600 \times 7.10)$ nests/mile/ft $= 124$ nests/square mile.

The density estimator expressed in terms of an estimated pdf, evaluated at zero, is convenient, as a large statistical literature exists on the subject of estimating a pdf. Thus, a large body of general knowledge can be brought to bear on this specific problem.

2.2.2 Point transects

In traditional circular plot sampling, k areas each of size πw^2 are censused and all n objects within the k plots are enumerated. By definition, density is the number per unit area, thus

$$\hat{D} = \frac{n}{k\pi w^2}$$

In point transect sampling, only a proportion of the objects in each sampled area is detected. Again, let this proportion be P_a. Then the estimator of density is

$$\hat{D} = \frac{n}{k\pi w^2 \hat{P}_a} \tag{2.6}$$

The unconditional probability of detecting an object that is in one of the k circular plots is

$$P_a = \int_0^w \frac{2\pi r g(r) dr}{\pi w^2} \tag{2.7}$$

$$= \frac{2}{w^2} \int_0^w r g(r) dr$$

Substituting Equation 2.7 into Equation 2.6 and cancelling the w^2 terms, the estimator of density is

$$\hat{D} = \frac{n}{2k\pi \int_0^w r\hat{g}(r) dr} \tag{2.8}$$

Defining $\nu = 2\pi \int_0^w r g(r) dr$

then

$$\hat{D} = n/k\hat{\nu}$$

Clearly, ν is the critical quantity to be estimated from the distance data for a point transect survey.

40

2.2.3 Summary

The statistical problem in the estimation of density of objects is the estimation of μ or ν. Then the estimator of density for line transect sampling is

$$\hat{D} = n/2L\hat{\mu}$$

where $\quad \mu = \int_0^w g(x)\,dx$

The estimator of density for point transect surveys can be given in a similar form:

$$\hat{D} = n/k\hat{\nu}$$

where $\quad \nu = 2\pi \int_0^w rg(r)\,dr$

This, then, entails careful modelling and estimation of $g(y)$. Good statistical theory now exists for these general problems. Finally, we note that the estimator of density from strip transect sampling is also similar:

$$\hat{D} = n/2wL$$

where $P_a = 1$ and, by assumption, n is the count from a complete census of each strip.

2.3 Philosophy and strategy

The true detection function $g(y)$ is not known. Furthermore, it varies due to numerous factors (Section 1.7). Therefore, it is important that strong assumptions about the shape of the detection function are avoided. In particular, a flexible or 'robust' model for $g(y)$ is essential.

The strategy used here is to select a few models for $g(y)$ that have desirable properties. These models are selected *a priori*, and without particular reference to the given data set. This class of models excludes those that are not robust, have restricted shapes, or have inefficient estimators. Because the estimator of density is closely linked to $g(y)$, it is of critical importance to select models for the detection function carefully. Three properties desired for a model for $g(y)$ are, in order of importance, **model robustness**, a **shape criterion**, and **efficiency**.

2.3.1 Model robustness

The most important property of a model for the detection function is **model robustness**. This means that the model is a general, flexible function that can take the variety of shapes that are likely for the true detection function. In general, this property excludes single parameter models; experience has shown that models with two or three parameters are frequently required. Most of the models recommended have a variable number of parameters, depending on how many are required to fit the specific data set. These are sometimes called semiparametric models.

The concept of **pooling robustness** (Burnham *et al.* 1980) is included here under model robustness. Models of $g(y)$ are pooling robust if the data can be pooled over many factors that affect detection probability (Section 1.7) and still yield a reliable estimate of density. Consider two approaches: stratified estimation $\hat{D}_{st}$ and pooled estimation $\hat{D}_p$. In the first case, the data could be stratified by factors affecting detectability (e.g. three observers and four habitat types) and an estimate of density made for each stratum. These separate estimates could be combined into an estimate of average density $\hat{D}_{st}$. In the second case, all data could be pooled, regardless of any stratification (e.g. the data for the three observers and four habitat types would be pooled) and a single estimate of density computed, $\hat{D}_p$. A model is **pooling robust** if $\hat{D}_{st} \doteq \hat{D}_p$. **Pooling robustness** is a desirable property. Only models that are linear in the parameters satisfy the condition with strict equality, although general models that are **model robust**, such as those recommended in this book, approximately satisfy the **pooling robust** property.

2.3.2 Shape criterion

Theoretical considerations and the examination of empirical data suggest that the detection function should have a 'shoulder' near the line or point. That is, detection remains nearly certain at small distances from the line or point. Mathematically, the derivative $g'(0)$ should be zero. This **shape criterion** excludes functions that are spiked near zero distance. Frequently, a histogram of the distance data will not reveal the presence of a shoulder, particularly if the histogram classes are large (Fig 2.3), or if the data include several large values (a long tail). Generally, good models for $g(y)$ will satisfy the **shape criterion** near zero distance. The **shape criterion** is especially important in the analysis of

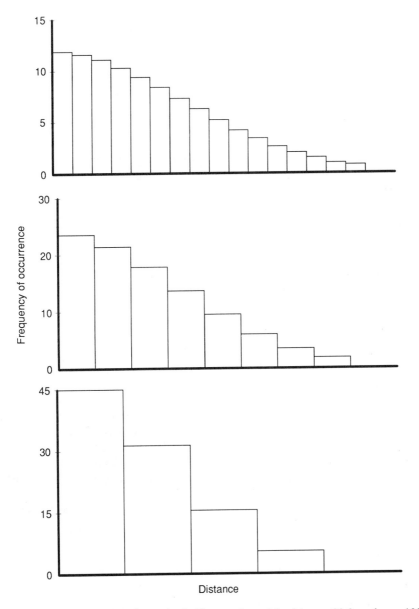

Fig. 2.3. Data ($n = 100$) from the half-normal model with $\sigma = 33.3$ and $w = 100$ shown with three different sets of group interval. As the group interval increases, the data appear to become more spiked. Adapted from Burnham *et al.* (1980).

data where some heaping at zero distance is suspected. This occurs most frequently when small sighting angles are rounded to zero, and gives

43

rise to histograms that show no evidence of a shoulder, even when the true detection function has a substantial shoulder.

2.3.3 Efficiency

Other things being equal, it is desirable to select a model that provides estimates that are relatively precise (i.e. have small variance). We recommend maximum likelihood methods, which have many good statistical properties, including that of asymptotic minimum variance. Efficient estimation is of benefit only for models that are model robust and have a shoulder near zero distance; otherwise, estimation might be precise but biased.

2.3.4 Model fit

Ideally, there would be powerful statistical tests of the fit of the model for $g(y)$ to the distance data. The only simple omnibus test available is the χ^2 goodness of fit test based on grouping the data. This test compares the observed frequencies n_i (based on the grouping selected) with the estimated expected frequencies under the model, $\hat{E}(n_i)$, in the usual way:

$$\chi^2 = \sum_{i=1}^{u} \frac{[n_i - \hat{E}(n_i)]^2}{\hat{E}(n_i)}$$

is approximately χ^2 with $u - m - 1$ degrees of freedom, where u is the number of groups and m is the number of parameters estimated. In isolation, this approach has severe limitations for choosing a model for $g(y)$, given a single data set (Fig. 2.4).

Generally, as the number of parameters in a model increases, the bias decreases but the sampling variance increases. A proper model should be supported by the particular data set and thus have enough parameters to avoid large bias but not so many that precision is lost (the **Principle of Parsimony**). Likelihood ratio tests (Lehmann 1959; Hogg and Craig 1970) are used in selecting the number of model parameters that are appropriate in modelling $f(y)$. The relative fit of alternative models may be evaluated using Akaike's Information Criterion (Akaike 1973; Sakamoto et al. 1986; Burnham and Anderson 1992). These technical subjects are presented in the following chapters.

2.3.5 Test power

The power of the goodness of fit test is quite low and, therefore, of little use in selecting a good model of $g(y)$ for the analysis of a particular

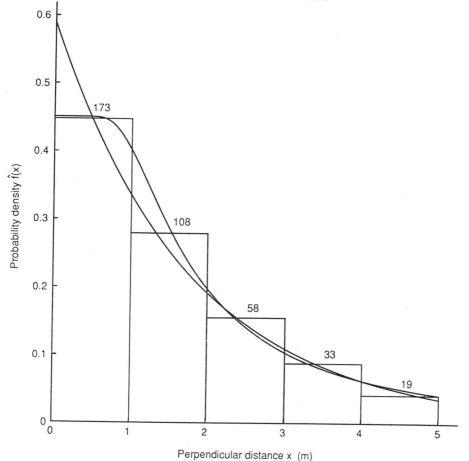

Fig. 2.4. The distance data are often of little help in testing the relative fit among models. Here, fits of the negative exponential model and the hazard-rate model to a line transect data set are shown. Both models provide an excellent fit ($\chi^2 = 0.49$, 3 df, $p = 0.92$, and $\chi^2 = 0.33$, 2 df, $p = 0.85$, respectively), even though the estimates of $f(0)$ are quite different ($f(0) = 0.589$ and 0.450, respectively).

data set. In particular, this test is incapable of discriminating between quite different models near the line or point, the most critical region (Fig. 2.4). In addition, grouping data into fewer groups frequently diminishes the power of the test still further and may give the visual impression that the data arise from a spiked distribution such as the negative exponential, when the true detection function has a shoulder (Fig. 2.3).

While goodness of fit test results should be considered in the analysis of distance data, they will be of limited value in selecting a model. Thus, a class of reliable models is recommended here, based on the three properties: **model robustness,** the **shape criterion** and **estimator efficiency**.

2.4 Robust models

Several models of $g(y)$ are recommended for the analysis of line or point transect data. These models, as implemented in program DISTANCE, have the three desired properties of **model robustness, shape criterion** and **estimator efficiency**. Following Buckland (1992a), the modelling process can be conceptualized in two steps. First, a '**key function**' is selected as a starting point, possibly based on visual inspection of the histogram of distances, after truncation of obvious outliers. Often, a simple key function is adequate as a model for $g(y)$, especially if the data have been properly truncated. Two key functions should probably receive initial consideration: the uniform and the half-normal (Fig. 2.5a). The uniform key function has no parameters, whereas the half-normal key has one unknown parameter to be estimated from the distance data. In some cases, the hazard-rate model (Fig. 2.5b) could be considered as a key function, although it requires that two key parameters be estimated.

Second, a flexible form, called a '**series expansion**', is used to adjust the key function, using perhaps one or two more parameters, to improve the fit of the model to the distance data. Conceptually, the detection function is modelled in the following general form:

$$g(y) = \text{key}(y) \left[1 + \text{series}(y) \right]$$

The key function alone may be adequate for modelling $g(y)$, especially if sample size is small or the distance data are easily described by a simple model. Theoretical considerations often suggest a series expansion appropriate for a given key. Three series expansions are considered here: (1) the cosine series, (2) simple polynomials, and (3) Hermite polynomials (Stuart and Ord 1987: 220–7). All three expansions are linear in their parameters. Thus, some generally useful models of $g(y)$ are:

Key function	*Series expansion*
Uniform, $1/w$	Cosine, $\sum_{j=1}^{m} a_j \cos\left(\dfrac{j\pi y}{w}\right)$
Uniform, $1/w$	Simple polynomial, $\sum_{j=1}^{m} a_j \left(\dfrac{y}{w}\right)^{2j}$

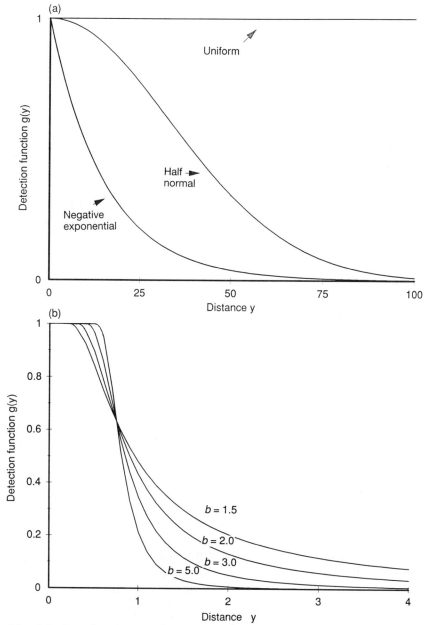

Fig. 2.5. Key functions useful in modelling distance data: (a) uniform, half-normal and negative exponential, and (b) hazard-rate model for four different values of the shape parameter b.

Half-normal, $\exp(-y^2/2\sigma^2)$ Cosine, $\sum_{j=2}^{m} a_j \cos\left(\dfrac{j\pi y}{w}\right)$

Half-normal, $\exp(-y^2/2\sigma^2)$ Hermite polynomial, $\sum_{j=2}^{m} a_j H_{2j}(y_s)$
where $y_s = y/\sigma$

Hazard-rate, $1 - \exp(-(y/\sigma)^{-b})$ Cosine, $\sum_{j=2}^{m} a_j \cos\left(\dfrac{j\pi y}{w}\right)$

Hazard-rate, $1 - \exp(-(y/\sigma)^{-b})$ Simple polynomial, $\sum_{j=2}^{m} a_j \left(\dfrac{y}{w}\right)^{2j}$

The uniform + cosine expression is the Fourier series model of Crain *et al.* (1979) and Burnham *et al.* (1980). This is an excellent omnibus model and has been shown to perform well in a variety of situations. The uniform + simple polynomial model includes the models of Anderson and Pospahala (1970), Anderson *et al.* (1980), and Gates and Smith (1980).

It may be desirable to use the half-normal key function with either a cosine expansion or Hermite polynomials. Because histograms of distance data often decline markedly with distance from the line, the half-normal may often represent a good choice as a key function. Similarly, the uniform key and one cosine term will often provide a good standard for possible further fitting with series adjustment terms. Theoretical reasons suggest the use of the Hermite polynomial in conjunction with the half-normal key, especially for the untruncated case. This is a minor point, and the reader should think of this as only an alternative form of a polynomial.

The final two models listed above use Buckland's (1985) hazard-rate as a two parameter key function and use cosine and simple polynomial expansions for additional fitting, if required. The hazard-rate model is a derived model in contrast to the others, which are proposed shapes. That is, the shape of this family of models is the result of *a priori* assumptions about the detection process. The hazard-rate model has been shown to possess good properties, especially for data that are genuinely spiked (as distinct from spuriously spiked, as a result of rounding). In addition, this model can have a marked shoulder that can be nearly flat for some distance from the line or point. Even for data appearing to be spiked, this model can fit a flat shoulder, yet provide a good fit.

These series-expansion models are non-parametric in the sense that the number of parameters used is data-dependent. The estimation theory for these models, including rules to select the number of parameters to use, is given in the following chapter. Typically, given suitable distance truncation, an adequate model for $g(y)$ will include only one or two parameters, sometimes three. Sometimes the key function by itself will be adequate, with no terms in the series expansion. We emphasize that truncation will often be required as part of the modelling, especially if the data are ungrouped. Outlier observations provide relatively little information about density, but are often difficult to model, so that proper truncation should always be considered in modelling $g(y)$. Program DISTANCE allows the combination of any of the key functions with any of the series expansions as a model for $f(y)$. Some models have appeared in the literature that assume $g(y) = 1$ for some considerable distance from the line or point; the models suggested above do not impose this assumption.

Only these general models are emphasized as state-of-the-art, general approaches at this time. Program DISTANCE allows any key function to be used with any series expansion; however, the combinations listed above should be satisfactory for general use. Further effort directed at model evaluation and development might now be better directed at survey design and data collection techniques to meet critical assumptions.

The exponential + simple polynomial is available for the salvage analysis of poorly collected data where there is strong reason to believe that the distance data are truly spiked. It has the form:

$$\exp(- y/\lambda) \cdot [1 + \sum_{j=1}^{m} a_j \, (y/\lambda)^{2j}]$$

Use of this approach should be accompanied by adequate justification and we recommend its use only in unusual circumstances. Every consideration should be given to the use of the hazard-rate model for distance data that appear spiked because this model enforces the shape criterion, offers greater flexibility in fitting a spike, and gives a more realistic (larger) variance when the data are inadequate for reliable modelling.

2.5 Some analysis guidelines

Distance sampling represents a broad area and includes many types of application and degree of complexity of design and data. Thus, specific 'cookbook' procedures for data analysis cannot be given safely. Instead, we will suggest a useful strategy that could be considered when planning the analysis of a data set. In this section we will consider only a simple survey and will not address stratification and other complications given in later chapters.

2.5.1 Exploratory phase

The exploratory phase of the analysis involves the preparation of histograms of the distance data under several groupings. Sometimes it is effective to partition the data into 10–20 groups to get a fine-grained picture of the distance data. Examination of such histograms can provide insight into the presence of heaping, evasive movement, outliers and the occasional gross error. Prominent heaps can be mitigated by judicious grouping or splitting prior to further analysis. Evasive movement is problematic, but it is important to know that movement is present (movement toward the line or point generally cannot be detected from the distance data alone). Some truncation of the distance data is nearly always suggested, even if no obvious outlier is noticed. We frequently recommend that 5–10% of the largest observations be truncated. A more refined rule of thumb is to truncate the data when $g(x) \doteq 0.15$ for line transects or 0.10 for point transects. If the data were taken as grouped data in the field, then options for further truncation are more limited. Some liberal truncation is generally recommended. Empirical estimates of var(n) can be computed and compared with the variance under the Poisson assumption (i.e. $\widehat{\text{var}}(n) = n$). One can examine the stability of the ratio $\widehat{\text{var}}(n)/n$ over various design features. If the data are from a clustered population, plots of s or $\log_e(s)$ vs x or r should be made and examined. Of course, data entry errors and other anomalies should be screened and corrected. This analysis phase is open-ended but the analyst is encouraged to begin to understand the data and possible violations of the assumptions. Chatfield (1988, 1991) offered some general practical advice relevant here. Program DISTANCE allows substantial exploratory options.

2.5.2 Model selection

Model selection cannot proceed until proper truncation and, where relevant, grouping have been tentatively addressed. Thus, this phase begins once a data set has been properly prepared. Several robust models should be considered (e.g. those in Section 2.4). The following chapters will introduce and demonstrate the use of likelihood ratio tests, goodness of fit tests and Akaike's Information Criterion (AIC; Akaike 1973) as aids in objective model selection. Here it might be appropriate to remind the analyst that it is the fit of the model to the distance data near the line or point that is most important (unless there is thought to be heaping at zero distance). Usually analysis will suggest additional exploratory work, so that the process is iterative. For example, it may become apparent that the fit of one or more models could be improved by selecting a different truncation point w, or by grouping ungrouped data, or by changing the choice of group intervals for grouped data. Further, if data are available over several years, taken in the same habitat type by the same observer, then it might be prudent to pool the data for the estimation of $f(0)$ or $h(0)$, but to use the year-specific sample sizes n_i, where i is year, to estimate annual abundance. The validity of this approach must then be assessed, for example using a likelihood ratio test to determine whether a common value for $f(0)$ or $h(0)$ can be assumed.

2.5.3 Final analysis and inference

At some point the analyst selects a model believed to be the best for the data set under consideration. In some cases, there may be several competing models that seem equally good. In most cases, there will be a subset of models that can be excluded from final consideration because they perform poorly relative to other models. Often, if two or three models seem to fit equally well to a data set, estimation of density D and mean cluster size $E(s)$ under these models will be quite similar (see examples in Chapter 8).

Once a single model has been selected, the analyst can address further issues. Thus, one might consider bootstrapping to obtain improved estimates of precision, or carry out a Monte Carlo study to understand further the effect of some assumption failure (e.g. overestimation of a significant proportion of detection distances in an aerial survey, due to the aircraft flying too low at times). Finally, estimates of density or abundance and their precision are made, and qualifying statements presented, such as discussion of the effects of failures of assumptions.

The above guidelines give a broad indication of how the analyst might proceed. They will be developed in the following chapters, both to give substance to the theory required at each step, and to show how the philosophy for analysis is implemented in real examples.

3

Statistical theory

3.1 General formula

The analysis methods for distance sampling described here model measured or estimated distances from a line or point so that density of objects in a study area may be estimated. Conceptually, object density varies spatially, and lines or points are placed at random or systematically in the study area to allow mean density to be estimated.

Suppose in a given survey that objects do not occur in clusters and that distances are only recorded out to a distance w from the line or point, or equivalently that recorded distances are truncated at distance w. Suppose further that the true density is D objects per unit area. Let the area covered by the survey within distance w of the line or point be a, and let the probability of detection for an object within this area, unconditional on its actual position, be P_a. Then the expected number of objects detected within distance w, $E(n)$, is equal to the expected number of animals in the surveyed area, $D \cdot a$, multiplied by the probability of detection, P_a, so that

$$D = \frac{E(n)}{a \cdot P_a}$$

If objects occur in clusters, so that $E(n)$ is the expected number of clusters, then the above equation should be multiplied by $E(s)$, the expectation of cluster size for the population:

$$D = \frac{E(n) \cdot E(s)}{a \cdot P_a}$$

Although the result is then perfectly general, it is convenient to modify the definitions of a and P_a to show explicitly two components of the general formula that are implicit in the above form of the equation.

52

First, if a is defined to be $2wL$ for a line transect survey, where L is the total transect length, or $a = k\pi w^2$, where k is the number of circular plots in a point transect survey, then the area surveyed within a distance w of the line or point can be expressed as $c \cdot a$. Usually $c = 1.0$, but if for example only one side of a line transect is counted, then $c = 0.5$. Similarly, if only an angle of ϕ radians ($\phi < 2\pi$) is counted in a point transect survey, then $c = \phi/2\pi$. This factor is required for example in the cue counting method (Section 6.10), in which the sector counted is of angle ϕ.

Second, a basic assumption of the standard line and point transect methods is that the probability of detection at zero distance $g(0)$ is unity. In surveys of inconspicuous objects or, for example, of whales, this assumption may be unreasonable. It may be possible to estimate $g(0)$, in which case it is convenient to rescale the detection function $g(y)$ such that $g(0) = 1.0$, and to define the probability of detection on the line or at the point to be g_0. The unconditional probability of detection of an object (or cluster) in the surveyed area can then be factorized into $g_0 \cdot P_a$. This yields the general equation

$$D = \frac{E(n) \cdot E(s)}{c \cdot a \cdot P_a \cdot g_0} \tag{3.1}$$

Estimation of g_0 is generally problematic, so that if at all possible, surveys should be designed such that all or almost all objects on or close to the line or point are detected. Further discussion of this issue is reserved for Chapter 6.

Replacing parameters in Equation 3.1 by their estimators gives

$$\hat{D} = \frac{n \cdot \hat{E}(s)}{c \cdot a \cdot \hat{P}_a \cdot \hat{g}_0} \tag{3.2}$$

The variance of $\hat{D}$ may be approximated using the delta method (Seber 1982: 7–9). Assuming correlations between the four estimation components are zero, the variance estimate is then:

$$\widehat{\text{var}}(\hat{D}) = \hat{D}^2 \cdot \left\{ \frac{\widehat{\text{var}}(n)}{n^2} + \frac{\widehat{\text{var}}[\hat{E}(s)]}{[\hat{E}(s)]^2} + \frac{\widehat{\text{var}}(a \cdot \hat{P}_a)}{(a \cdot \hat{P}_a)^2} + \frac{\widehat{\text{var}}[\hat{g}_0]}{[\hat{g}_0]^2} \right\} \tag{3.3}$$

The assumption of no correlation is a mild one in the sense that estimation is usually done in a way that ensures it holds. Because P_a is estimated conditional on n, no correlation term exists between n and

$\hat{P}_a$ if we can assume that $E(\hat{P}_a|n) = E(\hat{P}_a)$, independent of n. This assumption holds if $(\hat{P}_a|n)$ is unbiased for P_a:

$$\text{cov}[n,(\hat{P}_a|n)] = E[n \cdot \hat{P}_a|n] - E(n) \cdot E(\hat{P}_a|n)$$

$$= E_n[E[\{n \cdot (\hat{P}_a|n)\}|n]] - E(n) \cdot P_a$$

$$= E_n[n \cdot E(\hat{P}_a|n)] - E(n) \cdot P_a$$

$$= E(n) \cdot P_a - E(n) \cdot P_a$$

$$= 0$$

When sample size is adequate, $(\hat{P}_a|n)$ is approximately unbiased.

Similarly, $E(s)$ may be estimated conditional on n and the detection distances, rendering $\hat{E}(s)$ uncorrelated with n or $\hat{P}_a$. Estimation of g_0, if required, is usually based on additional, independent data.

Although area $a \to \infty$ as $w \to \infty$, the product $a \cdot \hat{P}_a$ remains finite, so that all three equations hold when there is no truncation. To estimate $a \cdot P_a$, a form must be specified, explicitly or implicitly, for the detection function $g(y)$, which represents the probability of detection of an object or object cluster at a distance y from the line or point. The simplest form is that of the Kelker strip: the truncation point w is selected such that it is reasonable to assume that $g(y) = 1.0$ for $0 \leq y \leq w$. More generally it seems desirable that the detection function has a 'shoulder'; that is, $g'(0)$ should be zero, so that the detection function is flat at zero. This is the shape criterion defined by Burnham et al. (1980). The detection function should also be non-increasing, and have a tail that goes asymptotically to zero.

The relationship between the detection function and the probability density function of distances, $f(y)$, is different for line and for point transects. We use the notation y to represent either x, the perpendicular distance of an object from the centreline in line transect sampling, or r, the distance of an object from the observer in point transect sampling.

For line transect sampling, the relationship between $g(x)$ and $f(x)$ is particularly simple. Intuitively, because the area of a strip of incremental width dx at distance x from the line is independent of x, it seems reasonable that the density function should be identical in shape to the detection function, but rescaled so that it integrates to unity. This result may be proven as follows. Suppose for the moment that w is finite.

$$f(x)dx = \text{pr}\{\text{object is in } (x, x + dx) \mid \text{object detected}\}$$

$$= \frac{\text{pr}\{\text{object is in } (x, x + dx) \text{ and object is detected}\}}{\text{pr}\{\text{object is detected}\}}$$

$$= \frac{\text{pr}\{\text{object is detected}\,|\,\text{object is in } (x, x + dx)\} \cdot \text{pr}\{\text{object is in } (x, x + dx)\}}{P_a}$$

$$= \frac{g(x) \cdot \dfrac{dx \cdot L}{w \cdot L}}{P_a}$$

Thus $\quad f(x) = \dfrac{g(x)}{w \cdot P_a}$

It is convenient to define $\mu = w \cdot P_a$, so that

$$f(x) = \frac{g(x)}{\mu}$$

Because $\int_0^w f(x)\,dx = 1$, it follows that $\mu = \int_0^w g(x)\,dx$. Figure 3.1 illustrates the result that P_a, the probability of detecting an object given that it is within w of the centreline, is $\mu = \int_0^w g(x)\,dx$ (the area under the curve) divided by $1.0w$ (the area of the rectangle); that is, $w \cdot P_a = \mu$, which is well-defined even when w is infinite.

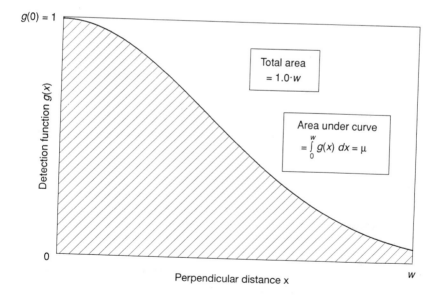

Fig. 3.1. The unconditional probability that an animal within distance w of the line is detected is the area under the detection function μ divided by the area of the rectangle $1.0\ w$.

55

The parameter μ is often termed the effective strip width (or more strictly, the effective strip half-width); if all objects were detected out to a distance μ on either side of the transect, and none beyond, then the expected number of objects detected would be the same as for the actual survey.

Let the total length of transect be L. Then the area surveyed is $a = 2wL$, and $a \cdot P_a = 2\mu L$. Since $\mu = g(x)/f(x)$ and $g(0) = 1.0$ after rescaling, if necessary, by the factor g_0, then $\mu = 1/f(0)$, and so for line transects, Equation 3.1 becomes:

$$D = \frac{E(n) \cdot f(0) \cdot E(s)}{2cLg_0} \qquad (3.4)$$

The parameter $f(0)$ is statistically well-defined and is estimable from the perpendicular distances $x_1, \ldots, x_n$ in a variety of ways.

The derivation for point transects is similar, but the relationship between $g(r)$ and $f(r)$ is less simple. The area of a ring of incremental width dr at distance r from the observer is proportional to r. Thus we might expect that $f(r)$ is proportional to $r \cdot g(r)$; using the constraint that $f(r)$ integrates to unity, $f(r) = 2\pi r g(r)/v$, where $v = 2\pi \int_0^w rg(r)dr$. A more rigorous proof follows.

$f(r)dr = \text{pr}\{\text{object is in the annulus } (r, r + dr) \mid \text{object detected}\}$

$= \dfrac{\text{pr}\{\text{object is in } (r, r + dr) \text{ and object is detected}\}}{\text{pr}\{\text{object is detected}\}}$

$= \dfrac{\text{pr}\{\text{object is detected} \mid \text{object is in } (r, r + dr)\} \cdot \text{pr}\{\text{object is in } (r, r + dr)\}}{P_a}$

$= \dfrac{g(r) \cdot \dfrac{2\pi r \, dr}{\pi w^2}}{P_a}$

so that $\quad f(r) = \dfrac{2\pi r \cdot g(r)}{\pi w^2 \cdot P_a}$

To be a valid density function, $\int_0^w f(r)dr = 1$, so that

$$f(r) = \frac{2\pi r \cdot g(r)}{v}, \text{ with } v = 2\pi \int_0^w rg(r)dr = \pi w^2 \cdot P_a$$

This result also holds for infinite w. Analogous to μ, v is sometimes called the effective area of detection.

Let there be k points, so that the surveyed area is $a = k\pi w^2$. The probability of detection of an object, given that it is within a distance w of the observer, is now $P_a = v/(\pi w^2)$, so that $a \cdot P_a$ becomes kv; again this holds as $w \to \infty$. Since $v = 2\pi r g(r)/f(r)$ and $g(0) = 1.0$ (after rescaling if necessary), then for point transects Equation 3.1 becomes:

$$D = \frac{E(n) \cdot h(0) \cdot E(s)}{2\pi c k g_0} \qquad (3.5)$$

where $h(0) = \lim_{r \to 0} f(r)/r = 2\pi/v$

Note that $h(0)$ is merely the derivative of the probability density $f(r)$ evaluated at $r = 0$; alternative notation would be $f'(0)$. It is thus estimable from the detection distances $r_1, \ldots, r_n$. Whereas $f(x)$ and $g(x)$ have identical shapes in line transect sampling, for point transects, $g(r)$ is proportional to $f(r)/r$. The constant of proportionality is $1/h(0)$.

Results equivalent to Equations 3.2 and 3.3 follow in the obvious way. Note that for both line and point transects, behaviour of the probability density function at zero distance is critical to object density estimation.

Burnham *et al.* (1980: 195) recommended that distances in point transect sampling should be transformed to 'areas' before analysis. Thus, the ith recorded area would be $u_i = \pi r_i^2$, $i = 1, \ldots, n$. If $f_u(u)$ denotes the probability density function of areas u_i, it may be shown that $f_u(u) = f(r)/(2\pi r) = g(r)/v$. The advantage of this transformation should now be apparent; the new density is identical in form to that for perpendicular distances in line transect sampling (where $f(x) = g(x)/\mu$), so line transect software may be used to analyse the data. Further, if r is allowed to tend to zero, then $f_u(0) = h(0)/(2\pi)$, and the development based on areas is therefore equivalent to that based on distances. This seems to suggest that modelling of areas rather than distances is preferable. However, as noted by Buckland (1987a), the transformation to area appreciably alters the shape of the detection function, and it is no longer clear that a model for area should satisfy the shape criterion. For example the half-normal model for distances, which satisfies the shape criterion, transforms to the negative exponential model for areas, which does not satisfy the shape criterion, whereas the hazard-rate model of Hayes and Buckland (1983) retains both its parametric form and a shoulder under the transformation, although the shoulder becomes narrower. We now recommend modelling the untransformed distance data, because line transect detection functions may then be more safely carried across to point transects, thus allowing the focus of analysis to be the detection function in both cases.

3.2 Hazard-rate modelling of the detection process

There are many possible models for the detection function that may fit any given data set well. If all give similar estimates, then model selection may not be critical. However, when the observed data exhibit little or no 'shoulder', it is not uncommon that one model yields an estimated density around double that for another model. Although the development of robust, flexible models allows workers to obtain good fits to most of their data sets, it does not guarantee that the resulting density estimators have low bias. There is some value therefore in attempting to model the detection process, to provide both some insight to the likely form of the detection function and a parametric model that might be expected to fit real data well. Hazard-rate methods have proved particularly useful for this purpose, and have been developed by Schweder (1977), Butterworth (1982a), Hayes and Buckland (1983) and Buckland (1985) for line transect sampling, and by Ramsey et al. (1979) and Buckland (1987a) for point transect sampling. We consider only continuous hazard-rate models at this stage; discrete hazard-rate models are described in Chapter 6.

3.2.1 Line transect sampling

At any one point in time, there is a 'hazard' that an object will be detected by the observer, which is a function of the distance r separating the object and observer. If the object is on the line, the observer will be moving directly towards it, so that r decreases quite quickly. The farther the object is from the line, the slower the rate of decrease in distance r, so that the observer has more time to detect the object at larger distances. Hazard-rate analysis models this effect, and also allows the hazard to depend on the angle of the object from the observer's direction of travel.

Suppose an object is at perpendicular distance x from the transect line, and let the length of transect line between the observer and the point of closest approach to the object be z, so that r, the distance between the observer and the object, satisfies $r^2 = x^2 + z^2$ (Fig. 1.5). Suppose also that the observer approaches from a remote point on the transect so that z may be considered to decrease from ∞ to 0, and assume for simplicity that the object cannot be detected once the observer has passed his/her point of closest approach. Let

$$h(z, x)dz = \text{pr}\{\text{object sighted while observer is in } (z, z - dz) | \text{not}$$
$$\text{sighted while observer is between } \infty \text{ and } z\}$$

and

$p(z, x)$ = pr{object not sighted while observer is between ∞ and z}

where both probabilities are conditional on the perpendicular distance x. Solving the forwards equations

$$p(z - dz, x) = p(z, x)\{1 - h(z, x)dz\}$$

for $p(z, x)$, and setting $p(\infty, x) = 1$, yields

$$p(z, x) = \exp\left\{-\int_z^\infty h(v, x)dv\right\}$$

so that

$$g(x) = 1 - p(0, x)$$

$$= 1 - \exp\left\{-\int_0^\infty h(z, x)dz\right\}, \quad 0 \leq x < \infty$$

Changing the variable of integration from z to r gives

$$g(x) = 1 - \exp\left\{-\int_x^\infty \frac{r}{\sqrt{(r^2 - x^2)}} k(r, x)dr\right\}$$

where $k(r, x) \equiv h\{\sqrt{(r^2 - x^2)}, x\}$

Time could be incorporated in the model, but for a continuous hazard-rate process, there is little value in doing so provided that the speed of the observer is not highly variable. Otherwise the development has been general up to this point. To progress further, it is necessary to restrict the form of the hazard. A plausible hazard should satisfy the following conditions:

1. $k(0, 0) = \infty$;
2. $k(\infty, x) = 0$;
3. $k(r, x)$ is non-increasing in r for any fixed x.

For example suppose that the hazard belongs to the family defined by:

$$\int_x^\infty \frac{r}{\sqrt{(r^2 - x^2)}} k(r, x)dr = (x/\sigma)^{-b} \quad \text{for some } \sigma \text{ and } b \quad (3.6)$$

Hayes and Buckland (1983) give two hazards from this family. In the first, the hazard of detection is a function of r alone:

$$k(r, x) = cr^{-d}, \; r \geqslant x$$

so that $b = d - 1$ and $\sigma = \left\{ \dfrac{c\Gamma[(d - 1)/2]\Gamma(0.5)}{2\Gamma(d/2)} \right\}^{1/(d-1)}$

The second hazard function allows the hazard of detection to be greater for objects directly ahead of the observer than for objects to the side. In practice this may arise if an object at distance r is more likely to flush when the observer moves towards it, or if the observer concentrates search effort in the forward direction. The functional form of the second hazard is:

$$k(r, x) = c \cdot r^{-d}\cos \theta, \qquad \text{where } \sin \theta = x/r$$

so that $b = d - 1$ and $\sigma = \left\{ \dfrac{c}{d - 1} \right\}^{1/(d-1)}$

The family of hazards defined by Equation 3.6, to which the above two belong, yields the detection function

$$g(x) = 1 - \exp[- (x/\sigma)^{-b}] \tag{3.7}$$

This is the hazard-rate model derived by Hayes and Buckland (1983) and investigated by Buckland (1985), although the above parameterization is slightly different from theirs, and has better convergence properties. The parameter b is a shape parameter, whereas σ is a scale parameter. The model should provide a good representation of the 'true' detection function when the hazard process is continuous, sighting (or auditory) conditions are homogeneous, and visibility (or sound) falls off with distance according to a power function, although it appears to be robust when these conditions are violated. It may be shown that $g'(0) = 0$ for $b > 0$, which covers all parameter values for which the detection function is a decreasing function. Hence the above two hazards which are sharply 'spiked' (the derivative of the hazard with respect to r, evaluated at $r = 0$, is infinite) give rise to a detection function that always satisfies the shape criterion of Burnham et al. (1980). For untruncated data the detection function integrates to a finite value only if $b > 1$. For truncated data, the model has a long tail and a narrow shoulder if $b < 1$, and convergence problems may be encountered for extreme data sets. These problems may be avoided and analyses are more robust when the constraint $b \geqslant 1$ is imposed (Buckland 1987b). Equation 3.7 is plotted for a range of values for the shape parameter likely to be encountered in real data sets in Fig. 2.5b.

Although in this book we describe the model of Equation 3.7 as 'the' hazard-rate model, any detection function may be described as a hazard-rate model in the sense that a (possibly implausible) hazard exists from which the detection function could be derived. Equation 3.7 is sometimes referred to as the complementary log-log model, a label which is both more accurate and more cumbersome.

3.2.2 Point transect sampling

For point transect sampling, the hazard-rate formulation is simpler, since there is only one distance, the sighting distance r, to model. The probability of detection is no longer a function of distance moved along the transect, but of time spent at the point. Define the hazard function $k(r, t)$ to be such that

$$k(r, t)dt = \text{pr}\{\text{an object at distance } r \text{ is detected during } (t + dt) | \text{it is not detected during } (0, t)\}$$

Then the detection function becomes:

$$g(r) = 1 - \exp\left[-\int_0^T k(r, t)dt\right]$$

where T is the recording time at each point. If the observer is assumed to search with constant effort during the recording period, then $k(r, t) = k(r)$, independent of t, so that

$$g(r) = 1 - \exp[- k(r)T] \tag{3.8}$$

If the hazard is assumed to be of the form $k(r) = cr^{-d}$, then

$$g(r) = 1 - \exp[- (r/\sigma)^{-b}]$$

where $b = d$ and $\sigma = (cT)^{1/b}$. The effect of increasing the time spent at each point is therefore to increase the scale parameter. This widens the shoulder on the detection function, making it easier to fit. Scott and Ramsey (1981) plotted the changing shape of a detection function as time spent at the point increases from four to 32 minutes. The disadvantages of choosing T large are that assumptions are more likely to be violated (Section 5.9), and after a few minutes, the number of new detections per minute will become small.

The parametric form of the above detection function is identical to that derived for line transects (Equation 3.7). Moreover, if sightings are squared prior to analysis, the parametric form remains unaltered, so that maximum likelihood estimation is invariant to the transformation to squared distances. This property is not shared by other widely used models for the detection function. Burnham *et al.* (1980: 195) suggested squaring distances, to allow standard line transect software to be used for analysing point transect data, but we now advise against this strategy (Buckland 1987a).

3.3 The key function formulation for distance data

Most formulations proposed for the probability density of distance data from line or point transects may be categorized into one of two groups. If there are theoretical reasons for supposing that the density has a given parametric form, then parametric modelling may be carried out. Otherwise, robust or non-parametric procedures such as Fourier series, splines, kernel methods or polynomials might be preferred. In practice it may be reasonable to assume that the true density function is close to a known parametric form, yet systematic departures can occur in some data sets. In this instance, a parametric procedure may not always give an adequate fit, yet a non-parametric method may be too flexible, perhaps giving very different fits to two related data sets from a single study, due to small random fluctuations in the data. An example of the latter occurs when a one term Fourier series model is selected for one data set and a two term model for a second. The second data set might be slightly larger, or show a slightly smaller shoulder; both increase the likelihood of rejecting the one term fit. Bias in estimation of $f(0)$ can be a strong function of the number of Fourier series terms selected (Buckland 1985), so that comparisons across data sets may be misleading. The technique described by Buckland (1992a, b) and summarized below incorporates knowledge of the likely shape of the density function, whether theoretical or from past experience, and allows polynomial or Fourier series adjustments to be made, to ensure a good fit to the data.

Simple polynomials have been used for fitting line transect data by Anderson and Pospahala (1970). However, low order simple polynomials may have unsuitable shapes. By taking the best available parametric form for the density, $\alpha(y)$, and multiplying it by a simple polynomial, this shortcoming is removed. We call $\alpha(y)$ the **key function**. If it is a good fit, it needs no adjustment; the worse the fit, the greater the adjustments required.

When the key function is the untruncated normal (or half-normal) density, Hermite polynomials (Stuart and Ord 1987: 220–7) are orthogonal with respect to the key, and may therefore be preferred to simple polynomials. Hermite polynomials are usually fitted by the method of moments, leading to unstable behaviour when the number of observations is not large or when high order terms are included. Buckland (1985) overcame these difficulties by using numerical techniques to obtain maximum likelihood estimates of the polynomial coefficients. These procedures have the further advantage that the Gram-Charlier type A and the Edgeworth formulations yield identical curves; for the method of moments, they do not (Stuart and Ord 1987: 222–5).

Let the density function be expressed as

$$f(y) \doteq \frac{\alpha(y)}{\beta} \cdot \left[1 + \sum_{j=1}^{m} a_j \cdot p_j(y_s) \right]$$

where $\alpha(y)$ is a parametric key, containing k parameters (usually 0, 1 or 2);

$$p_j(y_s) = \begin{cases} y_s^j, & \text{if a simple polynomial is desired, or} \\ H_j(y_s), & \text{the } j\text{th Hermite polynomial, } j = 1, \ldots, m, \text{ or} \\ \cos(j\pi y_s), & \text{if a Fourier series is desired;} \end{cases}$$

y_s is a standardized y value (see below);

$$a_j \begin{cases} = 0 \text{ if term } j \text{ of } p_j(y_s) \text{ is not used in the model, or} \\ \text{is estimated by maximum likelihood;} \end{cases}$$

β is a normalizing function of the parameters (key parameters and series coefficients) alone.

It is necessary to scale the observed distances. For the simple polynomial formulation, estimation is invariant to choice of scale, but the operation is still necessary to avoid numeric problems when fitting the model. If the key function is parameterized such that a single key parameter, σ say, is a scale parameter, y_s may be found as y/σ for each observation. If the parameters of the key function are fully integrated into the estimation routine, σ can be estimated by maximum likelihood (see below). Otherwise the key function may be fitted by maximum likelihood in the absence of polynomial adjustments, and subsequent fitting of polynomial terms can be carried out conditional on those key parameter estimates. For the Fourier series formulation, analyses are conditional on w, the truncation point, and $y_s = y/w$. In practice, it is simpler to use this standardization for all models, a strategy used in DISTANCE.

For line transect sampling, the standard form of the Fourier series model is obtained by setting the key function equal to the uniform distribution, so that $\alpha(y) = 1/w$. Used in conjunction with simple polynomials, this key gives the method of Anderson and Pospahala (1970). The standard form of the Hermite polynomial model arises when the key function is the half-normal. Point transect keys are found by multiplying their line transect counterparts by y (or, equivalently, $2\pi y$). The key need not be a valid density function. For the half-normal line transect key, define $\alpha(y) = \exp[-(y/\sigma)^2/2]$, and absorb the denominator of the half-normal density, $\sqrt{(\pi\sigma^2/2)}$, into β. In general, absorb any part of the key that is a function of the parameters alone into β.

For line transect sampling, the detection function is generally assumed to be symmetric about the line. Similarly for point transect sampling, detection probability is assumed to be independent of angle. The detection function may be envisaged as a continuous function on $(-w, +w)$; for line transects, negative distances would correspond say to sightings to the left of the line and positive to the right, and for point transects, this function can be thought of as a section through the detection 'dome', passing through the centre or point. The function is assumed to be symmetric about zero (although analyses are robust to this assumption). Hence only cosine terms are used for the Fourier series model, and only polynomials of even order for polynomial models, so that the detection function is an even function. In the case of the Hermite polynomial model, the parameter of the half-normal key corresponds to the second moment term, so that the first polynomial to be tested is of order four if terms are tested for inclusion in a sequential manner. The first adjustment to the half-normal fit therefore adjusts for kurtosis. It may be that kurtosis for the true detection function is close to that for the normal distribution, but a higher order moment may be very different. In this case it may be more profitable to test for inclusion of terms in a stepwise manner: select all terms of even order up to an arbitrary order, say 10, and include at the first step that term which gives the greatest increase in the value of the likelihood. Next include the term that gives the greatest improvement when fitted simultaneously with the first term selected. Continue until a likelihood ratio test indicates that no significant improvement in the fit has been achieved.

When the key function is not normal and testing is sequential, it is less clear which polynomial term should be tested first. Any key will contain a parameter, or a function of parameters, that corresponds to scale, so a possible rule is to start with the term of order four, whatever the key. An alternative rule is to start with the term of order $2 \cdot (k + 1)$, where k is the number of key parameters. We advise against the use of keys with more than two parameters. For stepwise testing, all even terms down to order two and up to an arbitrary limit can be included.

3.4 Maximum likelihood methods

We concentrate here on the likelihood function for the detection distances, $y_i, i = 1, \ldots, n$, conditional on n. If the full data set was to be modelled in a comprehensive way, then the probability of realizing the data $\{n, y_1, \ldots, y_n, s_1, \ldots, s_n\}$ might be expressed as

$$\Pr(n, y_1, \ldots, y_n, s_1, \ldots, s_n) = \Pr(n) \cdot \Pr(y_1, \ldots, y_n, s_1, \ldots, s_n | n)$$

$$= \Pr(n) \cdot \Pr(y_1, \ldots, y_n | n) \cdot \Pr(s_1, \ldots, s_n | n, y_1, \ldots, y_n)$$

Thus, inference on the distances y_i can be made conditional on n, and inference on the cluster sizes s_i can be conditional on n and the y_i. This provides the justification for treating estimation of D (with $g_0 = 1$) as a series of three univariate problems.

Rao (1973) and Burnham *et al.* (1980) present maximum likelihood estimation methods for both grouped and ungrouped distance data. Applying those techniques to the key formulation of Section 3.3 yields the following useful results.

3.4.1 Ungrouped data

Define $\mathcal{L}(\underline{\theta}) = \prod\limits_{i=1}^{n} f(y_i)$

where y_i is the ith recorded distance, $i = 1, \ldots, n$

$\theta_1, \ldots, \theta_k$ are the parameters of the key function

$\theta_{k+j} = a_j, j = 1, \ldots, m$, are the parameters (coefficients) of the adjustment terms.

Then $\log_e[\mathcal{L}(\underline{\theta})] = l = \sum\limits_{i=1}^{n} \log_e[f(y_i)] = \sum\limits_{i=1}^{n} \log_e[f(y_i) \cdot \beta] - n \cdot \log_e \beta$

Hence
$$\frac{\partial l}{\partial \theta_j} = \sum_{i=1}^{n} \frac{\partial \log_e[f(y_i)]}{\partial \theta_j}$$

$$= \sum_{i=1}^{n} \left\{ \frac{1}{f(y_i) \cdot \beta} \cdot \frac{\partial [f(y_i) \cdot \beta]}{\partial \theta_j} \right\} - \frac{n}{\beta} \cdot \frac{\partial \beta}{\partial \theta_j}, \, j = 1, \ldots, k + m$$

where

$$\frac{\partial [f(y_i) \cdot \beta]}{\partial \theta_j} = \begin{cases} \alpha(y_i) \cdot \left[\sum_{j'=1}^{m} a_{j'} \cdot \frac{\partial p_{j'}(y_{is})}{\partial y_{is}} \right] \cdot \frac{\partial y_{is}}{\partial \theta_j} + \left[1 + \sum_{j'=1}^{m} a_{j'} \cdot p_{j'}(y_{is}) \right] \cdot \frac{\partial \alpha(y_i)}{\partial \theta_j}, \\ \qquad\qquad\qquad\qquad\qquad\qquad\qquad\qquad\qquad\qquad\qquad 1 \leqslant j \leqslant k \quad (3.9) \\ \alpha(y_i) \cdot p_{j-k}(y_{is}), \text{ for all } j > k \text{ for which } a_{j-k} \text{ is non-zero} \end{cases}$$

$$\frac{\partial p_{j'}(y_{is})}{\partial y_{is}} = \begin{cases} j \cdot p_{j-1}(y_{is}), \text{ with } p_0(y_{is}) = 1, \\ \quad \text{for simple and Hermite polynomials} \\ -j\pi \cdot \sin(j\pi y_s), \text{ for the Fourier series model} \end{cases}$$

When $k = 1$ and $y_{is} = y_i/\theta_1$, $\frac{\partial y_{is}}{\partial \theta_1} = -y_i/\theta_1^2$; when $k = 1$ and $y_{is} = y_i/w$,

$$\frac{\partial y_{is}}{\partial \theta_1} = 0$$

The equations $\partial l/\partial \theta_j = 0, j = 1, \ldots, k + m$, may be solved using for example Newton–Raphson or a simplex procedure. To change between simple and Hermite polynomials, it is merely necessary to redefine $p_j(y_s)$, $j = 1, \ldots, m$; to change between polynomial and Fourier series adjustments, the derivative of $p_j(y_s)$ with respect to y_s must also be redefined. If a different key $\alpha(y)$ is required, the only additional algebra needed to implement the method is to find $\partial \alpha(y)/\partial \theta_j$ and $\partial y_s/\partial \theta_j$, $1 \leqslant j \leqslant k$; β and $\partial \beta/\partial \theta_j$ are evaluated by numerical integration.

The Fisher information matrix per observation may be estimated by the Hessian matrix $H(\hat{\theta})$, with jhth element

$$H_{jh}(\hat{\theta}) = \frac{1}{n} \cdot \left[\sum_{i=1}^{n} \frac{\partial \log_e[f(y_i)]}{\partial \hat{\theta}_j} \cdot \frac{\partial \log_e[f(y_i)]}{\partial \hat{\theta}_h} \right]$$

This may be formed from quantities already calculated. If a function of the parameters, $g(\underline{\theta})$, is to be estimated by $g(\hat{\underline{\theta}})$, then

$$\widehat{\mathrm{var}}\{g(\hat{\underline{\theta}})\} = \frac{1}{n} \cdot \left[\frac{\partial g(\hat{\underline{\theta}})}{\partial \underline{\hat{\theta}}}\right]' [H(\hat{\underline{\theta}})]^{-1} \left[\frac{\partial g(\hat{\underline{\theta}})}{\partial \underline{\hat{\theta}}}\right]$$

3.4.2 Grouped data

Suppose the observations y are grouped, the ith group spanning the interval (c_{i1}, c_{i2}), $i = 1, \ldots, u$. In general, the data may be truncated at either or both ends. For line and point transects, it is usual that $c_{i1} = 0$ (no left truncation) and $c_{i2} = c_{i+1,1}$, $i = 1, \ldots, u - 1$. The likelihood function is now multinomial. Let the group frequencies be $n_1, \ldots, n_u$, with cell probabilities

$$\pi_i = \int_{c_{i1}}^{c_{i2}} f(y) \, dy$$

Then

$$\mathcal{L}(\underline{\theta}) = \frac{n!}{n_1! \ldots n_u!} \prod_{i=1}^{u} \pi_i^{n_i}, \text{ with } n = \sum_{i=1}^{u} n_i$$

$$\log_e[\mathcal{L}(\underline{\theta})] = l = \sum_{i=1}^{u} n_i \cdot \log_e(\pi_i) + a \text{ constant}$$

$$\frac{\partial l}{\partial \theta_j} = \sum_{i=1}^{u} \frac{n_i}{\pi_i} \cdot \frac{\partial \pi_i}{\partial \theta_j}, \quad j = 1, \ldots, k + m$$

Define $P_i = \pi_i \cdot \beta$, so that

$$\frac{\partial \pi_i}{\partial \theta_j} = \frac{1}{\beta} \cdot \left[\frac{\partial P_i}{\partial \theta_j} - \frac{\partial \beta}{\partial \theta_j} \cdot \pi_i\right]$$

Then if P_i and $\partial P_i/\partial \theta_j$, $j = 1, \ldots, k + m$, $i = 1, \ldots, u$, can be found,

$$\beta = \sum_{i=1}^{u} P_i \qquad \text{and} \qquad \frac{\partial \beta}{\partial \theta_j} = \sum_{i=1}^{u} \frac{\partial P_i}{\partial \theta_j}$$

Given parameter estimates, the P_i may be evaluated by numerically integrating the numerator, $f(y) \cdot \beta$, of the density function:

$$P_i = \pi_i \cdot \beta = \int_{c_{i1}}^{c_{i2}} f(y) \cdot \beta \, dy$$

Similarly,

$$\frac{\partial P_i}{\partial \theta_j} = \int_{c_{i1}}^{c_{i2}} \frac{\partial \{ f(y) \cdot \beta \}}{\partial \theta_j} dy$$

and may be found using numerical integration on

$$\frac{\partial [f(y) \cdot \beta]}{\partial \theta_j} = \begin{cases} \alpha(y) \cdot \left[\sum\limits_{j'=1}^{m} a_{j'} \cdot \dfrac{\partial p_{j'}(y_s)}{\partial y_s} \right] \cdot \dfrac{\partial y_s}{\partial \theta_j} \\[2ex] + \left[1 + \sum\limits_{j'=1}^{m} a_{j'} \cdot p_{j'}(y_s) \right] \cdot \dfrac{\partial \alpha(y)}{\partial \theta_j}, & 1 \leq j \leq k \\[2ex] \alpha(y) \cdot p_{j-k}(y_s), & \text{for all } j > k \text{ for which } a_{j-k} \text{ is non-zero} \end{cases} \tag{3.10}$$

The implications of changing between simple and Hermite polynomials or between polynomial and Fourier series adjustments, and of changing the key function, are identical to the case of ungrouped data.

Again, a robust iterative procedure is required to maximize the likelihood. Variances follow as for ungrouped data, except that the information matrix per observation, $I(\theta)$, now has jhth element

$$I_{jh}(\underline{\theta}) = \sum_{i=1}^{u} \frac{1}{\pi_i} \cdot \frac{\partial \pi_i}{\partial \theta_j} \cdot \frac{\partial \pi_i}{\partial \theta_h}, \quad j, h = 1, \ldots, k + m$$

All of these quantities are now available, and so a function of the parameters $g(\theta)$ is estimated by $g(\hat{\theta})$ with variance

$$\widehat{\text{var}} \{ g(\hat{\underline{\theta}}) \} = \frac{1}{n} \cdot \left[\frac{\partial g(\hat{\underline{\theta}})}{\partial \hat{\underline{\theta}}} \right]' [I(\hat{\underline{\theta}})]^{-1} \left[\frac{\partial g(\hat{\underline{\theta}})}{\partial \hat{\underline{\theta}}} \right]$$

If data are analysed both grouped and ungrouped, and the respective maxima of the likelihood functions are compared, the constant combinatorial term in the likelihood for grouped data should be omitted. As the number of groups tends to infinity and interval width tends to zero, the likelihood for grouped data tends to that for ungrouped data, provided the constant is ignored.

3.4.3 Special cases

Suppose no polynomial or Fourier series adjustments are required. The method then reduces to a straightforward fit of a parametric density. The above results hold, except the range of j is now from 1 to k, and for ungrouped data Equation 3.9 reduces to

$$\frac{\partial[f(y_i) \cdot \beta]}{\partial\theta_j} = \frac{\partial\alpha(y_i)}{\partial\theta_j}, j = 1, \ldots, k$$

For grouped data, Equation 3.10 reduces to the above, with the suffix i deleted from y.

For the Hermite polynomial model, it is sometimes convenient to fit the half-normal model as described in the previous paragraph, and then to condition on that fit when making polynomial adjustments. For the standard Fourier series model, the key is a uniform distribution on $(0, w)$, where w is the truncation point, specified before analysis. In each of these cases, the adjustment terms are estimated conditional on the parameters of the key. Thus Equation 3.9 reduces to

$$\frac{\partial[f(y_i) \cdot \beta]}{\partial\theta_j} = \alpha(y_i) \cdot p_{j-k}(y_{is}), \text{ for non-zero } a_{j-k} \text{ and } k < j \leqslant k + m$$

Equation 3.10 reduces similarly, but with suffix i deleted; otherwise results follow through exactly as before, but with j restricted to the range $k + 1$ to $k + m$. This procedure is necessary whenever the uniform key is selected. For keys that have at least one parameter estimated from the data, the conditional maximization is useful only if simultaneous maximization across all parameters fails to converge.

A third option that is sometimes useful is to refit the key, conditional on polynomial or Fourier series adjustments. Equation 3.9 then becomes

$$\frac{\partial[f(y_i) \cdot \beta]}{\partial\theta_j} = \alpha(y_i) \cdot \left[\sum_{j'=1}^{m} a_{j'} \cdot \frac{\partial p_{j'}(y_{is})}{\partial y_{is}} \right] \cdot \frac{\partial y_{is}}{\partial\theta_j}$$

$$+ \left[1 + \sum_{j'=1}^{m} a_{j'} \cdot p_{j'}(y_{is}) \right] \cdot \frac{\partial\alpha(y_i)}{\partial\theta_j}, \ 1 \leqslant j \leqslant k$$

and similarly for Equation 3.10, but minus the suffix i throughout. The range of j is from 1 to k; otherwise results follow exactly as before.

3.4.4 The half-normal detection function

If the detection function is assumed to be half-normal, and the data are both ungrouped and untruncated, the above approach leads to closed form estimators and a particularly simple analysis for both line and point transect sampling. Suppose the detection function is given by $g(y) = \exp(-y^2/2\sigma^2)$, $0 \leq y < \infty$. We consider the derivation for line transects $(y = x)$ and point transects $(y = r)$ separately.

(a) *Line transects* With no truncation, the density function of detection distances is $f(x) = g(x)/\mu$, where

$$\mu = \int_0^\infty g(x)dx = \int_0^\infty \exp(-x^2/2\sigma^2)dx = \sqrt{\frac{\pi\sigma^2}{2}}$$

Given n detections, the likelihood function is

$$\mathcal{L} = \prod_{i=1}^n \{g(x_i)/\mu\} = \left\{ \prod_{i=1}^n \exp(-x_i^2/2\sigma^2) \right\}/\mu^n$$

so that $l = \log_e(\mathcal{L}) = -\sum_{i=1}^n \{x_i^2/2\sigma^2\} - n \cdot \log_e\{\sqrt{(\pi\sigma^2/2)}\}$

Differentiating l with respect to σ^2 (i.e. $k = 1$, $\theta_1 = \sigma^2$ and $m = 0$ in terms of the general notation) and setting the result equal to zero gives:

$$\frac{dl}{d\sigma^2} = \sum_{i=1}^n x_i^2/2\sigma^4 - n/2\sigma^2 = 0$$

so that $\hat{\sigma}^2 = \sum_{i=1}^n x_i^2/n$

Then $\hat{f}(0) = 1/\hat{\mu} = \sqrt{\{2/(\pi\hat{\sigma}^2)\}}$

By evaluating the Fisher information matrix, we get

$$\operatorname{var}(\hat{\sigma}^2) = \frac{2\sigma^4}{n}$$

from which

$$\text{var}\{\hat{f}(0)\} = \frac{1}{n\pi\sigma^2} = \frac{\{f(0)\}^2}{2n}$$

Equation 3.4, with each of $E(s)$, c and g_0 set equal to one, yields

$$D = \frac{E(n) \cdot f(0)}{2L}$$

from which

$$\hat{D} = \frac{n \cdot \hat{f}(0)}{2L} = \left[2\pi L^2 \sum_{i=1}^{n} x_i^2/n^3 \right]^{-0.5}$$

The methods of Section 3.7 yield an estimated variance for $\hat{D}$.

Quinn (1977) investigated the half-normal model, and derived an unbiased estimator for $f(0)$.

(b) *Point transects* The density function of detection distances is given by $f(r) = 2\pi r \cdot g(r)/v$. For the half-normal detection function,

$$v = 2\pi \int_0^w r \cdot g(r)dr = 2\pi \int_0^w r \cdot \exp(-r^2/2\sigma^2)dr$$

$$= [-2\pi\sigma^2 \cdot \exp(-r^2/2\sigma^2)]_0^w = 2\pi\sigma^2\{1 - \exp(-w^2/2\sigma^2)\}$$

Because there is no truncation, $w = \infty$, so that $v = 2\pi\sigma^2$. Note that if we substitute $w = \sigma$ into this equation, then the expected proportion of sightings within σ of the point transect is $2\pi\sigma^2\{1 - \exp(-0.5)\}/v = 39\%$. This compares with 68% for line transects; thus for the half-normal model, nearly 70% of detections occur within one standard deviation of the observer for line transects, whereas less than 40% occur within this distance for point transects. This highlights the fact that expected detection distance is greater for point transects than for line transects, a difference which is even more marked if the detection function is long-tailed.

If n detections are made, the likelihood function is given by

$$\mathcal{L} = \prod_{i=1}^{n} \{2\pi r_i \cdot g(r_i)/v\} = \left\{ \prod_{i=1}^{n} r_i \cdot \exp(-r_i^2/2\sigma^2) \right\}/(\sigma^{2n})$$

so that $l = \log_e(\mathcal{L}) = \sum_{i=1}^{n} \{\log_e(r_i) - r_i^2/2\sigma^2\} - n \cdot \log_e(\sigma^2)$

This is maximized by differentiating with respect to σ^2 and setting equal to zero:

$$\frac{dl}{d\sigma^2} = \sum_{i=1}^{n} r_i^2/2\sigma^4 - n/\sigma^2 = 0$$

so that $\hat{\sigma}^2 = \sum_{i=1}^{n} r_i^2/2n$

It follows that $\hat{h}(0) = 2\pi/\hat{v} = 1/\hat{\sigma}^2$. Equation 3.5, with each of $E(s)$, c and g_0 set equal to one, gives

$$D = \frac{E(n) \cdot h(0)}{2\pi k}$$

so that

$$\hat{D} = \frac{n \cdot \hat{h}(0)}{2\pi k} = \frac{n^2}{\pi k \sum_{i=1}^{n} r_i^2}$$

The maximum likelihood method yields $\mathrm{var}[\hat{h}(0)]$. The half-normal detection function has just one parameter (σ^2), so that the information matrix is a scalar. It yields

$$\mathrm{var}(\hat{\sigma}^2) = \frac{2\sigma^4}{n}$$

and

$$\mathrm{var}\{\hat{h}(0)\} = \frac{2}{n\sigma^4} = \frac{2\{h(0)\}^2}{n}$$

Estimation of $\mathrm{var}(n)$ and $\mathrm{var}(\hat{D})$ is covered in Section 3.7.

3.4.5 Constrained maximum likelihood estimation

The maximization routine used by DISTANCE allows constraints to be placed on the fitted detection function. In all analyses, the constraint $\hat{g}(y) \geq 0$ is imposed. In addition, $\hat{g}(y)$ is evaluated at ten y values, y_1 to y_{10}, and the non-linear constraint $\hat{g}(y_i) \geq \hat{g}(y_{i+1})$, $i = 1, \ldots, 9$, is enforced. The user may override this constraint, or replace it by the weaker constraint that $\hat{g}(0) \geq \hat{g}(y_i)$, $i = 1, \ldots, 10$. If the same data set is analysed by DISTANCE and by TRANSECT (Laake *et al.* 1979), different estimates may be obtained; TRANSECT does not impose constraints, and in addition does not fit the Fourier series model by maximum likelihood.

DISTANCE warns the user when a constraint has caused estimates to be modified. In these instances, the analytic variance of $\hat{f}(0)$ or $\hat{h}(0)$ may be unreliable, and we recommend that the bootstrap option for variance estimation is selected.

3.5 Choice of model

The key + adjustment formulation for line and point transect models outlined above has been implemented in DISTANCE, so that a large number of models are available to the user. Although this gives great flexibility, it also creates a problem of how to choose an appropriate model. We consider here criteria that models for the detection function should satisfy, and methods that allow selection between contending models.

3.5.1 Criteria for robust estimation

Burnham *et al.* (1979, 1980: 44) identified four criteria that relate to properties of the assumed model for the detection function. In order of importance, they were model robustness, pooling robustness, the shape criterion and estimator efficiency.

(a) *Model robustness* Given that the true form of the detection function is not known except in the case of computer simulations, models are required that are sufficiently flexible to fit a wide variety of shapes for the detection function. An estimator based on such a model is termed model robust. The adoption of the key + series expansion formulation means that any parametric model can yield model robust estimation, by allowing its fit to be adjusted when the data dictate. A model of this type is sometimes called 'semiparametric'.

(b) *Pooling robustness* Probability of detection is a function of many factors other than distance from the observer or line. Weather, time of day, observer, habitat, behaviour of the object, its size and many other factors influence the probability that the observer will detect it. Conditions will vary during the course of a survey, and different objects will have different intrinsic detectabilities. Thus the recorded data are realizations from a heterogeneous assortment of detection functions. A model is pooling robust if it is robust to variation in detection probability for any given distance y. A fuller definition of this concept is given by Burnham *et al.* (1980: 45).

(c) *Shape criterion* The shape criterion can be stated mathematically as $g'(0) = 0$. In words, it states that a model for the detection function should have a shoulder. The restriction is reasonable given the nature of the sighting process. Note that the hazard-rate derivations of Section 3.2 gave rise to detection functions which possess a shoulder for all parameter values, even though sharply spiked hazards with infinite slope at zero distance were assumed. If the shape criterion is violated, robust estimation of object density is problematic if not impossible.

(d) *Estimator efficiency* Estimators that have poor statistical efficiency (i.e. that have large variances) should be ruled out. However, an estimator that is highly efficient should be considered only if it satisfies the first three criteria. High estimator efficiency is easy to achieve at the expense of bias, and the analyst should be satisfied that an estimator is unbiased, or at least that there is no reason to suppose it might be more biased than other robust estimators, before selecting on the basis of efficiency.

3.5.2 *The likelihood ratio test*

The requirement for adjustment terms to a given key function can be judged using likelihood ratio tests. Suppose that a fitted model has m_1 adjustment terms (Model 1). A likelihood ratio test allows an assessment of whether the addition of another m_2 term improves the adequacy of a model significantly. The null hypothesis is that Model 1, with m_1 adjustment terms, is the true model, whereas the alternative hypothesis is that Model 2 with all $m_1 + m_2$ adjustment terms is the true model. The test statistic is

$$\chi^2 = -2\log_e(\mathscr{L}_1/\mathscr{L}_2)$$

$$= -2[\log_e(\mathcal{L}_1) - \log_e(\mathcal{L}_2)]$$

where $\mathcal{L}_1$ and $\mathcal{L}_2$ are the maximum values of the likelihood functions for Models 1 and 2 respectively. If Model 1 is the true model, the test statistic follows a χ^2 distribution with m_2 degrees of freedom.

The usual way to use likelihood ratio tests for line and point transect series expansion models is to fit the key function, and then fit a low order adjustment term. The adjustment would normally be a polynomial of order four or the first term in a cosine series. If it provides no significant improvement as judged by the above test, the fit of the key alone is taken. If the adjustment term does improve the fit, the next term is added (usually the polynomial of order six, or the second term of a cosine series), and a likelihood ratio test is again carried out. The process is repeated until the test is not significant, or until a maximum number of terms has been attained. This method is therefore sequential, and is the default method used by DISTANCE. The conventional significance level is 5% ($\alpha = 0.05$), so that the most recently added term is retained if the likelihood ratio statistic exceeds $\chi^2_{0.05} = 1.96^2 = 3.84$. Unless sample sizes are large, the test has rather low power, and the sequential method risks biased estimation of density and underestimation of variance. We suggest that $\alpha = 0.15$ be adopted, in which case the value 3.84 is replaced by $\chi^2_{0.15} = 2.07$.

Terms may be added in a stepwise manner, as in regression. For forward stepping, that term not yet in the model for which the χ^2 statistic from the likelihood ratio test is largest is included next, provided its test statistic is significant at the selected level. For backward stepping, the term already in the model with the smallest test statistic is dropped, unless it is significant at the $\alpha\%$ level.

The likelihood ratio test requires that Model 1 is a special case of Model 2. The models are said to be nested or hierarchical. The following procedure is similar in character, but allows the user to select between non-hierarchical models.

3.5.3 Akaike's Information Criterion

Akaike's Information Criterion (AIC) provides a quantitative method for model selection, whether or not models are hierarchical (Akaike 1973). It treats model selection within an optimization rather than a hypothesis testing framework. Burnham and Anderson (1992) illustrated the application of AIC, and Akaike (1985) presented the theory underlying the method. AIC is defined as

$$\text{AIC} = -2 \cdot \log_e(\mathcal{L}) + 2q$$

where $\log_e(\mathcal{L})$ is the log-likelihood function evaluated at the maximum likelihood estimates of the model parameters and q is the number of parameters in the model. The first term, $-2 \cdot \log_e(\mathcal{L})$, is a measure of how well the model fits the data, while the second term is a penalty for the addition of parameters (i.e. model complexity). For a given data set, AIC is computed for each candidate model and the model with the lowest AIC is selected. Thus, AIC attempts to identify a model that fits the data well and does not have too many parameters (the principle of parsimony). For the special case of nested models and $m_2 = 1$, model selection based on AIC is exactly equivalent to a likelihood ratio test with $\chi_\alpha^2 = 2.0$, which corresponds to $\alpha = 0.157$, close to the value of 0.15 recommended above for the likelihood ratio test.

For analyses of grouped data, DISTANCE omits the constant term from the multinomial likelihood when it calculates the AIC. This ensures that the AIC tends to the value obtained from analysis of ungrouped data as the number of groups tends to infinity, where each interval length tends to zero.

3.5.4 Goodness of fit

Goodness of fit can be a useful tool for model selection. Suppose the n distance data from line or point transects are split into u groups, with sample sizes $n_1, n_2, \ldots, n_u$. Let the cutpoints between groups be defined by $c_0, c_1, \ldots, c_u = w$ ($c_0 > 0$ corresponds to left truncation of the data). Suppose a model with q parameters is fitted to the data, so that the area under the estimated density function between cutpoints c_{i-1} and c_i is $\hat{\pi}_i$. Then

$$\chi^2 = \sum_{i=1}^{u} \frac{(n_i - n \cdot \hat{\pi}_i)^2}{n \cdot \hat{\pi}_i}$$

has a χ^2 distribution with $u - q - 1$ degrees of freedom if the fitted model is the true model.

Although a significantly poor fit need not be of great concern, it provides a warning of a problem in the data or the selected detection model structure, which should be investigated through closer examination of the data or by exploring other models and fitting options. Note that it is the fit of the model near zero distance that is most critical; none of the model selection criteria of goodness of fit statistics, AIC and likelihood ratio tests give special emphasis to this region. A possible criterion for selecting between models is to calculate the χ^2 goodness of fit statistic divided by its degrees of freedom for each model, and to select the model which gives the smallest value. A disadvantage of this approach is that the value of the χ^2 statistic depends on arbitrary decisions about the number of groups into which the data are divided and on where to place the cutpoints between groups. For several reasons, we prefer the use of AIC. However, a significant goodness of fit statistic is a useful warning that the model might be poor, or that an assumption might be seriously violated.

3.6 Estimation for clustered populations

Although the general formula of Section 3.1 incorporates the case in which the detections are clusters of objects, estimation of the expected cluster size $E(s)$ is often problematic. The obvious estimator, the average size of detected clusters, may be subject to size bias; if large clusters are detectable at greater distances than small clusters, mean size of detected clusters will be biased upwards.

3.6.1 Truncation

The simplest solution is to truncate clusters that are detected far from the line. The truncation distance need not be the same as that used if the detection function is fitted to truncated perpendicular distance data; if size bias is potentially severe, truncation should be greater. To be certain of eliminating the effects of size bias, the truncation distance should correspond roughly to the width of the shoulder of the detection function. Then $E(s)$ is estimated by $\bar{s}$, the mean size of the n clusters detected within the truncation distance. Generally, a truncation distance v corresponding to an estimated probability of detection $\hat{g}(v)$ in the range 0.6 to 0.8 ensures that bias in this estimate is small. Variance of $\bar{s}$ is estimated by:

$$\widehat{\operatorname{var}}(\bar{s}) = \frac{\sum_{i=1}^{n} (s_i - \bar{s})^2}{n(n-1)}$$

where s_i denotes the size of cluster i. This estimator remains unbiased when the individual s_i have different variances.

3.6.2 Weighted average of cluster sizes and stratification

Truncation may prove unsatisfactory if sample size is small. Quinn (1979) considered both post-stratifying detections by cluster size and pooling across cluster size in line transect sampling. He showed that estimation of the detection function, and hence of abundance of clusters, is not compromised by pooling the data. He noted the size bias in detected clusters, and proposed the estimator

$$\hat{E}(s) = \frac{\sum\limits_{v} n_v s_v \hat{f}_v(0 \mid s = s_v)}{\sum\limits_{v} n_v \hat{f}_v(0 \mid s = s_v)}$$

where summation is over the recorded cluster sizes. Thus there are n_v detections of clusters of size s_v, and the effective strip width for these clusters is $1/f_v(0 \mid s = s_v)$. The estimate is therefore the average size of detected clusters, weighted by the inverse of the effective strip width at each cluster size. For point transect sampling, $h_v(0 \mid s = s_v)$ would replace $\hat{f}_v(0 \mid s = s_v)$. As Quinn noted, if data are pooled with respect to cluster size, the $f_v(0 \mid s = s_v)$ are not individually estimated. He suggested that the effective strip width might be assumed to be proportional to the logarithm of cluster size, so that

$$\hat{E}(s) = \frac{\sum\limits_{v} n_v s_v / \log_e(s_v)}{\sum\limits_{v} n_v / \log_e(s_v)}$$

This method is used in the procedures developed by Holt and Powers (1982) for estimating dolphin abundance in the eastern tropical Pacific. If it is adopted, the recommendation of Quinn (1985) should be implemented: plot mean perpendicular distance as a function of cluster size to assess the functional relationship between cluster size and effective strip width. The method should not be used in conjunction with truncation of clusters at larger distances, because cluster size is then underestimated. The purpose of truncation is to restrict the mean cluster size calculation to those clusters that are relatively unaffected by size bias, so effective strip width of the retained clusters cannot be assumed proportional to the logarithm of cluster size. Clusters beyond the truncation distance are larger than average when size bias is present, so that the above weighted mean, if applied after truncating distant clusters, corrects for the effects of size bias twice.

Quinn (1985) examined further the method of post-stratifying by cluster size. He showed that the method necessarily yields a higher coefficient of variation for abundance of clusters than the above method in which data are pooled across cluster size, but found that the result

78

does not extend to estimates of object abundance. For his example, he concludes that the method of pooling is superior for estimating cluster abundance, and the method of post-stratification for estimating object abundance. This conclusion is likely to be true more generally. To apply the method of post-stratification, cluster size intervals should be defined so that sample size is adequate to allow estimation of $f(0)$ in each stratum. Stratification strategies relevant to this issue are discussed in more detail in Section 3.8.

3.6.3 Regression estimators

The solution of plotting mean distance y against cluster sizes was proposed by Best and Butterworth (1980), who predicted mean cluster size at zero distance, using a weighted linear regression of cluster size on distance. This suffers from the difficulty that, if the detection function has a shoulder, mean cluster size is not a function of distance until distance exceeds the width of the shoulder. Sample size is seldom sufficient to determine that a straight line fit is inadequate, so that estimated mean cluster size at zero distance is biased downwards. Because this is assumed to be an unbiased estimate of mean size of all clusters in the population, population abundance is underestimated. A solution to this problem is to replace detection distance y_i for the ith detection by $\hat{g}(y_i)$ in the regression, where $\hat{g}(y)$ is the detection function estimated from the fit of the selected model to the pooled data, and to predict mean cluster size when detection is certain ($\hat{g}(y) = 1.0$). Thus if there are n detections, at distances y_i and of sizes s_i, if $E_d(s|y)$ denotes the expected size of detected clusters at distance y, and $E(s)$ denotes the expected size of all clusters, whether detected or not (assumed independent of y), we have:

$$\hat{E}_d(s|y) = a + b \cdot \hat{g}(y)$$

where a and b are the intercept and slope respectively of the regression of s on $\hat{g}(y)$. Then

$$\hat{E}(s) = \hat{E}_d(s|y = 0) = a + b$$

and $\quad \widehat{\mathrm{var}}[\hat{E}(s)] = \left[\dfrac{1}{n} + \dfrac{(1 - \bar{g})^2}{\displaystyle\sum_{i=1}^{n} \{\hat{g}(y_i) - \bar{g}\}^2} \right] \cdot \hat{\sigma}^2$

with $\hat{\sigma}^2$ = residual mean square

and $\quad \bar{g} = \dfrac{\sum\limits_{i=1}^{n} \hat{g}(y_i)}{n}$

A further problem of the regression method occurs when cluster size is highly variable, so that one or two large clusters might have large influence on the fit of the regression line. Their influence may be reduced by transformation, for example to $z_i = \log_e(s_i)$. Suppose a regression of z_i on $\hat{g}(y_i)$ yields the equation $\hat{z} = a + b \cdot \hat{g}(y)$. Thus at $\hat{g}(y) = 1.0$, mean log cluster size is estimated by $a + b$ and $E(s)$ is estimated by

$$\hat{E}(s) = \exp(a + b + \widehat{\mathrm{var}}(\hat{z})/2)$$

where $\widehat{\mathrm{var}}(\hat{z}) = \left[1 + \dfrac{1}{n} + \dfrac{(1 - \bar{g})^2}{\sum\limits_{i=1}^{n} \{\hat{g}(y_i) - \bar{g}\}^2} \right] \cdot \hat{\sigma}^2$

$\hat{\sigma}^2$ is the residual mean square, and $\bar{g}$ is as above. Further,

$$\widehat{\mathrm{var}}\{\hat{E}(s)\} = \exp\{2(a + b) + \widehat{\mathrm{var}}(\hat{z})\} \cdot \{1 + \widehat{\mathrm{var}}(\hat{z})/2\} \cdot \widehat{\mathrm{var}}(\hat{z})/n$$

3.6.4 Use of covariates

The pooling method, with calculation of a weighted average cluster size, may be improved upon theoretically by incorporating cluster size as a covariate in the model for the detection function. Drummer and McDonald (1987) considered replacing detection distance y in a parametric model for the detection function by y/s^γ, where s is size of the cluster recorded at distance y and γ is a parameter to be estimated. Although their method was developed for line transect sampling, it can also be implemented for point transects. Ramsey et al. (1987) included covariates for point transect sampling by relating the logarithm of effective area searched to a linear function of covariates, one of which could be cluster size; this is in the spirit of general linear models. The same approach might be applied to effective strip width in line transect sampling, although the logarithmic link function might no longer be appropriate. These methods are discussed further in Section 3.8. Quang (1991) developed a method of modelling the bivariate detection function $g(y,s)$ using Fourier series.

3.6.5 *Replacing clusters by individual objects*

The problems of estimating mean cluster size can sometimes be avoided by taking the sampling unit to be the object, not the cluster. Even when detected clusters show extreme selection for large clusters, this approach can yield an unbiased estimate of object abundance, provided all clusters on or near the line are detected. The assumption of independence between sampling units is clearly violated, so robust methods of variance estimation that are insensitive to failures of this assumption should be adopted. Use of resampling methods allows the line, line segment or point to be the sampling unit instead of the object, so that valid variance estimation is possible. Under this approach, results from goodness of fit tests, likelihood ratio tests and AIC should not be used for model selection, since they will yield many spurious significant results. One solution is to select a model based on an analysis of clusters, then to refit the model, with the same number of adjustment terms, to the data recorded by object. If the number of clusters detected is small, if cluster size is highly variable, or if mean cluster size is large, the method may perform poorly.

3.6.6 *Some basic theory for size-biased detection of objects*

We present here some basic theoretical results when detection of clusters is size-biased. In this circumstance it is necessary to distinguish between the probability distribution of cluster sizes in the population from which the sample is taken from the distribution of s in the sample. Some of these results are in the literature (e.g. Quinn 1979; Burnham *et al.* 1980; Drummer and McDonald 1987; Drummer 1990; Quang 1991).

Let the probability distribution of cluster sizes in the region sampled be $\pi(s)$, $s = 1, 2, 3, \ldots$. This distribution applies to all the clusters, not to the detected sample. If there is size bias, then the sample of detected clusters has a different probability distribution, say $\pi^*(s)$, $s = 1, 2, 3, \ldots$. Consider first line transect sampling. Let the conditional detection function be $g(x|s) =$ probability of detection at perpendicular distance x given that the cluster is of size s, and let the detection function unconditional on cluster size be $g(x)$. Denote the corresponding probability density functions (pdf) by $f(x|s)$ and $f(x)$ respectively. The conditional pdf at $x = 0$ is

$$f(0|s) = \frac{1}{\displaystyle\int_0^w g(x|s)\,dx}$$

where w need not be finite. Note also the results

$$g(x|s) = \frac{f(x|s)}{f(0|s)} \quad \text{and} \quad g(x) = \frac{f(x)}{f(0)}$$

which are useful in derivations of results below.

For any fixed s, cluster density is given by the result for object density in the case without clusters:

$$D(s) = \frac{E[n(s)] \cdot f(0|s)}{2L}$$

where $n(s)$ is the number of detections of clusters of size s and $D(s)$ is the true density of clusters of size s. We need not assume that c and g_0 from Equation 3.1 equal one; however, the complication of g_0 varying by s is not considered here.

The key to deriving results is to realize that

$$\pi(s) = \frac{D(s)}{D}$$

and

$$\pi^*(s) = \frac{E[n(s)]}{E(n)}$$

where

$$D = \sum_{s=1}^{\infty} D(s) \quad \text{and} \quad n = \sum_{s=1}^{\infty} n(s)$$

By substituting the results for D and $D(s)$ into the first equation and using the result of the second, we derive

$$\pi^*(s) = \left[\frac{f(0)}{f(0|s)}\right] \cdot \pi(s)$$

Note that well-defined marginal probabilities and distributions exist; for example,

$$g(x) = \sum_{s=1}^{\infty} g(x|s) \cdot \pi(s)$$

82

from which

$$f(0) = \frac{1}{\int_0^w g(x)dx} \quad \text{and} \quad D = \frac{E(n) \cdot f(0)}{2L}$$

Given the above it is just a matter of using algebra to derive results of interest. Some key results are:

$$\pi^*(s) = \frac{\left[\int_0^w g(x|s)dx \right] \cdot \pi(s)}{\sum_{s=1}^{\infty} \left[\int_0^w g(x|s)dx \right] \cdot \pi(s)} = \frac{\frac{\pi(s)}{f(0|s)}}{\sum_{s=1}^{\infty} \frac{\pi(s)}{f(0|s)}}$$

By definition, $E(s) = \sum_{s=1}^{\infty} s \cdot \pi(s)$, so that

$$E(s) = \frac{\sum_{s=1}^{\infty} f(0|s) \cdot s \cdot \pi^*(s)}{\sum_{s=1}^{\infty} f(0|s) \cdot \pi^*(s)} = \frac{\sum_{s=1}^{\infty} f(0|s) \cdot s \cdot E[n(s)]}{\sum_{s=1}^{\infty} f(0|s) \cdot E[n(s)]}$$

The validity of Quinn's (1979) estimator, given in Section 3.6.2, is now apparent. The marginal pdf satisfies

$$f(0) = \sum_{s=1}^{\infty} f(0|s) \cdot \pi^*(s) = \frac{1}{\sum_{s=1}^{\infty} \frac{\pi(s)}{f(0|s)}}$$

These results are consistent with the formulae for density of individuals either as

$$D = \frac{E(n) \cdot f(0) \cdot E(s)}{2L}$$

or as

83

$$D = \frac{\sum_{s=1}^{\infty} E[n(s)] \cdot f(0|s) \cdot s}{2L}$$

A bivariate approach involves modelling the bivariate detection function $g(x, s)$, perhaps using generalized linear or non-linear regression. Adopting a univariate approach, we can estimate $E(s)$ in a linear regression framework. The key quantity needed here for theoretical work is the conditional probability distribution of detected cluster size given that detection was at perpendicular distance x, symbolized $\pi^*(s|x)$. Alternative representations are

$$\pi^*(s|x) = \frac{g(x|s) \cdot \pi(s)}{\sum_{s=1}^{\infty} g(x|s) \cdot \pi(s)} \equiv \frac{g(x|s) \cdot \pi(s)}{g(x)}$$

or

$$\pi^*(s|x) = \frac{f(x|s) \cdot \pi^*(s)}{\sum_{s=1}^{\infty} f(x|s) \cdot \pi^*(s)} \equiv \frac{f(x|s) \cdot \pi^*(s)}{f(x)}$$

If $T(s)$ represents any transformation of s, then we can compute conditional (on x) properties of $T(s)$, for example

$$E[T(s)|x] = \frac{\sum_{s=1}^{\infty} T(s) \cdot g(x|s) \cdot \pi(s)}{\sum_{s=1}^{\infty} g(x|s) \cdot \pi(s)} = \sum_{s=1}^{\infty} T(s) \cdot \pi^*(s|x)$$

$$= \frac{\sum_{s=1}^{\infty} T(s) \cdot f(x|s) \cdot \pi^*(s)}{\sum_{s=1}^{\infty} f(x|s) \cdot \pi^*(s)}$$

In particular, to evaluate the reasonableness of the regression estimator of $\log_e(s)$ on $\hat{g}(x)$, we can plot $E[\log_e(s)|x]$ against $\hat{g}(x)$ or otherwise explore this relationship, including computing $\text{var}[\log_e(s)|x]$.

Similar formulae exist for point transect sampling; in fact, many are the same. The relationship between the conditional detection function $g(r|s)$ and the corresponding pdf of distances to detected clusters is now

$$f(r|s) = \frac{r \cdot g(r|s)}{\int_0^w r \cdot g(r|s)\,dr}$$

and

$$h(0|s) = \lim_{r \to 0} \frac{f(r|s)}{r} = \frac{1}{\int_0^w r \cdot g(r|s)\,dr}$$

from which

$$D(s) = \frac{E[n(s)] \cdot h(0|s)}{2\pi k}$$

Univariate results are

$$g(r) = \sum_{s=1}^{\infty} g(r|s) \cdot \pi(s)$$

$$f(r) = \frac{r \cdot g(r)}{\int_0^w r \cdot g(r)\,dr}$$

$$h(0) = \frac{1}{\int_0^w r \cdot g(r)\,dr}$$

and

$$D = \frac{E(n) \cdot h(0)}{2\pi k}$$

85

Using these formulae we can establish that

$$\pi^*(s) = \frac{h(0)}{h(0\,|\,s)} \cdot \pi(s)$$

This is obtained from

$$\frac{D(s)}{D} = \frac{E[n(s)] \cdot h(0\,|\,s)}{E(n) \cdot h(0)}$$

The conditional and unconditional $h(\cdot)$ functions are related by

$$h(0) = \sum_{s=1}^{\infty} h(0\,|\,s) \cdot \pi^*(s) = \frac{1}{\displaystyle\sum_{s=1}^{\infty} \frac{\pi(s)}{h(0\,|\,s)}}$$

Also, analogous to the line transect case with $h(\cdot)$ in place of $f(\cdot)$, we have

$$\pi^*(s) = \frac{\left[\int_0^w r \cdot g(r\,|\,s)dr\right] \cdot \pi(s)}{\displaystyle\sum_{s=1}^{\infty}\left[\int_0^w r \cdot g(r\,|\,s)dr\right] \cdot \pi(s)} = \frac{\dfrac{\pi(s)}{h(0\,|\,s)}}{\displaystyle\sum_{s=1}^{\infty} \frac{\pi(s)}{h(0\,|\,s)}}$$

and

$$E(s) = \frac{\displaystyle\sum_{s=1}^{\infty} h(0\,|\,s) \cdot s \cdot \pi^*(s)}{\displaystyle\sum_{s=1}^{\infty} h(0\,|\,s) \cdot \pi^*(s)} = \frac{\displaystyle\sum_{s=1}^{\infty} h(0\,|\,s) \cdot s \cdot E[n(s)]}{\displaystyle\sum_{s=1}^{\infty} h(0\,|\,s) \cdot E[n(s)]}$$

In general, all the results for point transects can be obtained from the analogous results for line transects by making the following replacements: $r \cdot g(r)$ for $g(x)$, and $r \cdot g(r\,|\,s)$ for $g(x\,|\,s)$. In particular, note what happens to $\pi^*(s\,|\,r)$ in point transect sampling:

86

$$\pi^*(s|r) = \frac{r \cdot g(r|s) \cdot \pi(s)}{\sum_{s=1}^{\infty} r \cdot g(r|s) \cdot \pi(s)} \equiv \frac{r \cdot g(r|s) \cdot \pi(s)}{r \cdot g(r)}$$

$$\equiv \frac{g(r|s) \cdot \pi(s)}{\sum_{s=1}^{\infty} g(r|s) \cdot \pi(s)} \equiv \frac{g(r|s) \cdot \pi(s)}{g(r)}$$

This has exactly the same form as for line transects. An alternative expression is

$$\pi^*(s|r) = \frac{f(r|s) \cdot \pi^*(s)}{\sum_{s=1}^{\infty} f(r|s) \cdot \pi^*(s)} \equiv \frac{f(r|s) \cdot \pi^*(s)}{f(r)}$$

(defined to give continuity at $r = 0$), which looks structurally like the result for line transects. However, here the probability density function necessarily differs in shape from that for line transects, whereas the detection function of the previous expression might plausibly apply to both point and line transects.

3.7 Density, variance and interval estimation

3.7.1 Basic formulae

Substituting estimates into Equation 3.4, the general formula for estimating object density from line transect data is

$$\hat{D} = \frac{n \cdot \hat{f}(0) \cdot \hat{E}(s)}{2cL\hat{g}_0}$$

From Equation 3.3, the variance of $\hat{D}$ is approximately

$$\widehat{\text{var}}(\hat{D}) = \hat{D}^2 \cdot \left\{ \frac{\widehat{\text{var}}(n)}{n^2} + \frac{\widehat{\text{var}}[\hat{f}(0)]}{[\hat{f}(0)]^2} + \frac{\widehat{\text{var}}[\hat{E}(s)]}{[\hat{E}(s)]^2} + \frac{\widehat{\text{var}}[\hat{g}_0]}{[\hat{g}_0]^2} \right\}$$

Equivalent expressions for point transect sampling are

87

$$\hat{D} = \frac{n \cdot \hat{h}(0) \cdot \hat{E}(s)}{2\pi ck\hat{g}_0}$$

and

$$\widehat{var}(\hat{D}) = \hat{D}^2 \cdot \left\{ \frac{\widehat{var}(n)}{n^2} + \frac{\widehat{var}[\hat{h}(0)]}{[\hat{h}(0)]^2} + \frac{\widehat{var}[\hat{E}(s)]}{[\hat{E}(s)]^2} + \frac{\widehat{var}[\hat{g}_0]}{[\hat{g}_0]^2} \right\}$$

If $g_0 = 1$ (detection on the line or at the point is certain) or $E(s) = 1$ (no clusters), the terms involving estimates of these parameters are eliminated from the above equations. Generally, the constant $c = 1$, further simplifying the equations for $\hat{D}$.

To estimate the precision of $\hat{D}$, the precision of each component in the estimation equation must be estimated. Alternatively, resampling or empirical methods can be used to estimate $var(\hat{D})$ directly; some options are described in later sections. If precision is estimated component by component, then methods should be adopted for estimating mean cluster size and probability of detection on the line that provide variance estimates, $\widehat{var}[\hat{E}(s)]$ and $\widehat{var}[\hat{g}_0]$. Estimates of $f(0)$ or $h(0)$ and corresponding variance estimates are obtained from DISTANCE or similar software, using maximum likelihood theory. If objects are distributed randomly, then sample size n has a Poisson distribution, and $\widehat{var}(n) = n$. Generally, biological populations show some degree of aggregation, and Burnham et al. (1980: 55) suggested multiplication of the Poisson variance by two if no other approach for estimating $var(n)$ was available. If data are recorded by replicate lines or points, then a better method is to estimate $var(n)$ from the observed variation between lines and points. This method is described in the next section.

Having obtained $\hat{D}$ and $\widehat{var}(\hat{D})$, an approximate $100(1 - 2\alpha)\%$ confidence interval is given by

$$\hat{D} \pm z_\alpha \cdot \sqrt{\{\widehat{var}(\hat{D})\}}$$

where z_α is the upper $\alpha\%$ point of the $N(0,1)$ distribution. However, the distribution of $\hat{D}$ is positively skewed, and an interval with better coverage is obtained by assuming that $\hat{D}$ is log-normally distributed. Following the derivation of Burnham et al. (1987: 212), a $100(1 - 2\alpha)\%$ confidence interval is given by

$$(\hat{D}/C, \hat{D} \cdot C)$$

where

$$C = \exp[z_\alpha \cdot \sqrt{\{\widehat{\mathrm{var}}(\log_e \hat{D})\}}]$$

and

$$\widehat{\mathrm{var}}(\log_e \hat{D}) = \log_e \left[1 + \frac{\widehat{\mathrm{var}}(\hat{D})}{\hat{D}^2} \right]$$

This is the method used by DISTANCE, except z_α is replaced by a slightly better constant that reflects the actual finite and differing degrees of freedom of the variance estimates.

The use of the normal distribution to approximate the sampling distribution of $\log_e(\hat{D})$ is generally good when each component of $\widehat{\mathrm{var}}(\hat{D})$ (e.g. $\widehat{\mathrm{var}}(n)$ and $\widehat{\mathrm{var}}[\hat{f}(0)]$) is based on sufficient degrees of freedom (say 30 or more). However, sometimes the empirical estimate of $\mathrm{var}(n)$ in particular is based on less than 10 replicate lines, and hence on few degrees of freedom. When component degrees of freedom are small, it is better to replace z_α by a constant based on a t-distribution approximation. In this case we recommend an approach adapted from Satterthwaite (1946); see also Milliken and Johnson (1984) for a more accessible reference.

Adapting the method of Satterthwaite (1946) to this distance sampling context, z_α in the above log-based confidence interval is replaced by the two-sided alpha-level t-distribution percentile $t_{df}(\alpha)$ where df is computed as below. The coefficients of variation $\mathrm{cv}(\hat{D})$, $\mathrm{cv}(n)$, $\mathrm{cv}[\hat{f}(0)]$ or $\mathrm{cv}[\hat{h}(0)]$, and, where relevant, $\mathrm{cv}[\hat{E}(s)]$ and $\mathrm{cv}(\hat{g}_0)$ are required, together with the associated degrees of freedom. In general, if there are q estimated components in $\hat{D}$, then the computed degrees of freedom are

$$df = \frac{[\mathrm{cv}(\hat{D})]^4}{\sum\limits_{i=1}^{q} \dfrac{[\mathrm{cv}_i]^4}{df_i}} = \frac{\left[\sum\limits_{i=1}^{q} [\mathrm{cv}_i]^2 \right]^2}{\sum\limits_{i=1}^{q} \dfrac{[\mathrm{cv}_i]^4}{df_i}}$$

This value may be rounded to the nearest integer to allow use of tables of the t-statistic.

For the common case of line transect sampling of single objects using k replicate lines, the above formula for df becomes approximately

$$df = \frac{[\text{cv}(\hat{D})]^4}{\dfrac{[\text{cv}(n)]^4}{k-1} + \dfrac{\{\text{cv}[\hat{f}(0)]\}^4}{n}}$$

(The actual degrees of freedom for $\text{var}[\hat{f}(0)]$ are n minus the number of parameters estimated in $\hat{f}(x)$.) This Satterthwaite procedure is used by program DISTANCE, rather than just the first order z_α approximation. It makes a noticeable difference in confidence intervals for small k, especially if the ratio $\text{cv}(n)/\text{cv}[\hat{f}(0)]$ is greater than one; in practice, it is often as high as two or three.

3.7.2 Replicate lines or points

Replicate lines or points may be used to estimate the contribution to overall variance of the observed sample size. In line transects, the replicate lines may be defined by the design of the survey; for example if the lines are parallel and either systematically or randomly spaced, then each line is a replicate. Surveys of large areas by ship or air frequently do not utilize such a design for practical reasons. In this case, a 'leg' might be defined as a period of search without change of bearing, or all effort for a given day or watch period. The leg will then be treated as a replicate line. When data are collected on an opportunistic basis from, for example, fisheries vessels, an entire fishing trip might be considered to be the sampling unit.

Suppose the number of detections from line or point i is n_i, $i = 1, \ldots, k$, so that $n = \sum n_i$. Then for point transects (or for line transects when the replicate lines are all the same length), the empirical estimate of $\text{var}(n)$ is

$$\widehat{\text{var}}(n) = k \sum_{i=1}^{k} \left(n_i - \frac{n}{k} \right)^2 /(k-1)$$

For line transects, if line i is of length l_i and total line length $= L = \sum_{i=1}^{k} l_i$, then

$$\widehat{\text{var}}(n) = L \sum_{i=1}^{k} l_i \left(\frac{n_i}{l_i} - \frac{n}{L} \right)^2 /(k-1)$$

Encounter rate n/L is often a more useful form of the parameter than n alone; the variance of encounter rate is $\widehat{\text{var}}(n)/L^2$. There is a similarity here to ratio estimation in finite population sampling, except that we take all line lengths l_i, and hence L, to be fixed (as distinct from random) values. Consequently, the variance of a ratio estimator does not apply here, and our $\widehat{\text{var}}(n/L)$ is a little different from classical finite sampling theory.

If the same line or point is covered more than once, and an analysis of the pooled data is required, then the sampling unit should still be the line or point. That is, the distance data from repeat surveys over a short time period of a given line or point should be pooled prior to analysis. Consider point transects, in which point i is covered t_i times, and in total, n_i objects are detected. Then

$$\widehat{\text{var}}(n) = T \sum_{i=1}^{k} t_i \left(\frac{n_i}{t_i} - \frac{n}{T} \right)^2 / (k - 1)$$

where

$$T = \sum_{i=1}^{k} t_i$$

The formula for line transects becomes

$$\widehat{\text{var}}(n) = T_L \sum_{i=1}^{k} t_i \cdot l_i \left(\frac{n_i}{t_i \cdot l_i} - \frac{n}{T_L} \right)^2 / (k - 1)$$

· where

$$T_L = \sum_{i=1}^{k} t_i \cdot l_i$$

Generally, t_i will be the same for every point or line, in which case the above formulae simplify. The calculations may be carried out in DISTANCE by setting SAMPLE equal to t_i for point i (point transects) or $t_i \cdot l_i$ for line i (line transects).

The above provides empirical variance estimates for just one component of Equation 3.2, which may then be substituted in Equation 3.3. A more direct approach is to estimate object density for each replicate line or point. Define

$$\hat{D}_i = \frac{n_i \cdot \hat{E}_i(s)}{c \cdot a_i \cdot \hat{P}_{a_i} \cdot \hat{g}_{0i}}, \quad i = 1, \ldots, k$$

91

Then for point transects (and line transects when all lines are the same length),

$$\hat{D} = \left\{ \sum_{i=1}^{k} \hat{D}_i \right\} / k \qquad (3.11)$$

and

$$\widehat{\text{var}}(\hat{D}) = \left\{ \sum_{i=1}^{k} (\hat{D}_i - \hat{D})^2 \right\} / \{k(k-1)\} \qquad (3.12)$$

For line transects with replicate line i of length l_i,

$$\hat{D} = \left\{ \sum_{i=1}^{k} l_i \hat{D}_i \right\} / L \qquad (3.13)$$

and

$$\widehat{\text{var}}(\hat{D}) = \sum_{i=1}^{k} \{l_i(\hat{D}_i - \hat{D})^2\} / \{L(k-1)\} \qquad (3.14)$$

In practice, sample size is seldom sufficient to allow this approach, so that resampling methods such as the bootstrap and the jackknife are required.

3.7.3 The jackknife

Resampling methods start from the observed data and sample repeatedly from them to make inferences. The jackknife (Gray and Schucany 1972; Miller 1974) is carried out by removing each observation in turn from the data, and analysing the remaining data. It could be implemented for line and point transects by dropping each individual sighting from the data in turn, but it is more useful to define replicate points or lines, as above. The following development is for point transects, or line transects when the replicate lines are all of the same length.

First, delete all data from the first replicate point or line, so that sample size becomes $n - n_1$ and the number of points or lines becomes $k - 1$. Estimate object density using the reduced data set, and denote the estimate by $\hat{D}_{(1)}$. Repeat this step, reinstating the dropped point or line and removing the next, to give estimates $\hat{D}_{(i)}$, $i = 1, \ldots, k$. Now calculate the **pseudovalues:**

$$\hat{D}^{(i)} = k \cdot \hat{D} - (k - 1) \cdot \hat{D}_{(i)}, \; i = 1, \ldots, k \tag{3.15}$$

These pseudovalues are treated as k replicate estimators of density, and Equations 3.11 and 3.12 yield a jackknife estimate of density and variance:

$$\hat{D}_J = \left\{ \sum_{i=1}^{k} \hat{D}^{(i)} \right\} / k$$

and

$$\widehat{var}_J(\hat{D}_J) = \left\{ \sum_{i=1}^{k} (\hat{D}^{(i)} - \hat{D}_J)^2 \right\} / \{k(k - 1)\}$$

For line transects in general, Equation 3.15 is replaced by

$$\hat{D}^{(i)} = \{L \cdot \hat{D} - (L - l_i) \cdot \hat{D}_{(i)}\} / l_i, \; i = 1, \ldots, k$$

and the jackknife estimate and variance are found by substitution into Equations 3.13 and 3.14:

$$\hat{D}_J = \left\{ \sum_{i=1}^{k} l_i \hat{D}^{(i)} \right\} / L$$

and

$$\widehat{var}_J(\hat{D}_J) = \sum_{i=1}^{k} \{l_i(\hat{D}^{(i)} - \hat{D}_J)^2\} / \{L(k - 1)\}$$

An approximate $100(1 - 2\alpha)\%$ confidence interval for density D is given by

$$\hat{D}_J \pm t_{k-1}(\alpha) \cdot \sqrt{\{\widehat{var}_J(\hat{D}_J)\}}$$

where $t_{k-1}(\alpha)$ is from Student's t-distribution with $k - 1$ degrees of freedom.

This interval may have poor coverage when the number of replicate lines is small; Buckland (1982) found better coverage using

$$\hat{D} \pm t_{k-1}(\alpha) \cdot \sqrt{\{\widehat{var}_J(\hat{D})\}}$$

where $\hat{D}$ is the estimated density from the full data set and

$$\widehat{\text{var}}_J(\hat{D}) = \sum_{i=1}^{k} \{l_i(\hat{D}^{(i)} - \hat{D})^2\} / \{L(k-1)\}$$

The jackknife provides a strictly balanced resampling procedure. However there seems little justification for assuming that the pseudovalues are normally distributed, and the above confidence intervals may be poor when the number of replicate lines or points is small. Further there is little or no control over the number of resamples taken; under the above procedure, it is necessarily equal to the number of replicate lines or points k, and performance may be poor when k is small. Thirdly a resample can never be larger than the original sample, and will always be smaller unless there are no sightings on at least one of the replicate lines or points. The bootstrap therefore offers greater flexibility and robustness.

3.7.4 The bootstrap

The bootstrap (Efron 1979) provides a powerful yet simple method for variance and interval estimation. Consider first the non-parametric bootstrap, applied in the most obvious way to a line transect sample. Suppose the data set comprises n observations, $y_1, \ldots, y_n$, and the probability density evaluated at zero, $f(0)$, is to be estimated. Then a bootstrap sample may be generated by selecting a sample of size n **with replacement** from the observed sample. An estimate of $f(0)$ is found from the bootstrap sample using the same model as for the observed sample. A second bootstrap sample is then taken, and the process repeated. Suppose in total B samples are taken. Then the variance of $\hat{f}(0)$ is estimated by the sample variance of bootstrap estimates of $f(0)$, $\hat{f}_i(0)$, $i = 1, \ldots, B$ (Efron 1979). The percentiles of the distribution of bootstrap estimates give approximate confidence limits for $f(0)$ (Buckland 1980; Efron 1981). An approximate $100(1 - 2\alpha)\%$ central confidence interval is given by $[\hat{f}_{(j)}(0), \hat{f}_{(k)}(0)]$, where $j = (B + 1)\alpha$ and $k = (B + 1)(1 - \alpha)$ and $\hat{f}_{(i)}(0)$ denotes the ith smallest bootstrap estimate (Buckland 1984). To yield reliable confidence intervals, the number of bootstrap samples B should be at least 200, and preferably in the range 400–1000, although around 100 are adequate for estimating standard errors. The value of B may be chosen so that j and k are integer, or j and k may be rounded to the nearest integer values, or interpolation may be used between the ordered values that bracket the required percentile. Various modifications to the percentile method have been proposed, but the simple method is sufficient for our purposes.

The parametric bootstrap is applied in exactly the same manner, except that the bootstrap samples are generated by taking a random sample of size n from the fitted probability density, $\hat{f}(y)$.

If no polynomial or Fourier series adjustments are made to the fit of a parametric probability density, the above implementation of the boot-strap (whether parametric or non-parametric) yields variance estimates for $\hat{f}(0)$ close to those obtained using the information matrix. Since the bootstrap consumes considerably more computer time (up to B times that required by an analytical method), it would not normally be used in this case. When adjustments are made, precision as measured by the information matrix is conditional on the number of polynomial or Fourier series terms selected by the stopping rule (e.g. a likelihood ratio test). The Fourier series model in particular gives analytical standard errors that are strongly correlated with the number of terms selected (Buckland 1985). The above implementation of the bootstrap avoids this problem by applying the stopping rule independently to each bootstrap data set so that variation arising from estimating the number of terms required is accounted for (Buckland 1982).

In practice the bootstrap is usually more useful when the sampling unit is a replicate line or point, as for the jackknife method. The simplest procedure is to sample with replacement from the replicate lines or points using the non-parametric bootstrap. Unlike the jackknife, the sample need not be balanced, but a degree of balance may be forced by ensuring that each replicate line or point is used exactly B times in the B bootstrap samples (Davison et $al.$ 1986). Density D is estimated from each bootstrap sample, and the estimates are ordered, to give $\hat{D}_{(i)}$, $i = 1, \ldots, B$. Then

$$\hat{D}_B = \left\{ \sum_{i=1}^{B} \hat{D}_{(i)} \right\} / B$$

and

$$\widehat{\text{var}}_B(\hat{D}_B) = \left\{ \sum_{i=1}^{B} (\hat{D}_{(i)} - \hat{D}_B)^2 \right\} / (B - 1)$$

while a $100(1 - 2\alpha)\%$ confidence interval for D is given by $[\hat{D}_{(j)}, \hat{D}_{(k)}]$, with $j = (B + 1)\alpha$ and $k = (B + 1)(1 - \alpha)$ as above. (Note that the esti-mates do not need to be ordered if a confidence interval is not required.) The estimate based on the original data set, $\hat{D}$, is usually used in preference to the bootstrap estimate $\hat{D}_B$, with var$(\hat{D})$ estimated by $\widehat{\text{var}}_B(\hat{D}_B)$.

If an automated model selection procedure is implemented, for example using AIC, the bootstrap allows model selection to take place in each individual replicate. Thus the variability between bootstrap estimates of density reflects uncertainty due to having to estimate which model is appropriate. In other words, the bootstrap variance incorporates a component for model misspecification bias. By applying the full estimation procedure to each replicate, components of the variance for estimating the number of adjustment terms and for estimating $E(n)$, $E(s)$ and g_0 (where relevant) are all automatically incorporated. An example of such an analysis is given in Chapter 5.

A common misconception is that no model assumptions are made when using the non-parametric bootstrap. However, the sampling units from which resamples are drawn are assumed to be independently and identically distributed. If the sampling units are legs of effort, then each leg should be randomly located and independent of any other leg. In practice, this is seldom the case, but legs should be defined that do not seriously violate the assumption. For example, in marine line transect surveys, the sampling effort might be defined as all effort carried out in a single day. The overnight break in effort will reduce the dependence in the data between one sampling unit and the next, and the total number of sampling units should provide adequate replication except for surveys of short duration. It is wrong to break effort into small units and to bootstrap on those units. This is because the assumption of independence can be seriously violated, leading to bias in the variance estimate. If transect lines are designed to be perpendicular to object density contours, each line should be a sampling unit; subdivision of the lines may lead to overestimation of variance. In the case of point transects, if points are positioned along lines, then each line of points should be considered a sampling unit. If points are randomly distributed or evenly distributed throughout the study area, then individual points may be taken as sampling units. If a single line or point is covered more than once, and an analysis of the pooled data is required, the sampling unit should still be the line or point; it is incorrect to analyse the data as if different lines or points had been covered on each occasion. Analysis of such data is addressed in Section 3.7.2.

3.7.5 A finite population correction factor

We denote the size of the surveyed area, within distance w of the line or point, by a. If the size of the study area is A, a known proportion a/A is sampled. Moreover, a finite population of objects, N, exists in the area. Thus the question arises of whether a finite population correction (fpc) adjustment should be made to sampling variances. We give here a few thoughts on this matter.

Assume that there is no stratification (or that we are interested in results for a single stratum). Then for strip transect or plot sampling, fpc $= 1 - a/A$. The adjusted variance of $\hat{N}$ is

$$\mathrm{var}(\hat{N}) \cdot (1 - a/A)$$

where $\mathrm{var}(\hat{N})$ is computed from infinite population theory. In distance sampling, not all the objects are detected in the sampled area a, so that the fpc differs from $1 - a/A$. Also, no adjustment is warranted to $\mathrm{var}(\hat{P}_a)$ because this estimator is based on the detection distances, which conceptually arise from an infinite population of possible distances, given random placement of lines or points, or different choices of sample period.

Consider first the case where objects do not occur in clusters, and the following simple formula applies:

$$\hat{N} = A \cdot \frac{n}{a \cdot \hat{P}_a}$$

for which

$$[\mathrm{cv}(\hat{N})]^2 = [\mathrm{cv}(n)]^2 + [\mathrm{cv}(\hat{P}_a)]^2$$

The fpc is the same whether it is applied to coefficients of variation or variances. Heuristic arguments suggest that the fpc might be estimated by $1 - n/\hat{N}$ or by $1 - (a \cdot \hat{P}_a)/A$. These are clearly identical. In the case of a census of sample plots (or strips), $\hat{P}_a \equiv 1$ and the correct fpc is obtained. For the above simple case of distance sampling, $\mathrm{cv}(\hat{N})$ corrected for finite population sampling is

$$[\mathrm{cv}(\hat{N})]^2 = [\mathrm{cv}(n)]^2 \cdot \left[1 - \frac{a \cdot \hat{P}_a}{A} \right] + [\mathrm{cv}(\hat{P}_a)]^2$$

This fpc is seldom large enough to make any difference. When it is, then the assumptions on which it is based are likely to be violated. For the correction $1 - (a \cdot \hat{P}_a)/A$ to be valid, the surveyed areas within distance w of each line or point must be non-overlapping. Further, it must be assumed that an object cannot be detected from more than one line or point; if objects are mobile, the fpc $1 - n/\hat{N}$ is arguably inappropriate.

If objects occur in clusters, correction is more complicated. First consider when there is no size bias in the detection probability. The above result still applies to $\hat{N}_s$, the estimated number of clusters. However, the number of individuals is estimated as

$$\hat{N} = A \cdot \frac{n}{a \cdot \hat{P}_a} \cdot \bar{s}$$

Inference about N is limited to the time when the survey is done, hence to the actual individuals then present. If all individuals were counted ($P_a = 1$), var($\hat{N}$) should be zero; hence a fpc should be applied to $\bar{s}$ and conceptually, it should be $1 - (n \cdot \bar{s})/\hat{N} = 1 - (a \cdot \hat{P}_a)/A$. Thus for this case we have

$$[\mathrm{cv}(\hat{N})]^2 = \left[[\mathrm{cv}(n)]^2 + [\mathrm{cv}(\bar{s})]^2 \right] \cdot \left[1 - \frac{a \cdot \hat{P}_a}{A} \right] + [\mathrm{cv}(\hat{P}_a)]^2$$

Considerations are different for inference about $E(s)$. Usually one wants the inference to apply to the population in the (recent) past, present and (near) future, and possibly to populations in other areas as well. If this is the case, var($\bar{s}$) should not be corrected using the fpc.

Consider now the case of clusters with size-biased detection. The fpc applied to the number of clusters is as above. For inference about $\hat{N}$, the fpc applied to the variance of $\hat{E}(s)$ is still $1 - (n \cdot \bar{s})/\hat{N}$, which is now equal to

$$1 - \frac{\bar{s}}{\hat{E}(s)} \cdot \frac{a \cdot \hat{P}_a}{A}$$

Thus the adjusted coefficient of variation of $\hat{N}$ is given by

$$[\mathrm{cv}(\hat{N})]^2 = [\mathrm{cv}(n)]^2 \cdot \left[1 - \frac{a \cdot \hat{P}_a}{A} \right] + [\mathrm{cv}(\hat{P}_a)]^2 + [\mathrm{cv}\{\hat{E}(s)\}]^2 \cdot \left[1 - \frac{\bar{s}}{\hat{E}(s)} \cdot \frac{a \cdot \hat{P}_a}{A} \right]$$

in the case of size-biased detection of clusters.

We reiterate that these finite population corrections will rarely, if ever, be worth making.

3.8 Stratification and covariates

Two methods of handling heterogeneity in data, and of improving precision and reducing bias of estimates, are stratification and inclusion of covariates in the analysis. Stratification might be carried out by geographic region, environmental conditions, cluster size, time, animal behaviour, detection cue, observer, or many other factors. Different stratifications may be selected for different components of the estimation equation. For example, reliable estimation of $f(0)$ or $h(0)$ (or equivalently, effective strip width or effective area), and of g_0 where relevant, requires that sample size is quite large. Fortunately, it is often reasonable to assume that these parameters are constant across geographic strata. By contrast, encounter rate or cluster size may vary appreciably across strata, but can be estimated with low bias from small samples. In this case, reliable estimates can be obtained for each geographic stratum by estimating $f(0)$ or $h(0)$ from data pooled across strata and other parameters individually by stratum, although it may prove necessary to stratify by, say, cluster size or environmental conditions when estimating $f(0)$ or $h(0)$. In general, different stratifications may be needed for each component of Equation 3.2.

Post-stratification refers to stratification of the data after the data have been collected and examined. This practice is generally acceptable, but care must be taken. For example, if geographic strata are defined to separate areas for which encounter rate was high from those for which it was low, and estimates are given separately for these strata, there will be a tendency to overestimate density in the high encounter rate stratum, and underestimate density in the low encounter rate stratum. Variance will be underestimated in both strata. If prior to the survey, there is knowledge of relative density, geographic strata should be defined when the survey is designed, so that density is relatively homogeneous within each stratum. Survey effort should then be greater in strata for which density is higher (Section 7.2.3).

Variables such as environmental conditions, time of day, date or cluster size might enter the analysis as covariates rather than stratification factors. If the number of potential covariates is large, they might be reduced in some way, for example through stepwise regression or principal components regression. To carry out a covariate analysis, an appropriate model must be defined. For example, the scale parameter of a model for the detection function might be replaced by a linear function of parameters:

$$\beta_0 + \beta_1 \cdot X_{1i} + \beta_2 \cdot X_{2i} + \cdots$$

where X_{1i} might be sea state (Beaufort) at the time of detection i, X_{2i} might be cluster size for the detection, and so on.

3.8.1 Stratification

The simplest form of a stratified analysis is to estimate abundance independently within each stratum. A more parsimonious approach is to assume that at least one parameter is common across strata, or a subset of strata, an assumption that can be tested. Consider a point transect survey for which points were located in V geographic strata of areas A_v, $v = 1, \ldots, V$. Suppose we assume there is no size bias in detected clusters, and abundance estimates are required by stratum. Suppose further that data are sparse, so that $h(0)$ is estimated by pooling detection distances across strata. From Equation 3.2 with $c = 1$, $g_0 = 1$ and $a \cdot P_a = 2\pi k / h(0)$, we obtain

$$\hat{D}_v = \frac{n_v \cdot \hat{h}(0) \cdot \bar{s}_v}{2\pi k_v} = \frac{\hat{h}(0) \cdot \hat{M}_v}{2\pi} \quad \text{where } \hat{M}_v = \frac{n_v \bar{s}_v}{k_v}$$

for stratum v. Mean density $\hat{D}$ is then the average of the individual estimates, weighted by the respective stratum areas A_v:

$$\hat{D} = \frac{\sum_v A_v \hat{D}_v}{A} \quad \text{with } A = \sum_v A_v$$

The variance of any $\hat{D}_v$ may be found from Equation 3.3. However, to estimate var$(\hat{D})$, care must be taken, since one component of the estimation equation is common to all strata in a given year. The correct equation is:

$$\widehat{\text{var}}(\hat{D}) = \hat{D}^2 \cdot \left\{ \frac{\widehat{\text{var}}(\hat{M})}{\hat{M}^2} + \frac{\widehat{\text{var}}[\hat{h}(0)]}{[\hat{h}(0)]^2} \right\}$$

where $\quad \hat{M} = \dfrac{\sum_v A_v \hat{M}_v}{A}$

and $\quad \widehat{\text{var}}(\hat{M}) = \dfrac{\sum_v A_v^2 \cdot \widehat{\text{var}}(\hat{M}_v)}{A^2}$

with $\qquad \widehat{\mathrm{var}}(\hat{M}_v) = \hat{M}_v^2 \cdot \left\{ \dfrac{\widehat{\mathrm{var}}(n_v)}{n_v^2} + \dfrac{\widehat{\mathrm{var}}(\bar{s}_v)}{\bar{s}_v^2} \right\}$

Thus the estimation equation has been separated into two components, one of which ($\hat{M}_v$) is estimated independently in each stratum, and the other of which is common across strata. Population abundance is estimated by $\hat{N} = A \cdot \hat{D} = \sum A_v \hat{D}_v$, with $\widehat{\mathrm{var}}(\hat{N}) = A^2 \cdot \widehat{\mathrm{var}}(\hat{D})$. Further layers of strata might be superimposed on a design of this type. In the above example, each stratum might be covered by more than one observer, or several forests might be surveyed, and a set of geographic strata defined in each. Provided the principle of including each independent component of the estimation equation just once in the variance expression is adhered to, the above approach is easily generalized.

The areas A_v are weights in the above expressions. For many purposes, it may be appropriate to weight by effort rather than area. For example, suppose two observers independently survey the same area in a line transect study. Then density within the study area may be estimated separately from the data of each observer (perhaps with at least one parameter assumed to be common between the observers), and averaged by weighting the respective estimates by length of transect covered by the respective observers. Note that in this case, an average of the two abundance estimates from each stratum is required, rather than a total. If stratification is by factors such as geographic region, cluster size, animal behaviour or detection cue, then the strata correspond to mutually exclusive components of the population and the estimates should be summed, whereas if stratification is by factors such as environmental conditions, observer, time or date (assuming no migration), then each stratum provides an estimate of the whole population, so that an average is appropriate.

Note that the stratification factors for each component of estimation may be completely different provided the components are combined with care. As a general guide, stratification prior to estimation of $f(0)$ or $h(0)$ should only be carried out if there is evidence that the parameter varies between strata, and some assessment should be made of whether the number of strata can be reduced. This policy is recommended since estimation of the parameter is unreliable if sample size is not large. Encounter rate and mean cluster size on the other hand may be reliably estimated from small samples, so if there is doubt, stratification should be carried out. Further, if abundance estimates are required by stratum, then both encounter rate and mean cluster size should normally be estimated by stratum. If all parameters can be assumed common across strata, such as observers of equal ability covering a single study area at roughly the same time, stratification is of no benefit. Also, in the special case that strata correspond to different geographic regions, effort per unit area is the same in each region, the parameter $f(0)$ or $h(0)$ can be assumed constant across regions, and estimates are not required by region, stratification is unnecessary. Proration of a total abundance estimate by the area of each region is seldom satisfactory. An example for which stratification was used in a relatively complex way to improve abundance estimation of North Atlantic fin whales is given in Section 8.5.

Further parsimony may be introduced by noting that $\mathrm{var}(n_v)$ is a parameter to be estimated, and $b_v = \mathrm{var}(n_v)/n_v$ is often quite stable over strata. Especially if the n_v are small, it is useful to assess the assumption that $b_v = b$ for all v. If it is reasonable, the number of parameters is reduced. The parameter b can be estimated by

$$\hat{b} = \frac{\sum_v \widehat{\mathrm{var}}(n_v)}{n}$$

and $\mathrm{var}(n_v)$ is then more efficiently estimated as $\widehat{\mathrm{var}}_p(n_v) = \hat{b}/n_v$. This approach is described further in Section 6.3, and illustrated in Section 8.4. The same method might also be applied to improve the efficiency of $\widehat{\mathrm{var}}(\bar{s}_v)$.

3.8.2 Covariates

Several possibilities exist for incorporating covariates. Ramsey *et al.* (1987) used the effective area, v, as a scale parameter in point transect surveys, and related it to covariates using a log link function:

$$\log_e(v) = \beta_0 + \sum_j \beta_j \cdot X_j$$

where X_j is the jth covariate. Computer programs for implementing this approach for the case of an exponential power series detection function are available from the authors.

Drummer and McDonald (1987) considered a single covariate X, taken to be cluster size in their example, and incorporated it into detection functions by replacing y by y/X^γ, where γ is a parameter to be estimated. Thus the univariate half-normal detection function

$$g(y) = \exp\left[-\frac{y^2}{2\sigma^2}\right]$$

becomes the 'bivariate' detection function:

$$g(y \mid X) = \exp\left[-\frac{y^2}{2(\sigma X^\gamma)^2} \right]$$

The interpretation is now that $g(y \mid X)$ is the detection probability of a cluster at distance y, given that its size is X. Drummer and McDonald proposed the following detection functions as candidates for this approach: negative exponential, half-normal, generalized exponential, exponential power series and reversed logistic. They implemented the method for the first three, although their procedure failed to converge to plausible parameter values for the generalized exponential model for the data set they present. Their software (SIZETRAN) is available (Drummer 1991).

4

Line transects

4.1 Introduction

The purpose of this chapter is to illustrate the application of the theory of Chapter 3 to line transect data, and to present the strategies for analysis outlined in Section 2.5. In general, the principal parameter of a line transect analysis does not have a closed form estimator. Instead, numerical methods are required; it is generally not possible to substitute statistics computed from the data into formulae to estimate object density. Using pen and paper and a pocket calculator, a fairly simple analysis might take months. Instead, we rely on specialized computer software to analyse distance sampling data.

This chapter uses a series of examples in which complexities are progressively introduced. The examples come from simulated data for which the parameters are known; this makes comparisons between estimates and true parameters possible. However, in every other respect, the data are treated as any real data set undergoing analysis, where the parameters of interest are unknown. A simple data set, where each object represents an individual animal, plant, nest, etc., is first introduced. Truncation of the distance data, modelling the spatial variation of objects to estimate var(n), grouping of data, and model selection philosophy and methods are then addressed. Once an adequate model has been selected, we focus on statistical inference given that model, to illustrate estimation of density and measures of precision. Finally, the objects are allowed to be clusters (coveys, schools, flocks). Cluster size is first assumed to be independent of distance and then allowed to depend on distance.

The example data are chosen to be 'realistic' from a biological standpoint. The data (sample size, distances and cluster sizes) are generated stochastically and simulate the case where the assumptions of line transect sampling are true. Thus, no objects went undetected on the line ($g(0) = 1$), no movement occurred prior to detection, and data were free of measurement error (e.g. heaping at zero distance). In addition, the sample size n was adequate. The assumptions and survey design to ensure they are met are discussed in Chapter 7. Examples illustrating analysis of real data where some of these assumptions fail are provided in Chapter 8. The example data of this chapter are analysed using various options of program DISTANCE. In the penultimate section, some comparative analyses using program SIZETRAN (Drummer 1991) are carried out.

4.2 Example data

The example comprises an area of size A, whose boundary is well defined, and sampled by 12 parallel line transects $(l_1, l_2, \ldots, l_{12})$. The area is irregularly shaped, so that the lines running from boundary to boundary are of unequal length. We assume that no stratification is required and that the population was sampled once by a single observer to exacting standards; hence the key assumptions have been met. The distance data, recorded in metres, were taken without a fixed transect width (i.e. $w = \infty$), and ungrouped, to allow analysis of either grouped or ungrouped data. The detection function $g(x)$ was a simple half-normal with $\sigma = 10$ m, giving $f(0) = \sqrt{\{2/(\pi\sigma^2)\}} = 0.079788 \text{ m}^{-1}$. The n_i were drawn from a negative binomial distribution such that the spatial distribution of objects was somewhat clumped (i.e. $\text{var}(n) > n$). Specifically, $\text{var}(n_i|l_i) = 2E(n_i|l_i)$.

The total length $(L = \Sigma l_i)$ of the 12 transects was 48 000 m and $n = 105$ objects were detected. Their distances from the transect lines were measured carefully in metres. $E(n) = 96$, thus somewhat more were observed than expected (105 vs. 96). The true density is

$$D = \frac{N}{A} = \frac{E(n) \cdot f(0)}{2L} \text{ objects/m}^2$$

where all measurements are in metres. To convert density from numbers per m^2 into numbers per km^2, multiply by 1 000 000. The true density is known in this simulated example to be approximately 80 objects per km^2; the actual value is 79.788/km^2.

The first step is to examine the distance data by plotting histograms using various distance categories. It is often informative to plot a histogram with many fine intervals (Fig. 4.1). Here one can see the presence of a broad shoulder, no evidence of heaping, and no indication of evasive movement prior to detection; the data appear to be 'good', which we happen to know to be true here.

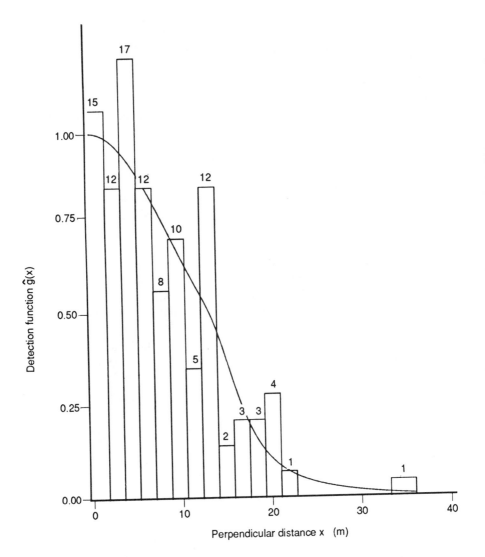

Fig. 4.1. Histogram of the example data using 20 distance categories. A fitted hazard-rate key with one cosine adjustment term is shown as a candidate model for the detection function, $g(x)$.

4.3 Truncation

Inspection of the histogram in Fig. 4.1 shows the existence of an extreme observation or 'outlier' at 35.82 m. A useful rule of thumb is to truncate at least 5% of the data; here the six most extreme distances are 19.27,

106

19.42, 19.44, 19.46, 21.21 and 35.82 m. Thus, w could initially be set for purposes of analysis at 19 m. An alternative is to fit a reasonable preliminary model to the data, compute $\hat{g}(x)$ to find the value of x such that $\hat{g}(x) = 0.15$, and use this value of x as the truncation point for further analysis.

As an illustration, the half-normal key function was fitted to the ungrouped, untruncated data and found to fit well. This approach suggested a truncation point of 19 m for the half-normal model, based on the criterion that $\hat{g}(x) \doteq 0.15$ (actually $\hat{g}(19) = 0.13$). The deletion of outliers is useful because these extreme observations provide little information for estimating $f(0)$, the density function at $x = 0$, but can be difficult to model. The series expansions require additional adjustment terms to fit the few data in the tail of the distance distribution, which may unnecessarily increase the sampling variance of the density estimate. In this example, both truncation rules suggest $w \doteq 19$, leaving $n = 99$ observations. For the rest of this chapter we will emphasize the results with $w = 19$ m, but estimates corresponding to no truncation will also be given and compared. The choice of truncation point is not a critical decision for these example data where all the assumptions are met and the true detection function is simple.

For the true model, the quantity $E(n) \cdot f(0)$ remains unchanged as the truncation point is varied. Consequently, for good data (i.e. data satisfying the assumptions) and a reasonable model for $g(x)$, the product $n \cdot \hat{f}(0)$ is quite stable over a range of truncation points. With increased truncation, n decreases, but $\hat{f}(0)$ increases to compensate. The estimate $n \cdot \hat{f}(0)$ under the half-normal model is 8.477 if data are truncated at 19 m, and 8.417 without truncation.

Truncation of the data at $w = 19$ m removed only six detections. If a series expansion model is used, up to three fewer parameters are required to model the truncated data than the untruncated data (Table 4.1). (Note that the truncation distance w supplied to DISTANCE must be finite; by 'untruncated data', we mean that w was at least as large as the largest recorded distance.) Outliers in the right tail of the distance distribution required additional adjustment parameters and the inclusion of such terms increased the sampling variance of $\hat{f}(0)$ and hence $\hat{D}$ when a robust but incorrect model was used (Table 4.2). If the correct model could somehow be known and used, then truncation is unimportant if the measurements are exact and no evasive movement prior to detection is present.

Truncation of the distance data for analysis deletes outliers and facilitates modelling of the data. However, as some data are discarded, one might ask if the uncertainty in $\hat{D}$ increases. First, this issue is examined when the true model is known and used (i.e. the half-normal in this case). The coefficient of variation increased about 1% when the data were truncated at 19 m relative to untruncated (Table 4.2). Thus, little precision was lost due to truncation if the data were analysed under the true model. Of course, one never knows the true detection function except for computer simulation examples.

Table 4.1 Summary of AIC values at two truncation values w for the example data analysed as ungrouped and at three different groupings (five groups of equal width, 20 groups of equal width, and five unequal groups such that the number detected was equal in each group). The models with minimum AIC values are indicated by an asterisk

Data type	Model (key + adjustment)	$w = 19$ m			$w =$ largest observation		
		No. of parameters			No. of parameters		
		Key	Adjust.	AIC	Key	Adjust.	AIC
Ungrouped	Uniform + cosine	0	1	562.98	0	2	636.48*
	Uniform + polynomial	0	1	563.28	0	4	638.18
	Half-normal + Hermite	1	0	562.60*	1	0	636.98
	Hazard-rate + cosine	2	0	565.22	2	0	639.16
Grouped	Uniform + cosine	0	1	300.91	0	3	224.77
(5 equal)	Uniform + polynomial	0	1	301.09	0	4	226.75
	Half-normal + Hermite	1	0	300.63*	1	0	222.13*
	Hazard-rate + cosine	2	0	303.18	2	0	224.21
Grouped	Uniform + cosine	0	1	563.58	0	2	520.88
(20 equal)	Uniform + polynomial	0	1	563.40	0	3	524.78
	Half-normal + Hermite	1	0	563.03*	1	0	520.31*
	Hazard-rate + cosine	2	0	565.80	2	1	523.53
Grouped	Uniform + cosine	0	1	323.05*	0	2	345.13
(5 unequal)	Uniform + polynomial	0	1	324.45	0	4	348.56
	Half-normal + Hermite	1	0	323.35	1	0	342.54*
	Hazard-rate + cosine	2	0	324.32	2	0	344.95

When series expansion models are used for the analysis of the example data, the uniform key function with either cosine or polynomial adjustments gives a smaller coefficient of variation when the data are truncated (Table 4.2). This small increase in precision is because only one parameter was required for a good model fit when $w = 19$ m, whereas two to four parameters were required to fit the untruncated data (Table 4.2). Precision was better for the untruncated data for the hazard-rate model.

The effect of truncation on the point estimates was relatively small, and estimates were not consistently smaller or larger than when data were untruncated (Table 4.2). The various density estimates ranged from 72.75 to 94.09, and their coefficients of variation ranged from 14.8% to 20.3%. The true parameter value was $D = 80$ objects/km^2.

In general, some truncation is recommended, especially for obvious outliers. Although some precision might be lost due to truncation, it is usually slight. Often, precision is increased because fewer parameters are required to model the detection function. Most importantly, truncation will often reduce bias in $\hat{D}$ or improve precision, or both, by making the data easier to model. Extreme observations in the right tail of the distribution may arise from a different detection process (e.g. a deer seen at some distance from the observer along a forest trail, or a whale breaching near the horizon), and are generally not informative, in addition to being difficult to model. Truncation is an important tool in the analysis of distance sampling data.

Table 4.2 Summary of estimated density $\hat{D}$ and coefficient of variation cv for two truncation values w for the example data. Estimates are derived for four robust models of the detection function. The data analysis was based on ungrouped data and three different groupings (five groups of equal width, 20 groups of equal width, and five unequal groups such that the number detected was nearly equal in each group)

Data type	Model (key + adjustment)	Truncation			
		$w = 19$ m		$w =$ largest obsn	
		$\hat{D}$	cv(%)	$\hat{D}$	cv(%)
Ungrouped	Uniform + cosine	90.38	15.9	80.52	16.8
	Uniform + polynomial	78.95	14.8	84.53	20.0
	Half-normal + Hermite	88.31	16.7	87.68	15.3
	Hazard-rate + cosine	84.23	18.4	72.75	15.6
Grouped (5 equal)	Uniform + cosine	88.69	15.9	94.09	16.7
	Uniform + polynomial	79.37	15.2	88.39	19.1
	Half-normal + Hermite	86.94	16.8	92.16	15.8
	Hazard-rate + cosine	84.49	19.6	80.80	16.7
Grouped (20 equal)	Uniform + cosine	89.95	15.8	80.06	15.1
	Uniform + polynomial	79.10	14.9	74.43	15.3
	Half-normal + Hermite	87.98	16.6	86.87	15.7
	Hazard-rate + cosine	85.81	19.2	84.06	18.1
Grouped (5 unequal)	Uniform + cosine	86.14	16.3	81.40	19.0
	Uniform + polynomial	78.60	15.8	86.91	17.8
	Half-normal + Hermite	85.12	17.0	88.84	16.3
	Hazard-rate + cosine	86.83	20.3	82.54	17.7

4.4 Estimating the variance in sample size

Before the precision of an estimate of density can be assessed, attention must be given to the spatial distribution of the objects of interest. If the n detected objects came from a sample of objects that were randomly (i.e. Poisson) distributed in space, then $\text{var}(n) = E(n)$ and $\widehat{\text{var}}(n) = n$.

Because most biological populations exhibit some degree of clumping, one expects $\text{var}(n) > E(n)$. Thus, empirical estimation of the sampling variance of n is recommended. This makes it nearly imperative to sample using several lines, l_i, such as the 12 used in the example. Variation in the number of detections found on each of the lines, n_i, provides a valid estimate of $\text{var}(n)$ without having to resort to the Poisson assumption and risk what may be a substantial underestimate of the sampling variance of the estimator of density.

After truncating at 19 m, the line lengths in km and numbers of detections (l_i, n_i) for the $k = 12$ lines were: (5, 14), (2, 2), (6, 8), (4, 8), (3, 3), (1, 4), (4, 10), (4, 8), (5, 17), (7, 20), (3, 0), and (4, 5). The estimator for the empirical variance of n is (from Section 3.7.2)

$$\widehat{\text{var}}(n) = L \sum_{i=1}^{k} l_i \left(\frac{n_i}{l_i} - \frac{n}{L} \right)^2 / (k - 1)$$

$$= 195.8$$

This estimate is based on $k - 1 = 11$ degrees of freedom. The ratio of the empirical variance to the estimated Poisson variance is $195.8/99 = 1.98$, indicating some spatial aggregation of objects. Equivalently, one can estimate the sampling variance of the encounter rate (n/L),

$$\widehat{\text{var}}(n/L) = \frac{\sum_{i=1}^{k} \frac{l_i}{L} \left(\frac{n_i}{l_i} - \frac{n}{L} \right)^2}{k - 1}$$

$$= 0.0850$$

$$\widehat{\text{se}}(n/L) = \sqrt{\widehat{\text{var}}(n/L)}$$

$$= 0.292$$

Then $\widehat{\text{se}}(n) = L \cdot \widehat{\text{se}}(n/L)$ and $\widehat{\text{var}}(n) = [\widehat{\text{se}}(n)]^2$. In most subsequent analyses of these data, we use the empirical estimate, $\widehat{\text{var}}(n) = 195.8$.

4.5 Analysis of grouped or ungrouped data

Analysis of the ungrouped data is recommended for the example because it is known that the assumptions of line transect sampling hold. General statistical theory and our experience indicate that little efficiency is lost by grouping data, even with as few as five or six well-chosen intervals. Grouping of the data can be used to improve robustness in the estimator of density in cases of heaping and movement prior to detection (Chapter 7).

For the example, changes in the estimates of density under a given model were in most cases slight (much smaller than the standard error) whether the analysis was based on the ungrouped data or one of the three sets of grouped data. This is a general result if the assumptions of distance sampling are met. If heaping, errors in measurement, or evasive movement prior to detection are present, then appropriate grouping will often lead to improved estimates of density and better model fit. Grouping the data is a tool for the analysis of real data to gain estimator robustness. When heaping occurs, cutpoints should be selected to avoid favoured rounding distances as far as possible. Thus, if values tend to be recorded to the nearest 10 m, cutpoints might be defined at 5 m, 15 m, 25 m, ... The first cutpoint is the most critical. If assumptions hold, the first interval should be relatively narrow, so that the first cutpoint is on the shoulder of the detection function. However, it is not unusual for 10% or more of detections to be recorded as on the centreline, especially when perpendicular distances are calculated from sighting distances and angles. In this circumstance, the width of the first interval should be chosen so that few detections are erroneously allocated to the first interval through measurement error, and in particular, through rounding a small sighting angle to zero.

4.6 Model selection

4.6.1 The models

Results for fitting the detection function are illustrated using the uniform, half-normal and hazard-rate models as key functions. Cosine and simple polynomial expansions are used with the uniform key, Hermite polynomials are used with the half-normal key, and a cosine expansion is used with the hazard-rate key. Thus, four models for $g(x)$ are considered for the analysis of these data. Modelling in this example can be expected to be relatively easy as the data are well behaved, exhibit a shoulder, and the sample size is relatively large ($n = 105$ before truncation). With such ideal data, the choice of model is unlikely to affect the abundance estimate much, whereas if survey design or data collection is poor, different models might yield substantially different estimates.

From an inspection of the data in Fig. 4.1, it is clear that the uniform key function will require at least one cosine or polynomial adjustment term. The data here were generated under a half-normal detection function so we might expect the half-normal key to be sufficient without any adjustment terms. However, the data were stochastically generated; the addition of a Hermite polynomial term is quite possible, although it would just fit 'noise'. The hazard-rate key has two parameters and seldom requires adjustment terms when data are good. In general, a histogram of the untruncated data using 15–20 intervals will reveal the characteristics of the data. Such a histogram will help identify outliers, heaping, measurement errors, and evasive movement prior to detection.

4.6.2 Likelihood ratio tests

The addition of adjustment terms to a given key function can be judged using likelihood ratio tests (LRTs). This procedure is illustrated using the example data, with $w = 19$, and ungrouped data. Assume the key function is the 1-parameter half-normal. This model is fitted to the distance data to provide the MLE of the parameter σ. Does an adjustment term significantly improve the fit of the model of $g(x)$ to the data? Let $\mathcal{L}_0$ be the value of the likelihood for the 1-parameter half-normal model and $\mathcal{L}_1$ be its value for the 2-parameter model (half-normal model plus one Hermite polynomial adjustment term). Then, the test statistic for this likelihood ratio test is

$$\chi^2 = -2 \log_e (\mathcal{L}_0/\mathcal{L}_1)$$

and is distributed asymptotically as χ^2 with 1 df if the 1-parameter model ($\mathcal{L}_0$) is the true model. In general, the degrees of freedom are calculated as the difference in the number of parameters between the two models being tested. This is a test of the null hypothesis that the 1-parameter model is the true model against the alternative hypothesis that the 2-parameter model is the true model. If the additional term makes a significant improvement in the fit, then the test statistic will be 'large'. For the example, $\log_e(\mathcal{L}_0) = -280.3000$ and $\log_e(\mathcal{L}_1) = -280.2999$. These are values of the log-likelihood function computed at the MLE values of the parameters. Then, the test statistic is

$$\chi^2 = -2 \log_e(\mathcal{L}_0/\mathcal{L}_1)$$
$$= -2[\log_e(\mathcal{L}_0) - \log_e(\mathcal{L}_1)]$$
$$= -2[-280.3000 - (-280.2999)]$$
$$= 0.0002$$

This test statistic has 1 df, so that $p = 0.988$, and there is no reason to add a Hermite polynomial adjustment term. This does not necessarily mean that the 1-parameter half-normal model is an adequate fit to the example data; it only informs us that the 2-parameter model is not a significant improvement over the 1-parameter model. If goodness of fit for example indicates that both models are poor, it is worth investigating whether a 3-parameter model is significantly better than the 1-parameter model. This may be done by setting the LOOKAHEAD option in DISTANCE to 2. Another solution is to try a different model.

A second example is that shown in Fig. 4.1, the untruncated example data modelled by a hazard-rate key and cosine adjustment terms. Let $\mathscr{L}_0$ be the likelihood under the 2-parameter hazard-rate model and $\mathscr{L}_1$ be the likelihood under this same model with one cosine adjustment term. MLEs of the parameters are found under both models with the resulting log-likelihood values: $\log_e(\mathscr{L}_0) = -259.898$ and $\log_e(\mathscr{L}_1) = -258.763$. Which is the better model of the data? Should the cosine term be retained? These questions are answered by the LRT statistic,

$$\chi^2 = -2\left[\log_e(\mathscr{L}_0) - \log_e(\mathscr{L}_1)\right]$$

$$= -2\left[-259.898 - (-258.763)\right]$$

$$= 2.27$$

Because $\mathscr{L}_0$ has two parameters and $\mathscr{L}_1$ has three parameters, the LRT has 1 df. Here, $\chi^2 = 2.27$, 1 df, $p = 0.132$. As noted in Section 3.5.2, use of $\alpha = 0.15$ instead of the conventional $\alpha = 0.05$ might be found useful as a rejection criterion. Thus, the 2-parameter model is rejected in favour of a 3-parameter model, with a single cosine adjustment to the hazard-rate key. The procedure is repeated to examine the adequacy of the 3-parameter model against a 4-parameter model with two cosine terms. Note that this illustration used the untruncated data; additional terms are frequently needed to model the right tail of the distance data if proper truncation has not been done.

If the LRT indicates that a further term is not required but goodness of fit (below) indicates that the fit is poor, the addition of two terms (using DISTANCE option LOOKAHEAD = 2) may provide a significantly better fit. If it is important to obtain the best possible fit, options SELECT = **forward** and SELECT = **all** of DISTANCE may prove useful.

4.6.3 *Akaike's Information Criterion*

The use of the Akaike's Information Criterion (AIC) provides an objective, quantitative method for model selection (see Burnham and Anderson (1992) for application of AIC and Akaike (1985) for theoretical synthesis). It is similar in character to a likelihood ratio test for hierarchical models, but is equally applicable to selection between non-hierarchical models. The criterion is

$$\text{AIC} = -2 \cdot [\log_e(\mathcal{L}) - q]$$

where $\log_e(\mathcal{L})$ is the value of the log-likelihood function evaluated at the maximum likelihood estimates of the model parameters and q is the number of parameters in the model (Section 3.5.3). AIC is computed for each candidate model, and that with the lowest AIC is selected for analysis and inference. Having selected a model, one should check that it fits as judged by the usual χ^2 goodness of fit statistics. Visual inspection of the estimated detection function plotted on the histogram is also informative because one can better judge the model fit near the line, and perhaps discount some lack of fit in the right tail of the data.

AIC was computed for the four models noted above for both grouped and ungrouped data with and without truncation (Table 4.1). For computational reasons, w was set equal to the largest observation in the case of no truncation. Three sets of cutpoints were considered for each model for illustration. Set 1 had five groups of equal width, set 2 had 20 groups of equal width, and set 3 had five groups whose width increased with distance, such that the number detected in each distance category was nearly equal. Note that AIC cannot be used to select between models if the truncation distances w differ, or, in the case of an analysis of grouped data, if the cutpoints differ.

AIC values in Table 4.1 indicate that the half-normal model is the best of the four models considered. Here, we happen to know that this is the true model. All four models have generally similar AIC values within any set of analyses of Table 4.1. Still, AIC selects the half-normal model in three of the four instances, both with truncation and without. Thus, the main analysis will focus on the ungrouped data, truncated at $w = 19$ m, under the assumption that $g(x)$ is well modelled by the half-normal key function with no adjustment parameters. The fit of this model is shown in Fig. 4.2. One might suspect that all four models would provide valid inference because of the similarity of the AIC values. Often, the AIC will identify a subset of models that are clearly inferior and these should be discarded from further consideration.

4.6.4 Goodness of fit

Goodness of fit is described briefly in Section 3.5.4, and is the last of the model selection criteria we consider here. Although it is the fit of

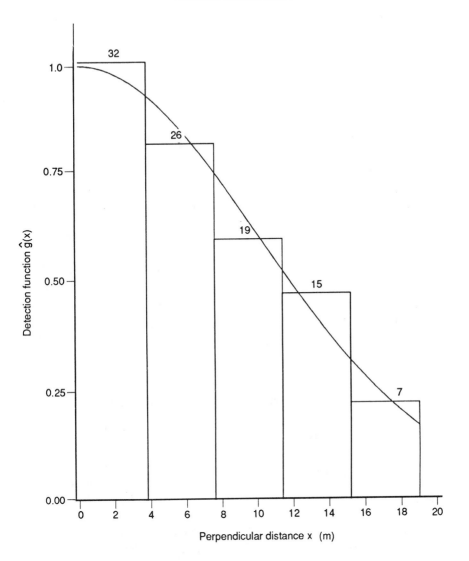

Fig. 4.2. Histogram of the example data using five distance categories. A half-normal detection function, fitted to the ungrouped data with $w = 19$ m, is shown and was used as a basis for final inference from these data.

the model near zero distance that is most critical, none of the model selection criteria of goodness of fit statistics, AIC and likelihood ratio tests give special emphasis to this region.

Some goodness of fit statistics for the example with $w = 19$ m and 20 groups are given in Table 4.3. These data were taken when all the assumptions were met; all four models fit the data well and yield similar estimates of density.

Table 4.3 Goodness of fit statistics for models fitted to the example data with $w = 19$ m and 20 groups

Model	χ^2	df	p
Uniform + cos	14.58	17	0.62
Uniform + poly	13.11	17	0.73
Half-normal + Hermite	13.63	17	0.69
Hazard-rate + cos	14.91	16	0.53

Real data are often heaped, so that no parsimonious model seems to fit the data well, as judged by the χ^2 test. Grouping can be carried out to smooth the distance data and, thus, obtain an improved fit. While grouping usually results in little change in $\hat{D}$, it provides a more acceptable assessment of the fit of the model to the data. If possible, groups should be selected so that there is one favoured rounding distance per interval, and it should occur at the midpoint of the interval. The grouped nature of the (rounded) data is then correctly recognized in the analysis. If cutpoints are badly chosen, heaping will generate spurious significant χ^2 values.

4.7 Estimation of density and measures of precision

4.7.1 The standard analysis

Preliminary analysis leads us to conclude that the half-normal model is an adequate model of the detection function, with truncation of the distance data at $w = 19$ m, fitted to ungrouped data, and using the empirical variance of n.

Replacing the parameters of Equation 3.4 by their estimators and simplifying under the assumptions that objects on the line are detected with certainty, detected objects are recorded irrespective of which side of the line they occur, and objects do not occur in clusters, estimated density becomes

$$\hat{D} = \frac{n \cdot \hat{f}(0)}{2L}$$

where n is the number of objects detected, L is the total length of transect line, and $\hat{f}(0)$ is the estimated probability density evaluated at zero perpendicular distance. For the example data, and adopting the preferred analysis, program DISTANCE yields $\hat{f}(0) = 0.08563$ with $\widehat{se}\{\hat{f}(0)\} = 0.007601$. The units of measure for $f(0)$ are 1/metres. Often, estimates of the effective strip width, $\mu = 1/f(0)$, are given in preference to $\hat{f}(0)$, since it has an intuitive interpretation. It is the perpendicular distance from the line for which the number of objects closer to the line that are missed is equal to the number of objects farther from the line (but within the truncation distance w) that are detected.

For the half-normal model, if data are neither grouped nor truncated, a closed form expression for $\hat{f}(0)$ exists (Chapter 3):

$$\hat{f}(0) = \sqrt{\{2/(\pi\hat{\sigma}^2)\}}$$

where $\qquad \hat{\sigma}^2 = \sum x_i^2/n$

Similarly, closed form expressions exist for the Fourier series estimator (uniform key + cosine adjustment terms) of $f(0)$ (Burnham *et al.* 1980: 56–61). However, generally the MLE of $f(0)$ must be computed numerically because no closed form equation exists.

The estimate of density for the example data truncated at 19 m, and using $\hat{f}(0)$ from DISTANCE, is

$$\hat{D} = n \cdot \hat{f}(0)/2L$$
$$= (99 \times 0.08563)/(2 \times 48)$$
$$= 0.0883$$

Since the units of $\hat{f}(0)$ are m and those for L are km, multiplying by 1000 gives

$$\hat{D} = 88.3 \text{ objects/km}^2$$

The estimator of the sampling variance of this estimate is

$$\widehat{var}(\hat{D}) = \hat{D}^2 \cdot \{[cv(n)]^2 + [cv\{\hat{f}(0)\}]^2\}$$

where

$$[cv(n)]^2 = \widehat{var}(n)/n^2 = 195.8/99^2 = 0.01998$$

and

117

$$[cv\{\hat{f}(0)\}]^2 = \widehat{var}\{\hat{f}(0)\}/\{\hat{f}(0)\}^2 = 0.007601^2/0.08563^2 = 0.007879$$

where $\widehat{var}\{\hat{f}(0)\}$ is based on approximately $n = 99$ degrees of freedom. Then

$$\widehat{var}(\hat{D}) = (88.3)^2 [0.01998 + 0.007879]$$

$$= 217.2$$

and

$$\widehat{se}(\hat{D}) = \sqrt{\widehat{var}(\hat{D})}$$

$$= 14.74$$

The coefficient of variation of estimated density is $cv(\hat{D}) = \widehat{se}(\hat{D})/\hat{D} = 0.167$, or 16.7%, which might be adequate for many purposes. An approximate 95% confidence interval could be set in the usual way as $\hat{D} \pm 1.96 \cdot \widehat{se}(\hat{D})$, resulting in the interval [59.4, 117.2]. Log-based confidence intervals (Burnham et al. 1987: 211–3) offer improved coverage by allowing for the asymmetric shape of the sampling distribution of $\hat{D}$ for small n. The procedure allows lower and upper 95% bounds to be computed as

$$\hat{D}_L = \hat{D}/C$$

and

$$\hat{D}_U = \hat{D} \cdot C$$

where

$$C = \exp \{1.96 \cdot \sqrt{\log_e(1 + [cv(\hat{D})]^2)}\}$$

This method gives the interval [63.8, 122.2], which is wider than the symmetric interval, but is a better measure of the precision of the estimate $\hat{D} = 88.3$. The imprecision in $\hat{D}$ is primarily due to the variance component associated with n; approximately 72% (i.e. $0.01998/(0.01998 + 0.007879)$) of $\widehat{var}(\hat{D})$ is due here to $\widehat{var}(n)$.

The use of 1.96 in constructing the above confidence intervals is only justified if the degrees of freedom of all variance components in

118

$\widehat{\text{var}}(\hat{D})$ are large, say greater than 30. In this example, the degrees of freedom for the component $\widehat{\text{var}}(\hat{n})$ are only 11. If there were only one variance component, it would be standard procedure to use the t-distribution on the relevant degrees of freedom, rather than the standard normal distribution, as the basis for a confidence interval. When the relevant variance estimator is a linear combination of variances, there is a procedure using an approximating t-distribution as the basis for the confidence interval. This more complicated procedure is explained in Section 4.7.4 below, and is used automatically by program DISTANCE.

The effective strip width is estimated by $\hat{\mu} = 1/\hat{f}(0) = 11.68$ m. The unconditional probability of detecting an object in the surveyed area, $a = 2wL$, is $P_a = \hat{\mu}/w = 0.61$, which is simply the ratio of the effective strip width to the truncation distance, $w = 19$ m. These estimates are MLE as they are one-to-one transformations of the MLE of $f(0)$.

In summary, we obtain $\hat{D} = 88.3$, $\widehat{\text{se}}(\hat{D}) = 14.7$, cv = 16.7%, with a 95% confidence interval of [63.8, 122.2]. Recalling that the true parameter $D = 80$, this particular estimate is a little high, largely because the sample size ($n = 105$, untruncated) happened to be above that expected ($E(n) = 96$). This is not unusual, given the large variability in n due to spatial aggregation of the objects, and the confidence interval easily covers the parameter. Some alternative analyses and issues and their consequences will now be explored.

4.7.2 Ignoring information from replicate lines

If the Poisson assumption ($\widehat{\text{var}}(n) = n$) had been used with $w = 19$ m and $L = 48$ km, then the estimate of density would not change, but $\widehat{\text{se}}(\hat{D})$ would be underestimated at 11.84, with 95% confidence interval of [67.98, 114.70]. While this interval happens to cover D, the method underestimates the uncertainty of the estimator $\hat{D}$; if many data sets were generated, the true coverage of the interval would be well below 95%. This procedure cannot be recommended; one should estimate the variance associated with sample size empirically from the counts on the individual replicate lines, including those lines with zero counts. For example, line 11 had no observations ($n_{11} = 0$), which must be included in the analysis as a zero count.

4.7.3 Bootstrap variances and confidence intervals

The selected model for $g(x)$ for the example data was the half-normal, with $w = 19$ m, fitted to ungrouped distance data. The MLE of $f(0)$ was 0.08563 with $\widehat{\text{se}} = 0.007601$. The bootstrap procedure (Section 3.7.4) can be used to obtain a more robust estimate of this standard error. The required number of series expansion terms can be estimated in each resample, and variance due to this estimation, ignored in the analytical method, is then a component of the bootstrap variance. As an illustration, 1000 bootstrap replications were performed, yielding an average $\hat{f}(0) = 0.08587$ with $\widehat{se} = 0.00748$. In this simple example where the true model is the fitted model without any adjustment terms, the two procedures yield nearly identical results.

A superior use of the bootstrap in line transect sampling is to sample with replacement from the replicate lines, until either the number of lines in the resample equals the number in the original data set, or the total effort in the resample approximately equals the total effort in the real data set. If the model selection procedure is automated, it can be applied to each resample, so that model misspecification bias can be incorporated in the variance estimate. Further, the density D may be estimated for each resample, so that robust standard errors and confidence intervals may be set that automatically incorporate variance in sample size (or equivalently, encounter rate) and cluster size if relevant, as well as in the estimate of $f(0)$. The method is described in Section 3.7.4, and an example of its application to point transect data is given in Section 5.7.2.

A possible analysis strategy is to carry out model selection and choice of truncation distance first, and then to evaluate bootstrap standard errors only after a particular model has been identified. Although model misspecification bias is then ignored, the bootstrap is computationally intensive, and its use at every step in the analysis will be prohibitive.

4.7.4 Satterthwaite degrees of freedom for confidence intervals

For the log-based confidence interval approach, there is a method to allow for the finite degrees of freedom of each estimated variance component in $\widehat{var}(\hat{D})$. This procedure dates from Satterthwaite (1946); a more accessible reference is Milliken and Johnson (1984). Assuming the log-based approach, $[\log_e(\hat{D}) - \log_e(D)]/cv(\hat{D})$ is well approximated by a t-distribution with degrees of freedom computed in the case of two variance components by the formula

$$df = \frac{[cv(\hat{D})]^4}{\dfrac{[cv(n)]^4}{k-1} + \dfrac{[cv\{\hat{f}(0)\}]^4}{n}}$$

where

$$[cv(\hat{D})]^2 = \frac{\widehat{var}(\hat{D})}{\hat{D}^2} \qquad [cv(n)]^2 = \frac{\widehat{var}(n)}{n^2} \qquad [cv\{\hat{f}(0)\}]^2 = \frac{\widehat{var}\{\hat{f}(0)\}}{\{\hat{f}(0)\}^2}$$

Given the computed degrees of freedom, one finds the two-sided $100(1 - 2\alpha)$ percentile of the t-distribution with these degrees of freedom; df is in general non-integer, but may be rounded to the nearest integer. Usually $\alpha = 0.025$, giving a 95% confidence interval. Then one uses the value of $t_{df}(0.025)$ in place of 1.96 in the confidence interval calculations, so that

$$\hat{D}_L = \hat{D}/C$$

and

$$\hat{D}_U = \hat{D} \cdot C$$

where

$$C = \exp\{t_{df}(0.025) \cdot \sqrt{\log_e(1 + [cv(\hat{D})]^2)}\}$$

This lengthens the confidence interval noticeably when the number of replicate lines is small.

We illustrate this procedure with the current example for which $[cv(n)]^2 = 0.01998$ on 11 degrees of freedom, and $[cv\{\hat{f}(0)\}]^2 = 0.007879$ on 99 degrees of freedom. Thus $[cv(\hat{D})]^2 = 0.027859$. The above formula for df gives

$$df = \frac{0.0007761}{\dfrac{0.0003992}{11} + \dfrac{0.00006208}{99}} = 21.02$$

which we round to 21 for looking up $t_{21}(0.025) = 2.08$ in tables. Using 2.08 rather than 1.96, we find that $C = 1.4117$, and the improved 95% confidence interval is [62.6, 124.7], compared with [63.8, 122.2] using $z = 1.96$. The Satterthwaite procedure is implemented in DISTANCE, so that it produces the improved interval.

This procedure for computing the degrees of freedom for an approximating t-distribution generalizes to the case of more than two components, for example when there are three parameter estimates, n, $\hat{f}(0)$ and $\hat{E}(s)$. Section 3.7.1 gives the general formula.

4.8 Estimation when the objects are in clusters

Often the objects are detected in clusters (flocks, coveys, schools) and further considerations are necessary in this case. The density of clusters (D_s), the density of individual objects (D), and average cluster size $E(s)$ are the biological parameters of interest in surveys of clustered populations, and several intermediate parameters are of statistical interest (e.g. $f(0)$). Here, we will assume that the clusters are reasonably well defined; populations that form loose aggregations of objects are more problematic.

It is assumed that the distance measurement is taken from the line to the geometric centre of the cluster. If a truncation distance w is adopted in the field (as distinct from in the analysis), then a cluster is recorded if its centre lies within distance w and a count made of **all** individuals within the cluster, including those individuals that are at distances greater than w. If the geometric centre of the cluster lies at a distance greater than w, then no measurement should be recorded, even if some individuals in the cluster are within distance w of the line. The sample size of detected objects n is the number of clusters, not the total number of individuals detected.

4.8.1 Observed cluster size independent of distance

If the size of detected clusters is independent of distance from the line (i.e. $g(x)$ does not depend on s), then estimation of D_s, D and $E(s)$ is relatively simple. The sample mean $\bar{s}$ is taken as an unbiased estimator of the average cluster size. Then $\hat{E}(s) = \bar{s} = \Sigma\, s_i/n$, where s_i is the size of the ith cluster. In general the density of clusters D_s and measures of precision are estimated exactly as given in Sections 4.3–4.7. Then, $\hat{D} = \hat{D}_s \cdot \bar{s}$; the estimator of the density of individuals is merely the product of the density of clusters times the average cluster size. Alternatively, the expression can be written

$$\hat{D} = \frac{n \cdot \hat{f}(0) \cdot \bar{s}}{2L}$$

The example data set used throughout this chapter is now reconsidered in view of the (now revealed) clustered nature of the population. The distribution of true cluster size in the population was simulated from a Poisson distribution and the size of detected clusters was independent of distance from the line. The value of s was simulated as $1 + $ a Poisson variable with a mean of two or, equivalently, $(s - 1) \sim \text{Poisson}(2)$, so that $E(s) = 3$. Theoretically, $\text{var}(s) = \text{var}(s - 1) = E(s - 1) = E(s) - 1 = 2$. Under the independence assumption, the sample mean $\bar{s}$ is an unbiased estimate of $E(s)$. The true density of individuals was 240. Estimates of D_s (called D in previous sections), $f(0)$, effective strip width and the various measures of precision are exactly those derived in Section 4.7.

The estimated average cluster size, $\bar{s}$, for the example data with $w = 19$ m is 2.859 ($\bar{s} = 283/99$) and the empirical sampling variance on $n - 1 = 98$ degrees of freedom is

$$\widehat{\text{var}}(\bar{s}) = \frac{\sum\limits_{i=1}^{n} (s_i - \bar{s})^2}{n(n - 1)}$$

$$= 0.02062$$

so that

$$\widehat{\text{se}}(\bar{s}) = \sqrt{0.02062}$$

$$= 0.1436$$

These empirical estimates compare quite well with the true parameters; $\text{var}(\bar{s}) = 2/n = 2/99 = 0.0202$, $\widehat{\text{se}}(\bar{s}) = 0.142$. If one uses the knowledge that cluster sizes were based on a Poisson process, one could estimate this true standard error as $\sqrt{\{(\bar{s} - 1)/n\}} = \sqrt{(1.859/99)} = 0.137$, which is also close to the true value. The point here is that the empirical estimate is quite good and can be computed when the Poisson assumption is false.

A plot of cluster size s_i vs. distance x_i (Fig. 4.3) provides only weak evidence of dependence ($r = 0.16, p = 0.10$). In this case, we take $\hat{E}(s) = \bar{s}$, the sample mean. Thus, the density of individuals is estimated as

$$\hat{D} = \hat{D}_s \cdot \bar{s}$$

$$= 88.3 \times 2.859$$

$$= 252.4 \text{ individuals/km}^2$$

Then, for large samples,

$$\widehat{\mathrm{var}}(\hat{D}) = \hat{D}^2 \cdot \{[\mathrm{cv}(n)]^2 + [\mathrm{cv}\{\hat{f}(0)\}]^2 + [\mathrm{cv}(\bar{s})]^2\}$$

$$= 252.4^2 \cdot [0.01998 + 0.007879 + 0.002523]$$

$$= 1936$$

so that

$$\widehat{\mathrm{se}}(\hat{D}) = \sqrt{1936} = 44.0$$

This gives $\hat{D} = 252.4$ individuals/km^2 with cv = 17.4%. The log-based 95% confidence interval, using the convenient multiplier $z = 1.96$, is [179.8, 354.3]. This interval is somewhat wide, due primarily to the spatial variation in n; $\widehat{\mathrm{var}}(n)$ makes up 66% of $\widehat{\mathrm{var}}(\hat{D})$, while $\widehat{\mathrm{var}}(\hat{f}(0))$ contributes 26% and $\widehat{\mathrm{var}}(\bar{s})$ contributes only 8% (e.g. 66% = {0.01998/(0.01998 + 0.007879 + 0.002523)} × 100).

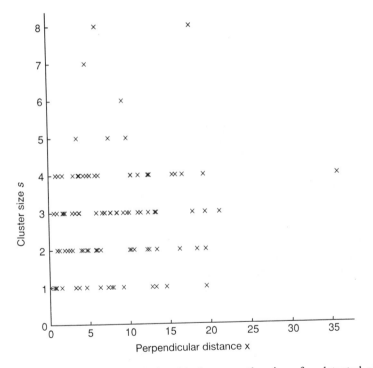

Fig. 4.3. Scatterplot of the relationship between the size of a detected cluster and the distance from the line to the geometric centre of the cluster for the example in which size and detection distance are independent. The correlation coefficient is 0.16.

A theoretically better confidence interval is one based not on the standard normal distribution (i.e. on $z = 1.96$) but rather on a t-distribution with degrees of freedom computed here as

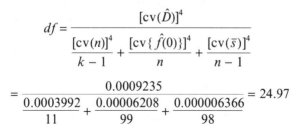

$$df = \frac{[cv(\hat{D})]^4}{\dfrac{[cv(n)]^4}{k-1} + \dfrac{[cv\{\hat{f}(0)\}]^4}{n} + \dfrac{[cv(\bar{s})]^4}{n-1}}$$

$$= \frac{0.0009235}{\dfrac{0.0003992}{11} + \dfrac{0.00006208}{99} + \dfrac{0.000006366}{98}} = 24.97$$

which we round to 25. Given this computed value for the degrees of freedom, one finds the two-sided $100(1 - 2\alpha)$ percentile $t_{df}(\alpha)$ of the t-distribution. In this example, we find $t_{25}(0.025) = 2.06$. Using 2.06 rather than 1.96 in the log-based confidence interval procedure, $C = 1.4282$, giving an improved 95% confidence interval of [176.8, 360.5]. It is this latter interval which DISTANCE computes, applying the procedure of Satterthwaite (1946) to $\log_e(\hat{D})$.

Plots and correlations should always be examined prior to proceeding as if cluster size and detection distance were independent. In particular, some truncation of the data will often have the added benefit of weakening the dependence between s_i and x_i. If truncation is appropriate, then $\hat{E}(s)$ should be based on only those clusters within $(0, w)$. Our experience suggests that data from surveys of many clustered populations can be treated under the assumption that s_i and x_i are independent. For small clusters (e.g. coveys of quail of 5–12 or family groups of antelope of 2–4), the independence assumption is likely to be reasonable. This allows the analysis of $f(0)$, D_s, and measures of their precision to be separated from the analysis of the data on cluster size and its variability. Then, estimation of the density of individuals is fairly simple.

4.8.2 Observed cluster size dependent on distance

The analysis of survey data where the cluster size is dependent on the detection distance is more complicated because of difficulties in obtaining an unbiased estimate of $E(s)$ (Drummer and McDonald 1987; Drummer et al. 1990; Otto and Pollock 1990). The dependence arises because large clusters might be seen at some distance from the line (near w), while small clusters might remain undetected. This phenomenon causes an overestimation of $E(s)$ because too few small clusters are detected (i.e. they are underrepresented in the sample). Thus, $\hat{D} = \hat{D}_s \cdot \bar{s}$ is also an overestimate. Another complication is that large clusters might be more easily detected near w than small clusters, but their size might be underestimated due to reduced detectability of individuals at long distances. This phenomenon has a counter-balancing effect on the estimates of $E(s)$ and D. The dependence of x_i on s_i is a case of size-biased sampling (Cox 1969; Patil and Ord 1976; Patil and Rao 1978; Rao and Portier 1985).

The analysis of sample data from clustered populations where a dependence exists between the distances x_i and the cluster sizes s_i can take several avenues. Some of these require that the detection function $g(x)$ is fitted unconditional on cluster size, using the robust models and model selection tools already discussed. The simplest method exploits the fact that size bias in detected clusters does not occur at distances from the line for which detection is certain. Hence, $E(s)$ may be estimated by the mean size of clusters detected within distance v, where $g(v)$ is reasonably close to one, say 0.6 or 0.8. In the second method, a cluster of size s_i at distance x_i from the line is replaced by s_i objects, each at distance x_i. Thus, the sampling unit is assumed to be the individual object rather than the cluster, and the issue of estimating true mean cluster size is side-stepped. For the third method, data are stratified by cluster size (Quinn 1979, 1985). The selected model is then fitted independently to the data in each stratum. If size bias is large or cluster size very variable, smaller truncation distances are likely to be required for strata corresponding to small clusters. The fourth method estimates cluster density D_s conventionally, as does the first. Then, given the x_i, $E(s)$ is estimated by some form of regression of s_i on the x_i (i.e. an appropriate model is identified for $E(s|x)$). This sequential procedure seems to have a great deal of flexibility. In the final approach considered here, a bivariate model of $g(x, s)$ is fitted to the data to obtain the estimates of D, D_s and $E(s)$ simultaneously. The first four approaches are illustrated in this section using program DISTANCE, and the fifth using program SIZETRAN (Drummer 1991).

The data used in this section are sampled from the same population as in earlier sections of this chapter (i.e. $L = 48$, $D_s = 80$ $f(0) = 0.079$, $E(s) = 3$ and $E(n) = 96$, so that $D = 240 = 3 \times 80$). The half-normal detection function was used, as before, but σ was allowed to be a function of cluster size:

$$\sigma(s) = \sigma_0 \left(1 + b \cdot \frac{s - E(s)}{E(s)}\right)$$

where $b = 1$ and $E(s) = 3$ in the population. Selecting $b = 1$ represents a strong size bias and corresponds to Drummer and McDonald's (1987) form with $\alpha = 1$. Cluster size in the entire population (detected or not) was distributed as $s \sim (1 + \text{Poisson})$. Given a cluster size s, the detection distance was generated from the half-normal detection function $g(x_i|s_i)$. Because of the dependence between cluster size and detection distance, the distance data differ from those in the earlier parts of this chapter. In particular, some large clusters were detected at greater distances (e.g. one detection of 5 objects at 50.9 m). Because of the dependence on cluster size, the bivariate detection function $g(x, s)$ is not half-normal. This detection function is monotone non-increasing in x and monotone non-decreasing in s. In addition, the detected cluster sizes do not represent a random sample from the population of cluster sizes, as small clusters tend to remain undetected except at short distances. Thus, the size of **detected** clusters is not any simple function of a Poisson variate. Generation of these data is a technical matter and is treated in Section 6.7.2.

A histogram of the distance data indicates little heaping and a somewhat long right tail (Fig. 4.4). Truncation at 20 m seemed reasonable and eliminated only 16 observations, leaving $n = 89$ (15% truncation). Truncation makes modelling of the detection function easier and always reduces, at least theoretically, the correlation between detection distance and cluster size. Three robust models were chosen as candidates for modelling $g(x)$: uniform + cosine, half-normal + Hermite, and hazard-rate + cosine. All three models fit the truncated data well. AIC suggested the use of the uniform + cosine model by a small margin (506.26 vs. 506.96 for the half-normal), and both models gave very similar estimates of density. The hazard-rate model (AIC = 509.209) provided rather high estimates of density with less precision, although confidence intervals easily covered the true parameter. The uniform + cosine model and the half-normal model both required only a single parameter to be estimated from the data, while the hazard-rate has two parameters. This may account for some of the increased standard error of the hazard-rate model, but the main reason for the high estimate and standard error is that the hazard-rate model attempts to fit the spike in the histogram of Fig. 4.4 in the first distance category. Because we know the true model in this case, we know the spike is spurious, and arises because for this data set, more simulated values occurred within 2.5 m than would be expected. Generally, if such a spike is real, the hazard-rate model yields lower bias (but also higher variance) than most series expansion models, whereas its performance is poor if the spike is spurious. Since AIC selected the uniform + cosine model, we use it below to illustrate methods of analysis of the example data.

The uniform + cosine model for the untruncated data required five cosine terms to fit the right tail of the data adequately (Fig. 4.4). Failure to truncate the data here would have resulted in lower precision, the model would have required five cosine terms instead of just one

127

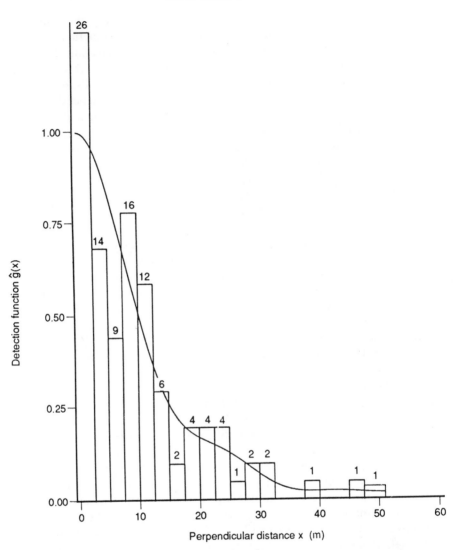

Fig. 4.4. Histogram of the example data using 20 distance categories for the case where cluster size and detection distance are dependent. The fit of a uniform + 5-term cosine model is shown.

(Fig. 4.5), and the mean cluster size would have been less reliably estimated (below).

The uniform key function and a 1-term cosine model fit the truncated data well (Fig. 4.5, $0.38 \leq p \leq 0.71$, depending on the grouping used for the χ^2 test). The estimated density of clusters was 81.56 ($\hat{se} = 12.43$),

128

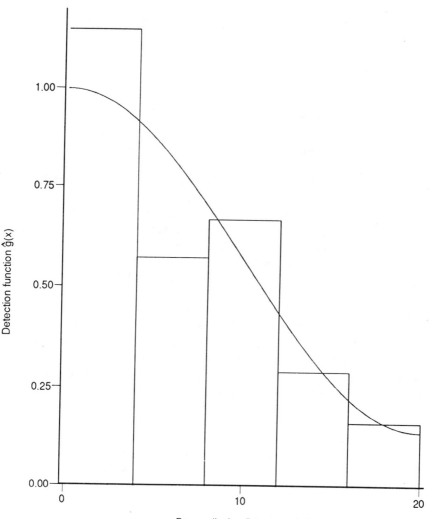

Fig. 4.5. Histogram of the example data using five distance categories and truncation at $w = 20$ m for the case where cluster size and detection distance are dependent. The fit of a uniform + 1-term cosine model is shown.

quite close to the true value (80). The mean cluster size from the sample data was 3.258 ($\widehat{se} = 0.134$) which is surely too high in view of the size-biased sampling caused by the correlation between cluster size and detection distance. However, truncation at $w = 20$ reduced this correlation from 0.485 to 0.224 so this uncorrected estimate of $E(s)$ may not be heavily biased. Multiplying the density of clusters by the uncorrected estimate of mean cluster size (3.258), the density of individuals is estimated as 265.7 ($\widehat{se}$ = 41.9; 95% CI = [195.4, 361.4]), which is a little high, but still a quite acceptable estimate for these data, for which actual density was 240 individuals/km^2.

(a) *Truncation* Observed mean cluster sizes and standard errors for various truncation distances are shown in Table 4.4. The detection function $g(x)$ was estimated using truncation distance w and mean cluster size was estimated after truncating data at distance v, $v \leqslant w$. The gain in precision by reducing the truncation distance from 51 m to 20 m arises because predominantly large clusters are truncated, and variability in the size of remaining clusters is reduced. It is clear from Table 4.4 that 20 m is too large a truncation distance for unbiased estimation of mean cluster size, since mean cluster size continues to fall when the truncation distance is reduced further. The choice of truncation distance is a compromise between reducing bias and retaining adequate precision. Here, mean cluster size appears to stabilize at a truncation distance of 10.5 m, for which $\hat{g}(10.5) = 0.5$. Thus, mean cluster size is estimated to be 3.116 with $\widehat{se} = 0.152$. Replacing the estimates $\bar{s} = 3.258$ and $\widehat{se} = 0.134$ by these values, density is estimated as 254.1 individuals/km^2, with $\widehat{se} = 40.7$ and approximate 95% confidence interval [186.1, 347.0] (based on $z = 1.96$ rather than Satterthwaite's correction). This estimate is closer to the true value of 240 individuals/km^2, and precision is almost unaffected, because the contribution to the overall variance due to variation in cluster size is slight.

Table 4.4 Observed mean cluster sizes and standard errors for various truncation distances v. Probability of detection $\hat{g}(v)$ at the truncation distance v for cluster size estimation was estimated from a uniform + 1-term cosine model with $w = 20$ m (Fig. 4.5) for $v \leqslant 20$ m, and from a uniform + 5-term cosine model with $w = 51$ m for $v = 51$ m

Truncation distance v(m)	n	$\bar{s}$	$\widehat{se}(\bar{s})$	$\hat{g}(v)$
51.0	105	3.581	0.150	0.02
20.0	89	3.258	0.134	0.14
13.6	80	3.188	0.144	0.30
10.5	69	3.116	0.152	0.50
9.1	61	3.098	0.166	0.60
7.6	49	3.061	0.192	0.70
6.0	44	3.114	0.206	0.80
4.1	37	3.081	0.214	0.90

Hence if sample size is large, one may select a truncation point for the estimation of $E(s)$ that is smaller than the truncation point for the estimation of $f(0)$ to reduce the size bias in $\hat{E}(s)$. For example, $E(s)$ might be estimated by the mean size of clusters detected within a distance x of the line, where $g(x) = 0.6$. Often bias reduction is more important than precision in estimating mean cluster size because its relative contribution to $\mathrm{var}(\hat{D})$ may be small, as in this example.

(b) *Replacement of clusters by individuals* If a cluster of size s_i is replaced by s_i objects at the same distance, the assumption that detections are independent is violated. This compromises analytic variance estimates and model selection procedures. The first difficulty may be overcome by using robust methods for variance estimation, but model selection is more problematic. If likelihood ratio tests are used to determine the number of terms, too many terms are fitted on average, since heaping in the data at distances where large clusters were recorded yield significant departures from a smooth detection function when observations are assumed to be independent. The effect may be reduced by imposing a monotonicity constraint (Section 3.4.5). Another option is to select a model taking clusters as the sampling unit, then refit the model (with the same series terms, if any) to the data with object as the sampling unit. Neither of these is entirely satisfactory. If both strategies are adopted in the same analysis, so that a uniform + 1-term cosine model is fitted to the distance data truncated at 20 m, the following estimates are obtained. Number of objects detected, $n = 290$. Estimated density, $\hat{D} = 255.6$ objects/km^2, with analytic $\widehat{se} = 38.3$ and 95% confidence interval [184.7, 353.8]. These estimates are very close to those obtained assuming cluster size is independent of distance, although the point estimate is rather closer to the true density of 240 objects/km^2. Average cluster size can be estimated by the ratio of estimated object density (255.6) to estimated cluster density (81.56), giving 3.134. The precision of this estimate could be quantified using the bootstrap. In each bootstrap resample, both densities, and hence their ratio, would be estimated, and a variance and confidence interval obtained as described in Section 4.7.3.

This procedure cannot generally be recommended. However, it may be useful if the population being sampled occurs in loose aggregations, rather than tight, easily defined clusters. The distance to each individual object should ideally be measured in this case, although it may be sufficient to record positions and sizes of smaller groups within a cluster. The method will often perform poorly unless sample size is fairly large.

(c) *Stratification* Choice of number of strata is determined largely by sample size. The more strata, the greater the reduction in size bias, but an adequate sample size for estimating $f(0)$ is required in each stratum (perhaps at least 20–30 per stratum). Defining two strata, corresponding to cluster sizes 1–3 and $\geqslant 4$, sample sizes before truncation are 52 and 53 respectively. If four strata are defined, for cluster sizes 1–2, 3, 4 and $\geqslant 5$, sample sizes before truncation are 29, 23, 26 and 27. The data were analysed for both choices of stratification.

Table 4.5 Summary of results for different stratification options. Model was uniform with cosine adjustments; distance data were truncated at $w = 20$ m, except for the stratum comprising clusters of size 5–9, for which $w = 35$ m

Cluster sizes	Sample size after truncation	Effective strip width (m)	$\hat{D}$	$\widehat{se}(\hat{D})$	95% CI for D
All	89	11.4	265.7	41.9	(195.4, 361.4)
1–3	51	11.0	96.9	17.6	
4–9	38	12.1	147.4	35.0	
All			244.3	39.2	(178.8, 333.8)
1–2	29	10.0	51.2	16.0	
3	22	13.7	50.1	16.3	
4	22	11.7	78.2	21.0	
5–9	24	22.6	53.3	18.0	
All			232.8	35.8	(172.5, 314.2)

Results are summarized in Table 4.5. In this case, no precision is lost by stratification, despite the small samples from which $f(0)$ was estimated, and the estimated densities were closer to the true value of 240 objects//km^2 than for the case without stratification. In our experience, loss of precision arising from stratification by cluster size is seldom large, provided sample size in each stratum does not fall below 20, and the method is a simple way of reducing the effects of size-biased sampling. Mean cluster size may be estimated by a weighted average of the mean size per stratum, with weights equal to the estimated density of clusters by stratum. Alternatively, $E(s)$ may be estimated as overall $\hat{D}$ from the stratified analysis divided by $\hat{D}_s$ from the unstratified analysis. For two strata, this yields $\hat{E}(s) = 244.3/81.56 = 2.995$, and for four strata, $\hat{E}(s) = 232.8/81.56 = 2.854$. Both estimates are close to the true mean cluster size of 3.0. The reader is referred to Drummer (1985) and Quinn (1985) for further information on stratification.

(d) *Regression estimator* The procedure we recommend in most cases is a regression of s_i or $\log_e(s_i)$ on $\hat{g}(x_i)$ (Section 3.6.3). This allows an

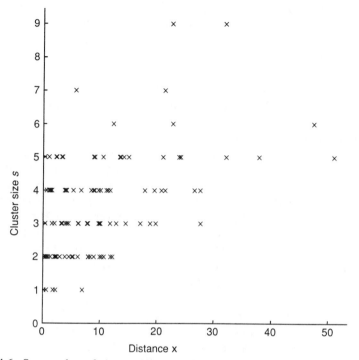

Fig. 4.6. Scatterplot of the relationship between the size of a detected cluster and the distance from the line to the geometric centre of the cluster for the example in which probability of detection is a function of cluster size. The correlation coefficient is 0.485 ($w = \infty$).

estimate of $E(s)$ at the point where $\hat{g}(x_i) = 1$; that is, the point at which detectability is certain, where size bias should not occur. Proper truncation of the distance data should be considered prior to the regression analysis (e.g. $g(x) \doteq 0.15$). Applying this method to the example, with dependent variable $\log_e(s_i)$, yields $\hat{E}(s) = 2.930$ and $\widehat{se}\{\hat{E}(s)\} = \sqrt{[\widehat{var}\{\hat{E}(s)\}]} = 0.139$, which is close to the true mean cluster size of 3.0 with good precision. The corresponding density estimate is 239.0 individuals/km^2, with $\widehat{se} = 38.1$, cv = 16.0% and 95% confidence interval [171.3, 333.5]. The regression approach allows $f(0)$ to be estimated using all the robust theory available and then treats the estimation of mean cluster size as a separate problem. The analyst has good control over the estimation under this procedure. A scatter plot of cluster size against distance or estimated detection probability can be used to investigate the form of the relationship, although the scatter can be wide (Fig. 4.6).

The regression estimate of $E(s)$ reduces the bias but some precision may be lost in correcting for the size-biased sampling of cluster size.

Often, var($\bar{s}$) is a small component of var($\hat{D}$), so that little precision is lost by applying an adjustment for size bias. Use of $\bar{s}$ as an estimate of $E(s)$ is not recommended if dependence between cluster size and detection distance is suspected.

(e) *Bivariate models for the detection function* This methodology, due to Drummer and McDonald (1987), relies on several parametric models, each incorporating a parameter (α) to reflect the size bias in the sample data. The data are transformed to x/s^{α} and the bivariate detection function is expressed as $g(x, s) = g(x/s^{\alpha})$; $\alpha = 0$ represents the special case where cluster size and detection distance are independent and no size bias exists (Otto and Pollock 1990). Program SIZETRAN was used for the analysis of the example data and four models were considered: the negative exponential, the half-normal, the reversed logistic, and the generalized exponential (Drummer and McDonald 1987, Table 1). These models incorporate the size bias by modelling the scale parameter as a simple increasing power function of observed cluster size, s^{α}. Under this approach, truncation is not an option with current software and separate consideration of $\hat{E}(s)$ is not necessary.

The choice of the best model might be guided by AIC and this leads to the half-normal model (Table 4.6). The generalized exponential model failed to converge and cannot be considered. The estimates under the half-normal model are quite good (Table 4.6), but the precision is poorer than under the sequential approach using regression (cv = 21.3% instead of 16.0%). This model fits the data well as judged by χ^2 goodness of fit tests ($p > 0.15$). The data were simulated from this model with $\alpha = 1$, so that it would be expected to fit well. Estimation under the other models seems less satisfactory; the estimated density of individuals is too high under both the negative exponential and the reversed logistic (Table 4.6). Still, the confidence intervals cover the true density, partially because the estimated standard errors are so large. In each case, there is clear evidence of a size-biased sample (i.e. α is significantly greater than zero, Table 4.6).

Table 4.6 Summary of results for three models allowing for dependence between cluster size and detection distance (standard errors). The parameter α is incorporated into these models to account for size-biased sampling (Drummer and McDonald 1987). Data are untruncated ($w = \infty$)

Model	AIC	$\hat{\alpha}$		$\hat{D}_s$		$\hat{D}$	
Negative exponential	674.4	1.113	(0.214)	140.7	(41.4)	397.6	(114.6)
Half-normal	668.9	1.092	(0.150)	87.1	(19.3)	247.5	(52.9)
Reversed logistic	671.1	1.076	(0.171)	95.2	(27.8)	271.4	(77.5)

If the size-biased nature of the data had been ignored (i.e. $\alpha = 0$), the point estimates would have been less satisfactory. In general, the point estimates of the density of individuals would have been too high (e.g. $\hat{D} = 385.0$ and 384.6 for the negative exponential and reversed logistic, respectively). In this example, AIC selected a good model and the point estimates were quite satisfactory. Again, this illustrates the importance of meeting the key assumptions and obtaining an adequate sample of quality data.

The bivariate approach is interesting and appealing from a statistical viewpoint, but precision seems poorer than for the sequential regression approach. Software development is required to address convergence problems, and to allow the user to specify a finite truncation width. Further study is needed to investigate the robustness of the approach. The method of Quang (1991), using Fourier series, may partially address this aspect.

4.9 Assumptions

Assumptions of line transect sampling are covered in detail in Section 2.1, and further discussion is given in Chapter 7. We outline a few issues here, to counter some of the more common misconceptions about what is assumed in deriving density or abundance estimates.

Most model selection procedures and some variance estimation procedures assume that objects are randomly and independently distributed throughout the study area. Provided lines are randomly located, or a systematic grid of lines is randomly positioned in the study area, the assumption is not required. If object density is highly variable, or dependence between detections is strong, then the possible effect on model selection should be borne in mind, and robust variance methods should be adopted, with care taken to ensure that the correct sampling unit is selected; whether a detection associated with one sampling unit is made should be largely independent of detections made in other sampling units. Often, all effort carried out by a single observer in a single session comprises a suitable sampling unit. Detections made within the unit might be highly dependent (e.g. if one bird calls in response to the calls of another, both might be detected by the observer, or encounter rate might be abnormally high on one leg of a marine mammal survey because of exceptional sighting conditions). Between units, dependence should be slight. If random lines are used, the appropriate sampling unit is the line, and all data associated with it. If the design comprises a systematic grid of lines, use of lines as sampling units should again prove satisfactory. These issues are discussed further in

Section 3.7. The most extreme departure from independent detections is when objects occur in clusters. Strategies for this case are outlined in Section 4.8.

Estimation of $g(0)$ for line transect surveys is considered in Chapter 6. There is currently no approach to this problem that is wholly satisfactory, so whenever possible, surveys should be designed to ensure that $g(0) = 1$. The solution of guarding the centreline can be counter-productive if this gives rise to two detection processes. If one process generates detections at large distances, but is such that $g(0)$ is appreciably less than one, and the other generates detections only at small distances, then the composite detection function will be impossible to model adequately. If such a field strategy is adopted, data from the two processes should be recorded and analysed separately, although problems are likely to remain. If it is suspected that $g(0)$ is less than one, methods that might increase it include using more observers to cover the line, travelling more slowly along the line, using only experienced observers, improving the training of observers, and upgrading optical aids. For terrestrial surveys in which animals are flushed, trained dogs can be an effective aid, allowing a wider area to be efficiently searched.

Random movement of objects before detection generates positive bias in estimates of object density. Hiby (1986) showed that bias is small provided that object movement is slow relative to that of the observer (up to around a third of the observer's speed). A strategy for line transect analysis of fast moving objects is outlined in Section 7.6. Movement in response to the observer is problematic, and is discussed in Section 2.1. From a practical viewpoint, field procedures should be developed that ensure that most detections occur at distances for which responsive movement is unlikely to have occurred. In other words, the observer should strive to detect the object before the object is able to move far from its initial position in response to the observer's presence. If this is not possible, the methods of Turnock and Quinn (1991) or Buckland and Turnock (1992), which use ancillary data to adjust for the effect of movement, might be attempted. The latter method is described in Section 6.4.

If the distance data appear to have a distinct mode away from the origin, the analysis is problematic. This might happen by chance, as a result of heaping, or through the presence of evasive movement prior to detection. Some robust models will attempt to fit the data near the origin, so that the mode of the density function is to the right of the origin. In these cases, it is often prudent to constrain the estimated detection function to be monotone non-increasing in an attempt to minimize bias. A weak constraint is to impose the condition $\hat{g}(x) \leqslant g(0) = 1$ for all $x > 0$. This condition is often sufficient to achieve a monotone non-increasing function and satisfactory estimates of $f(0)$. The alternative is a constraint forcing a strict non-increasing function such that $\hat{g}(x_i) \geqslant \hat{g}(x_{i+1})$, where $x_i < x_{i+1}$. The default option of DISTANCE imposes the strong constraint. To reduce the computational cost of applying the constraint, the estimated density is evaluated at just ten points, and the fit is modified if the constraint is not satisfied for all successive pairs of these points. The user may instead select the weak constraint, or override both constraints. DISTANCE warns the user if a constraint has caused the model fit to be modified. In this case, the bootstrap estimate of var($\hat{f}(0)$) is recommended, since the assumptions on which the analytic variance is based are violated.

Consistent bias in distance estimation should be avoided. If distances are overestimated by 10%, densities are underestimated by 9%; if they are underestimated by 10%, densities are overestimated by 11%. If on the other hand distance estimation is unbiased on average, measurement errors must be large to be problematic. For marine surveys, reticles or graticules (Section 7.4.2) are almost essential for accurate distance estimation. In terrestrial surveys, distances can often be measured, and if this is not practical, good range finders can be effective up to around 300 m. Distance categories can be accurately determined in aerial surveys by lining up markers on the windows with markers or streamers on the wing struts, although the height of the aircraft must be accurately measured and constant. In hilly terrain, perpendicular distances from the aircraft must be determined by other means.

4.10 Summary

Data analysis is relatively easy if the survey is well designed and the data properly collected. The analysis of small samples, especially where some assumptions have been violated, is more problematic. The analysis of 'good' data, such as here, is relatively easy using available software. Adequate analysis cannot be carried out without specialist software.

An objective strategy must be followed, such as that outlined in this chapter and Section 2.5. The data must be checked for recording or data-entry errors. Plotting the distance data as histograms will often reveal anomalies that must be further considered. Truncation of some observations in the right tail of the distance data should always receive consideration. Several candidate, robust models should be considered. The use of AIC and other criteria are helpful in selecting the best model, or a small subset of good models, for final analysis and inference. Once a model is selected, MLE is used to obtain parameter estimates and measures of their precision. With good data (adequate sample size and validity of the key assumptions), inference using two or three good, robust models is likely to yield similar estimates. This is reassuring because the methods to select the best model are subject to uncertainty.

If objects on the centreline are missed, $E(\hat{D})$ will be too low. If 20% of objects on the centreline are missed, the density estimate can be expected to show a negative bias of around 20%. Movement prior to detection is also problematic. Measurement errors, especially near the centreline, are more difficult to treat. If measurement errors are random, then the sampling variance may be somewhat inflated, but bias may be small. Systematic measurement errors invariably generate bias and should be avoided. Valid inference depends on field design and attention to the assumptions. While analysis procedures are robust to some types of assumption failure, there is no substitute for quality data taken carefully under the assumptions. Searching should be conducted such that the distance data have a broad shoulder. The presence of a shoulder makes model selection less important and improves the quality of inference. The reader is urged to study the material in Chapter 7 prior to the conduct of a survey involving distance sampling.

These strategies for analysis carry over to more complicated surveys involving stratification, surveys repeated in time using the same lines, multiple observers, aerial or underwater platforms, or samples of very large areas. Some of these issues are illustrated in Chapter 8 (and by Burnham *et al.* 1980: 41–55), and specialized theory is extended in Chapter 6.

Surveys of clustered populations require additional care in counting the number of individuals in each cluster detected and addressing the possible size-biased aspects of such sampling. Plotting the cluster sizes s_i against the x_i distances is always recommended. Our experience suggests that size bias is often a minor issue if cluster size is not too variable; proper truncation of perpendicular distance data can often allow simple models to provide valid inference concerning the density of clusters and individuals. However, if the largest cluster is, say, more than five times the size of the smallest, correction for possible size bias should be investigated. When cluster size is highly variable (e.g. from one or two individuals to many thousands, as in some species of marine mammals), then very careful modelling and analysis of the data is required.

Populations in large, loose aggregations, scattered around the sample area, are problematic. Theory and software are readily available for the analysis of sample data from populations of individuals randomly distributed in space, and the same is true of populations distributed under some regular stochastic process that generates some degree of spatial aggregation, by computing var(n) empirically. Good theory and software

138

now exist for the analysis of populations that are clustered in definable clusters where the cluster size is not too variable. Difficulties arise when populations are spatially distributed in loose clusters whose boundaries, and therefore size, must be determined subjectively. This situation is in need of additional research, but bootstrap methods may play an important role in the analysis of such data. If at all possible, the location of each individual object should be recorded in this circumstance, so that the method of Section 4.8.2(b) can be applied, but the cluster to which each individual belongs should also be noted, to allow comparative analyses of clusters. Populations in large or highly variable groups require great care in estimating $E(s)$ in ways that minimize or avoid bias. Estimation of average cluster size must receive special emphasis in the design of the survey and the pilot study (e.g. temporarily leaving the planned centreline in aerial surveys of cetaceans to count individuals more accurately).

The following is intended as a crude checklist of the stages required to carry out a full analysis of line transect data. Not all steps are necessarily required in any given analysis, especially if similar data sets have been analysed previously.

1. Key in and validate the data. The data should not be aggregated in any way prior to entry. Thus if distances are ungrouped, they should not be entered as grouped data, even if they are subsequently grouped for analysis. Distances should be entered by line, so that individual lines can be defined as the sampling units. For stratified designs, these lines should be allocated to their strata.

2. Plot histograms of the perpendicular distance data, using different choices for the cutpoints, and fit a preliminary model to the data. Examine the histograms for evidence of failure of assumptions. If data are ungrouped, assess whether they should be grouped before analysis, selecting group cutpoints to reduce the effect of heaping, or to alleviate the effects of a spurious spike in the data at zero distance (Section 4.5). If data are grouped, assess whether any groups should be amalgamated.

3. Identify a truncation point w for perpendicular distances, preferably such that $\hat{g}(w) \doteq 0.15$, although truncation of roughly 5% of observations is often satisfactory (Section 4.3). Assess from the histograms whether this truncation distance is reasonable; if not, select one or more alternatives. Try fitting a few models, possibly with different grouping options or different truncation points.

4. Where relevant, select an appropriate truncation point w and an appropriate choice of grouping (if any). Fit several models that satisfy the model robustness, shape and estimator efficiency criteria. We recommend some or all combinations of a half-normal, uniform

or hazard-rate key with simple or Hermite polynomial or cosine adjustments. Select a single model, for example using Akaike's Information Criterion, and assess its adequacy using goodness of fit (Section 4.6). If the fit is poor, investigate the reasons, and evaluate possible solutions. Assess the sensitivity of estimation to the model selected; if sensitivity is high (e.g. the detection curve is excessively spiked under one or more models), examine whether the estimates from the selected model should be replaced or supplemented by those from other models that yield adequate fits.

5. If the detections are of clusters of objects, assess whether there is evidence of size bias, and if necessary, try one or more of the methods of Section 4.8 to correct for it.

6. Having identified a model for the perpendicular distance data, review the options for variance estimation, for stratifying some or all components of estimation, and for including covariates. Select options that are likely to reduce bias; of the options remaining, select those that yield the most efficient estimation. Fit the data using the preferred model(s) and options.

5

Point transects

5.1 Introduction

Songbird surveys often utilize point transects rather than line transects for several reasons. Once the observer is at the point, he or she can concentrate solely on detecting, locating and identifying birds, without the need to traverse what may be difficult terrain; he or she can take the easiest route into and away from the point, whereas good line transect practice dictates that the observer follows routes determined in advance and according to a randomized design. Further, patchy habitats can be sampled more easily by point transects. Frequently, density estimates are required for each habitat type, or estimation is stratified by habitat to improve precision. Designing a point transect survey so that each habitat type is represented in the desired proportions is easier than for line transects, and describing the vegetation structure associated with a point is also easier than for a line. If line transects are used in patchy habitats, either each line traverses several habitat types, and data must be recorded separately for each section of line within a single habitat type, or the design comprises many short lines, so that end effects (e.g. objects detected behind the observer when he or she first starts a transect, or objects detected by the observer as he or she approaches the end of a transect, which are beyond the area to be surveyed) become problematic. Other advantages of point transects are that known distances from the points may be flagged, to aid distance estimation, and only the observer-to-object distance is required, which is easier to estimate than the perpendicular distance required in line transect sampling if the observer is far from that part of the line closest to the object.

At the time of writing, point transect sampling seems to be restricted to bird surveys, although the theory also applies to the cue count and trapping web methods described in Chapter 6. The disadvantages of point transect sampling that make it unsuitable for many purposes include the following. Objects may be disturbed or flushed by an observer approaching the point. It is difficult to determine which of these would have been detected from the point, but if they are ignored, density will be underestimated. The observer may detect many objects and waste much time while travelling between points; for line transects, a higher proportion of time in the field is spent surveying, and a higher proportion of detections is made while surveying. Thus, point transects may be inefficient for objects that occur at low densities.

This chapter illustrates point transect analysis through simulated data sets for which the parameters are known. As in Chapter 4, a simple data set is first introduced. Truncation of the distance data, modelling the spatial variation of objects to estimate var(n), grouping of data, and model selection philosophy and methods are then addressed. Having selected an appropriate model, estimation of density and measures of precision are discussed. In a final example, the objects are assumed to occur in clusters (e.g. family parties or flocks).

5.2 Example data

The example data were generated from a half-normal detection function, $g(r) = \exp(-r^2/2\sigma^2)$, $0 \leq r < \infty$ with $\sigma = 10$ m. There were $k = 30$ points, and the number of sightings per point followed a Poisson distribution with parameter $E(n_i) = 5$, $i = 1, \ldots, k$. Each sighting is of a single object. Thus $E(n) = 5 \times 30 = 150$, $h(0) = 1/\sigma^2 = 0.01$, and true density is

$$D = \frac{E(n) \cdot h(0)}{2\pi k} = 0.00796 \text{ objects/m}^2 = 79.6 \text{ objects/ha}$$

Untruncated data generated from this model, together with the fitted half-normal model, are shown in Fig. 5.1. Data truncated at 20 m and the corresponding fit of the half-normal are shown in Fig. 5.2. For comparison, the fit of the uniform + one term simple polynomial detection function is shown in Fig. 5.3.

The histograms of Figs 5.1 and 5.2 illustrate two methods of presenting point transect data. In Figs 5.1b and 5.2b, the frequency of distances is shown by distance interval, as for a conventional histogram. The curve is the fitted probability density function of recorded distances, with scale chosen to match that of the frequency data. The parameter $h(0)$ is estimated by the slope of this curve at distance $r = 0$. At small distances, the function increases because area surveyed at a given distance increases with distance from the point. For example, the area

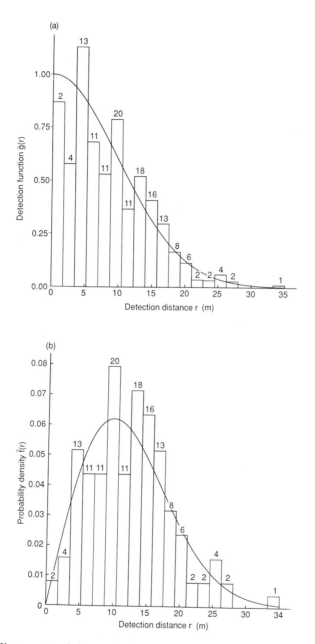

Fig. 5.1. Histograms of the example data using 20 distance categories. The fit of the half-normal detection function to untruncated data is shown in (a), in which frequencies are divided by detection distance, and the corresponding density function is shown in (b).

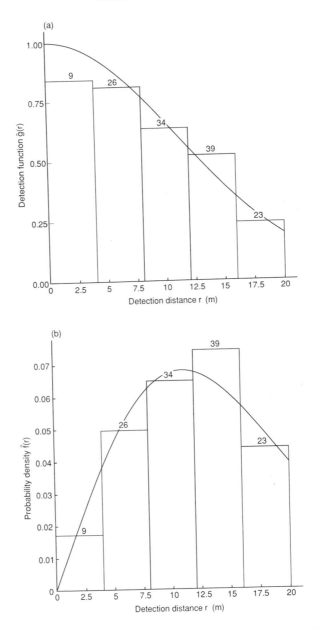

Fig. 5.2. Histograms of the example data, truncated at 20 m, using five distance categories. The half-normal model, fitted to the ungrouped data, is shown and was used for final analysis of these data. The fitted detection function is shown in (a), in which frequencies are divided by detection distance, and the corresponding density function is shown in (b).

144

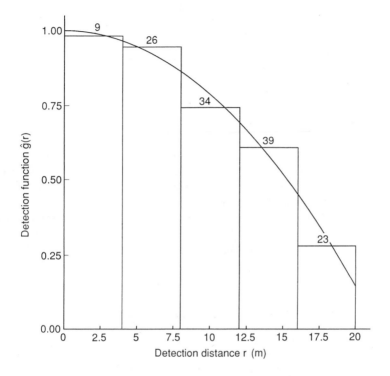

Fig. 5.3. The fit of the detection function using a uniform key with a single simple polynomial adjustment term to the example data, ungrouped and truncated at $w = 20$ m.

surveyed between r and $r + \delta$, where δ is small, is approximately $2\pi r\delta$, whereas the area between $2r$ and $2r + \delta$ is roughly $4\pi r\delta$. To correct for this increase in area, the ith distance r_i may be assigned a weight $1/r_i$. For each distance interval, these weights are summed across those observations falling within the interval. The sums are the 'corrected frequencies' of Figs 5.1a and 5.2a. To guard against infinite weights, program DISTANCE assigns weights to any zero distances equal to the weight for the smallest non-zero distance. If data are in frequency form, DISTANCE approximates the weights by the reciprocals of the mid-points of the groups. The fitted detection function, plotted so that its scale corresponds to that of the data, is also given in Figs 5.1a and 5.2a. Note that the detection function may sometimes appear to fit badly at small distances, as in Fig. 5.2; this is not a programming error, but arises because of the deceptive nature of point transect data. Relatively few distances are recorded close to the point, where area surveyed is small, so the fit of the model is not heavily influenced by distances close to zero, whereas the height of the first histogram bar is dominated by small distances. The corresponding density is plotted with untransformed frequencies, and should appear to fit the data well, as in Fig. 5.2b, provided heaping is not severe, adequate truncation is carried out, and an appropriate model is selected.

5.3 Truncation

The largest detection distance in the example data was 34.16 m, considerably greater than the second largest of 26.87 m (Fig. 5.1). Unless the true detection function is somehow known (as it is for this simulated example), large distances can prove difficult to model, and the extra terms required increase the variance in $\hat{h}(0)$. If the uniform + simple polynomial model is fitted to the untruncated data of Fig. 5.1, four polynomial terms are required, and a less plausible shape for the detection function is obtained than for the single term fit of Fig. 5.3. In Chapter 4, we suggested as rules of thumb that either roughly 5% of observations be truncated or truncation distance w be chosen such that $\hat{g}(w) \doteq 0.15$. These rules do not carry across to point transects, for which

Table 5.1. Summary of AIC values for two truncation values (w) for the example data analysed as ungrouped and three different groupings (five groups of equal width, 20 groups of equal width, and five unequal groups such that the number detected was nearly equal in each group). For each analysis, the model with the smallest AIC is indicated by an asterisk

Data type	Model (key + adjustment)	$w = 20$ m			$w =$ largest obsn		
		No. of parameters			No. of parameters		
		Key	Adjust.	AIC	Key	Adjust.	AIC
Ungrouped	Uniform + cosine	0	1	765.51	0	2	918.46*
	Uniform + polynomial	0	1	764.48	0	4	922.32
	Half-normal + Hermite	1	0	764.31*	1	0	919.16
	Hazard-rate + cosine	2	0	767.22	2	1	919.79
Grouped	Uniform + cosine	0	1	403.36	0	2	374.06*
(5 equal)	Uniform + polynomial	0	1	400.83*	0	4	377.78
	Half-normal + Hermite	1	0	401.97	1	1	374.22
	Hazard-rate + cosine	2	0	403.52	2	1	375.91
Grouped	Uniform + cosine	0	1	768.16	0	2	764.20*
(20 equal)	Uniform + polynomial	0	1	766.90	0	2	830.04
	Half-normal + Hermite	1	0	766.85*	1	0	764.34
	Hazard-rate + cosine	2	1	769.94	2	1	765.25
Grouped	Uniform + cosine	0	1	426.28	0	2	468.50
(5 unequal)	Uniform + polynomial	0	1	426.69	0	4	476.91
	Half-normal + Hermite	1	0	425.93*	1	0	467.17*
	Hazard-rate + cosine	2	0	428.04	2	0	472.39

a higher proportion of detections occurs in the tail of the detection function. This can be seen in Fig. 5.1; Fig. 5.1a shows that probability of detection is as small as 0.12 at a distance of 20 m, yet 13 of 144 observations (9%) lie beyond 20 m (Fig. 5.1b). We suggest that roughly 10% of observations should be truncated for point transects, or alternatively, w should be chosen such that $g(w) \doteq 0.1$, where $g(w)$ is estimated from a preliminary fit of a plausible model to the data. In this example, a truncation distance of 20 m roughly satisfies both criteria, and is used subsequently for the example data.

Truncation of the data at $w = 20$ m removed 13 detections. If a series expansion model is used, up to three fewer parameters are required to model the truncated data than the untruncated data (Table 5.1). Outliers in the right tail of the distance distribution required additional adjustment parameters. Except for the hazard-rate model, which performed relatively poorly on these data, density estimates varied more when data were untruncated (Table 5.2). The poor performance of the uniform + polynomial model when fitted to untruncated data divided into 20

Table 5.2 Summary of estimated density ($\hat{D}$) and coefficient of variation (cv) for two truncation values (w) for the example data. Estimates are derived for four robust models of the detection function. The data analysis was based on ungrouped data and three different groupings (five groups of equal width, 20 groups of equal width, and five unequal groups such that the number detected was equal in each group)

Data type	Model (key + adjustment)	Truncation			
		$w = 20$ m		$w =$ largest obsn	
		$\hat{D}$	cv (%)	$\hat{D}$	cv (%)
Ungrouped	Uniform + cosine	75.05	14.4	74.13	10.6
	Uniform + polynomial	60.76	12.1	70.88	18.0
	Half-normal + Hermite	70.82	15.7	79.62	12.6
	Hazard-rate + cosine	62.36	18.7	71.02	18.1
Grouped	Uniform + cosine	73.74	14.5	73.77	11.5
(5 equal)	Uniform + polynomial	62.01	12.7	70.14	25.7
	Half-normal + Hermite	69.06	16.0	64.53	26.3
	Hazard-rate + cosine	52.14	14.5	79.13	17.2
Grouped	Uniform + cosine	75.54	14.0	74.24	10.7
(20 equal)	Uniform + polynomial	61.09	12.2	42.26	9.0
	Half-normal + Hermite	71.30	15.6	80.25	12.3
	Hazard-rate + cosine	82.98	26.9	70.85	18.6
Grouped	Uniform + cosine	74.91	14.3	74.36	15.8
(5 unequal)	Uniform + polynomial	61.93	12.8	57.45	37.0
	Half-normal + Hermite	71.57	15.8	80.73	13.0
	Hazard-rate + cosine	84.76	37.2	57.50	13.9

groups was because DISTANCE failed to converge when attempting to fit a better model; convergence problems are encountered more commonly when appropriate truncation is not carried out. If the correct model is known and used, then truncation is not necessary provided the measurements are exact and no evasive movement prior to detection occurs. However, the true model is never known in field surveys.

Truncation of the distance data deletes outliers and facilitates model fitting. However, as some data are discarded, the uncertainty in $\hat{D}$ may increase. For the example data, when the true model was used (i.e. the half-normal), the coefficient of variation was around 3% higher for analyses of truncated data in three of the four analyses, but 10% lower (16.0%, compared with 26.3%) in one of the analyses of grouped data. When an incorrect model was fitted, the cv increased after truncation in eight analyses and decreased in four.

The true density in this example was 79.6 objects/ha. In exactly one half of the 16 analyses of Table 5.2, the estimate was closer to the true density after truncation than before. The case for truncation is therefore not compelling for these simulated data. However, real data tend to be less well behaved, and if no truncation is imposed in the field, truncation at the analysis stage is advisable.

5.4 Estimating the variance in sample size

If objects were known to be distributed at random, the distribution of sample size n would be Poisson with $\text{var}(n) = E(n)$, so that $\widehat{\text{var}}(n) = n$. Most biological populations exhibit some degree of clumping, so that $\text{var}(n) > E(n)$. If the survey is well designed so that points are spread either systematically or randomly throughout the study area, or within each stratum if the study area is divided into strata, then point transect methods are ideally suited to estimating $\text{var}(n)$ empirically, from the variability in sample size between individual points. For the example data, there were $k = 30$ points, and sample sizes n_i within the truncation distance of $w = 20$ m were 1, 1, 5, 6, 3, 8, 7, 5, 3, 4, 1, 8, 3, 1, 2, 7, 4, 4, 6, 7, 6, 8, 4, 3, 5, 2, 4, 2, 9 and 2.

From Section 3.7.2,

$$\widehat{\text{var}}(n) = k \sum_{i=1}^{k} (n_i - \bar{n})^2 / (k - 1)$$

with $k - 1$ degrees of freedom. All counts here are positive; had any been zero, they would be retained when calculating the variance. For the example, the above formula yields $\widehat{\text{var}}(n) = 172.7$, or $\widehat{\text{se}}(n) = 13.14$. Equivalently, the variance of the mean number of objects per point, $\bar{n} = n/k = 4.367$, may be estimated:

$$\widehat{\text{var}}(\bar{n}) = \sum_{i=1}^{k} (n_i - \bar{n})^2/\{k \cdot (k-1)\}$$

so that $\widehat{\text{var}}(\bar{n}) = 0.1919 = \widehat{\text{var}}(n)/k^2$. Since $n = 131$ after truncation, $\widehat{\text{var}}(n) > n$, indicating possible clumping of objects. However, the variance-mean ratio does not differ significantly from one ($p = 0.12$):

$$(k-1) \cdot \frac{\widehat{\text{var}}(n)}{n} = 38.2$$

which is a value from $\chi^2_{k-1} = \chi^2_{29}$ if the true distribution of n is Poisson. For this simulated example, we know the true distribution is indeed Poisson.

5.5 Analysis of grouped or ungrouped data

Because the assumptions of point transect sampling are known to hold for the example, analysis of ungrouped data is preferred. Generally, little efficiency is lost by grouping data prior to analysis, even with as few as five or six well-chosen intervals. If recorded distances tend to be rounded to favoured values (heaping), or if there is evidence of movement of objects in response to the observer before detection, appropriate grouping of data can lead to more robust estimation of density (Chapter 7). Often, there are sound practical reasons for recording data by distance group, instead of measuring each individual detection distance, in which case the field methods determine the analysis option.

For the example, estimated densities tended to be rather more variable between models when analysis was based on grouped data, although coefficients of variation were not consistently higher (Table 5.2). Provided distances can be measured accurately, and movement in response to the observer before detection is not a problem, we recommend that analysis should be of ungrouped data. Otherwise, data should be grouped. If heaping occurs, group cutpoints should be selected so that favoured distances for rounding tend to occur midway between cutpoints. Choice of group interval is often more critical than for line transect sampling, since a smaller proportion of detections occurs near zero distance, yet it is the value of a function at zero distance, $h(0)$, that must be estimated.

It is this difficulty that gives rise to the relatively large variability in density estimates in Table 5.2. On the other hand, although poor practice, it is not uncommon for 10% or more of perpendicular distances to be recorded as on the line in line transect sampling. It is rare for an object to be recorded as at the point ($r = 0$) in point transect sampling, so that spurious spikes in the detection function at small distances are uncommon.

5.6 Model selection

5.6.1 The models

The same four models for the detection function are considered as in Chapter 4. Thus, the uniform, half-normal and hazard-rate models are used as key functions. Cosine and simple polynomial expansions are used with the uniform key, Hermite polynomials are used with the half-normal key, and a cosine expansion is used with the hazard-rate key. The data were generated under a half-normal detection function so we might expect the half-normal key to be sufficient without any adjustment terms. However, the data were stochastically generated, so that the addition of a Hermite polynomial term in one analysis of Table 5.1 is not particularly surprising. A histogram of the data using 15–20 intervals, as in Fig. 5.1, tends to reveal the characteristics of the data, such as outliers, heaping, measurement errors, and evasive movement prior to detection.

5.6.2 Likelihood ratio tests

If default settings are accepted, DISTANCE determines the number of adjustment terms required to attain an adequate fit of the data using likelihood ratio tests. Consider the example data, ungrouped and with $w = 20$ m, analysed using the uniform key with a single polynomial adjustment (Table 5.1). How was it determined that a single adjustment was required for this model? Let $\mathcal{L}_0$ be the value of the likelihood for fitting a uniform key alone, let $\mathcal{L}_1$ be the maximum value of the likelihood when a single polynomial term is added, and $\mathcal{L}_2$ be the value after fitting two polynomial terms. Program DISTANCE gives $\log_e (\mathcal{L}_0) = -394.986$, $\log_e (\mathcal{L}_1) = -381.239$ and $\log_e (\mathcal{L}_2) = -381.061$. The likelihood ratio test of the hypothesis that the uniform key provides an adequate description of the data against the alternative that a single polynomial adjustment to the key provides a better fit is carried out by calculating

$$\chi^2 = -2 \log_e (\mathcal{L}_0 / \mathcal{L}_1)$$
$$= -2[\log_e (\mathcal{L}_0) - \log_e (\mathcal{L}_1)]$$
$$= -2[-394.986 + 381.239]$$
$$= 27.49$$

If the true model is the uniform key without adjustment, this statistic is distributed asymptotically as χ_1^2. In general, the *df* for this test statistic is the difference in the number of parameters between the two models being tested. A value of 27.49 is much larger than would be expected if the distribution really was χ_1^2 ($p < 0.001$), suggesting that a uniform detection function is not an adequate description of the data, a conclusion that is obvious from Fig. 5.3. Less obvious is whether an additional polynomial term should be fitted. The above test is now carried out, but with $\mathcal{L}_1$ replacing $\mathcal{L}_0$ and $\mathcal{L}_2$ replacing $\mathcal{L}_1$:

$$\chi^2 = -2[\log_e (\mathcal{L}_1) - \log_e (\mathcal{L}_2)]$$
$$= 0.36$$

Again comparing with χ_1^2, this test statistic is not significant ($p = 0.55$), so a further term does not improve the fit of the model significantly. Our experience suggests that a larger value than the conventional $\alpha = 0.05$ is often preferable for the size of the test, and we suggest $\alpha = 0.15$ (Section 3.5.2).

If the likelihood ratio test indicates that a further term is not required but goodness of fit (below) indicates that the fit is poor, the addition of two terms (using DISTANCE option LOOKAHEAD = 2) rather than just one may provide a significantly better fit. Another solution is to change the default setting of SELECT = **sequential** to SELECT = **forward** or SELECT = **all** in DISTANCE.

5.6.3 Akaike's Information Criterion

Akaike's Information Criterion (AIC) provides a quantitative method for model selection, whether models are hierarchical or not (Section 3.5.3). The adequacy of the selected model should still be assessed, for example using the usual χ^2 goodness of fit statistics and visual inspection of both the estimated detection function and the corresponding density plotted on histograms of the data, as shown in Figs 5.1 and 5.2. The plots allow the fit of the model near the point to be assessed; some lack of fit in the right tail of the data can be tolerated.

AIC was computed for the four models for both grouped and ungrouped data, with truncation distance *w* set first to 20 m (13 observations truncated) and then to the largest observation, selected so that no observations were truncated (Table 5.1). Three sets of cutpoints were considered for grouped analyses under each model. Set 1 had five equal groups, set 2 had 20 equal groups, and set 3 had five groups whose width varied, such that the number detected in each distance category was nearly equal. AIC cannot be used to select between models if the truncation distances w differ, or, in the case of an analysis of grouped data, if the cutpoints differ, so AIC values can only be compared within each of the eight sets of results in Table 5.1.

The AIC values in Table 5.1 select the half-normal (true) model in four of the eight sets of results. The uniform key with cosine adjustments is selected three times, and the uniform key with simple polynomial adjustments once. Since the half-normal model is selected for the preferred analysis of ungrouped data, truncated at 20 m, the main analysis will be based upon it. However, the AIC value for the uniform + polynomial model is almost the same as for the half-normal + Hermite model, and might equally well be adopted on this basis. We examine the consequences of selecting this model later. The only model that might reasonably be excluded from further consideration on the basis of its AIC value is the hazard-rate + cosine model.

Table 5.3 Goodness of fit statistics for models fitted to the example data with $w = 20$ m and 20 groups

Model	χ^2	df	p
Uniform + cosine	20.34	17	0.26
Uniform + polynomial	19.84	17	0.28
Half-normal + Hermite	19.20	17	0.32
Hazard-rate + cosine	20.32	16	0.21

5.6.4 Goodness of fit

Goodness of fit is another useful tool for model selection (Section 3.5.4). Goodness of fit statistics for the example data without grouping, with $w = 20$ m, and using 20 groups of equal width to evaluate the χ^2 statistic, are given in Table 5.3. These data were taken when all the assumptions were met, and all four models fit the data well. If a model was to be selected from these results, there might be a marginal preference for the half-normal + Hermite polynomial model, which we know to be the correct choice in this case. Heaping in real data sets generally means that fewer than 20 groups should be used, with perhaps six to eight usually being reasonable. If heaping is severe, fewer groups might be required, ideally with each preferred rounding distance falling near

152

the middle of each group. The grouped nature of the (rounded) data is then correctly recognized in the analysis. If cutpoints are badly chosen, heaping will lead to spurious significant χ^2 values. If data are collected as grouped, the group cutpoints are determined before analysis, although consecutive groups may be merged.

5.7 Estimation of density and measures of precision

5.7.1 The standard analysis

The preferred analysis from the above considerations comprises the fit of the half-normal key without adjustments to ungrouped data, truncated at $w = 20$ m. The variance of n is estimated empirically.

Replacing the parameters of Equation 3.5 by their estimators and simplifying under the assumptions that objects at zero distance are detected with certainty, detected objects are recorded irrespective of their angle from the observer, and objects do not occur in clusters, estimated density becomes

$$\hat{D} = \frac{n \cdot \hat{h}(0)}{2\pi k}$$

where n is the number of objects detected, k is the number of point transects sampled, and $\hat{h}(0)$ is the slope of the estimated density $\hat{f}(r)$ of observed detection distances evaluated at $r = 0$; $\hat{h}(0) = 2\pi/\hat{v}$, where $\hat{v}$ is the effective area of detection.

For the example data, and adopting the preferred analysis, program DISTANCE yields $\hat{h}(0) = 0.01019$, with $\widehat{se}\{\hat{h}(0)\} = 0.001233$ (based on approximately $n = 131$ degrees of freedom). The units of $\hat{h}(0)$ are m^{-2}. Thus

$$\hat{D} = \frac{131 \times 0.01019}{2\pi \times 30} = 0.00708 \text{ objects/m}^2 \text{ or } 70.8 \text{ objects/ha}$$

The estimator of the sampling variance of this estimate is

$$\widehat{var}(\hat{D}) = \hat{D}^2 \cdot \{[cv(n)]^2 + [cv\{\hat{h}(0)\}]^2\}$$

where $\qquad [cv(n)]^2 = \widehat{var}(n)/n^2 = 172.7/131^2 = 0.010065$

and $\qquad [cv\{\hat{h}(0)\}]^2 = \widehat{var}\{\hat{h}(0)\}/\{\hat{h}(0)\}^2 = 0.001233^2/0.01019^2 = 0.01464$

153

Then $$\widehat{\text{var}}(\hat{D}) = (70.8)^2\,[0.010065 + 0.01464]$$

$$= 123.84$$

and $$\widehat{\text{se}}(\hat{D}) = \sqrt{\widehat{\text{var}}(\hat{D})}$$

$$= 11.13$$

The coefficient of variation of estimated density is $\text{cv}(\hat{D}) = \widehat{\text{se}}(\hat{D})/\hat{D} = 15.7\%$, which is likely to be adequate for some purposes. Note that even with a sample size of $n = 131$ after truncation, the coefficient of variation is over 15%. A 95% confidence interval could be calculated as $\hat{D} \pm 1.96(\widehat{\text{se}}(\hat{D}))$, giving the interval [49.0, 92.6]. Log-based confidence intervals offer improved coverage by allowing for the asymmetric shape of the sampling distribution of $\hat{D}$ for small n. Applying the procedure of Section 3.7.1, the interval [52.1, 96.2] is obtained, which is wider than the symmetric interval, but is a better measure of the uncertainty in the estimate $\hat{D} = 70.8$. In line transect sampling, the variance of $\hat{D}$ is usually primarily due to the variance in n, but this is less often the case in point transect sampling, where precision in $\hat{h}(0)$ can be poor; here, variance in n accounts for 41% of the total variance estimate.

If the uniform model with polynomial adjustments is adopted, estimated density is 60.9 objects/ha, with 95% log-based confidence interval [48.2, 77.0]. The true parameter value, $D = 79.6$ objects/ha, lies above the upper limit of this interval. We return to this example later, to show how the bootstrap may be used to estimate variances and to determine confidence limits that incorporate model misspecification uncertainty.

For some purposes it is convenient to have a measure of detectability. For example, it may be useful to assess whether the detectability for a species is a function of habitat, which may have implications for survey design. The effective radius of detection $\rho = \sqrt{(v/\pi)}$, estimated by $\hat{\rho} = \sqrt{\{2/\hat{h}(0)\}}$, may be used for this purpose. For long-tailed detection functions, ρ may be considerably larger than intuition would suggest, because large numbers of objects are detected at far distances, where the area surveyed is great, relative to close distances, where the surveyed area is small. A parameter that is unaffected either by this phenomenon or by truncation is $r_{1/2}$, the distance at which the probability of detecting an object is one-half. For any fitted detection function $\hat{g}(r)$, it may be estimated by solving $\hat{g}(\hat{r}_{1/2}) = 0.5$ for $\hat{r}_{1/2}$. For the example with $w = 20$ m and $\hat{\rho} = 14.0$ m and $\hat{r}_{1/2} = 13.0$ m.

We know that the true detection function is half-normal for the example. Using that knowledge, closed form estimators are available and the analysis is simple to carry out by hand, provided the data are both ungrouped and untruncated. Using the results of Chapter 3, Section 3.4.4,

$$\hat{\sigma}^2 = \sum_{i=1}^{n} r_i^2/2n = 94.81 \text{ m}^2$$

It follows that

$$\hat{h}(0) = 2\pi/\hat{v} = 1/\hat{\sigma}^2 = 0.01055$$

and estimated density is

$$\hat{D} = 144 \times \hat{h}(0)/(2\pi \times 30) = 0.00806 \text{ objects/m}^2, \text{ or } 80.6 \text{ objects/ha}$$

The effective radius of detection is estimated as $\hat{\rho} = \sqrt{(2\hat{\sigma}^2)} = 13.8$ m, and the radius at which probability of detection is one-half is estimated by $\hat{r}_{1/2} = \sqrt{(2\hat{\sigma}^2 \log_e 2)} = 11.5$ m. These estimates are in excellent agreement with the true values of $D = 79.6$ objects/ha, $\rho = 14.1$ m, and $r_{1/2} = 11.8$ m.

The results of Section 3.4 also yield variance estimates for this special case:

$$\widehat{\text{var}}[\hat{h}(0)] = 4/\sum_{i=1}^{n} (r_i^2 - 2\hat{\sigma}^2)^2 = 8.850 \times 10^{-7}, \text{ or } \widehat{\text{se}}[\hat{h}(0)] = 9.407 \times 10^{-4}$$

Thus $[\text{cv}\{\hat{h}(0)\}]^2 = \widehat{\text{var}}\{\hat{h}(0)\}/\{\hat{h}(0)\}^2 = 0.0009407^2/0.01055^2 = 0.007951$

Also, $[\text{cv}(n)]^2 = 0.010065$ from above

so that $\widehat{\text{var}}(\hat{D}) = \hat{D}^2 \cdot \{[\text{cv}(n)]^2 + [\text{cv}\{\hat{h}(0)\}]^2\}$

$$= (80.6)^2 [0.010065 + 0.007951]$$

$$= 117.04$$

and $\widehat{\text{se}}(\hat{D}) = 10.82$

The 95% log-based confidence interval is then [62.0, 104.7] objects/ha.

5.7.2 Bootstrap variances and confidence intervals

The bootstrap is a robust method, based on resampling, for quantifying precision of estimates. One circumstance in which the bootstrap is likely to be preferred is when the user wishes to incorporate in the standard error the component of variation arising from estimating the number of

155

polynomial or Fourier series adjustments to be carried out. We recommend the following inplementation.

Generate a bootstrap sample by selecting points **with replacement** from the k points recorded until the bootstrap sample also comprises k points. Repeat until B bootstrap samples have been selected. Typically, B will be around 200 to 1000. Density D is estimated from each bootstrap sample, and the estimates are ordered, to give $\hat{D}_{(i)}$, $i = 1, \ldots, B$. Then

$$\hat{D}_B = \left\{ \sum_{i=1}^{B} \hat{D}_{(i)} \right\} / B$$

and

$$\widehat{\text{var}}_B (\hat{D}_B) = \left\{ \sum_{i=1}^{B} (\hat{D}_{(i)} - \hat{D}_B)^2 \right\} / (B - 1)$$

while a $100(1 - 2\alpha)\%$ confidence interval for D is given by $[\hat{D}_{(j)}, \hat{D}_{(j')}]$, with $j = (B + 1)\alpha$ and $j' = (B + 1)(1 - \alpha)$. It is convenient to select B so that j and j' are integer. Thus for $\alpha = 0.025$, one might select from the following values: 199, 239, 279, ..., 999. The estimate $\hat{D}$ calculated from the original data set is usually used in preference to the bootstrap estimate $\hat{D}_B$, with se$(\hat{D})$ estimated by $\sqrt{\{\widehat{\text{var}}_B(\hat{D}_B)\}}$. Applying this to the example with $B = 399$ (so that $(B + 1)\alpha = 10$, an integer, for $\alpha = 0.025$), we take a sample of 30 points at random and with replacement from the 30 in the example data set. Suppose this yields the following points: 1, 1, 3, 5, 6, 6, 6, 8, 10, 10, 11, 12, 15, 15, 17, 17, 17, 18, 18, 20, 21, 22, 22, 25, 26, 26, 26, 28, 30, 30. The bootstrap sample therefore comprises each detection distance recorded at points 6, 17 and 26 three times, each distance recorded at points 1, 10, 15, 18, 22 and 30 twice, and each distance from points 3, 5, 8, 11, 12, 20, 21, 25 and 28 once. Those for remaining points are excluded. This bootstrap sample is analysed in exactly the same way as the actual sample, to yield an estimate $\hat{D}_1$. The exercise is repeated 399 times. The sample variance of these bootstrap estimates was 159.8, giving $\widehat{\text{se}}(\hat{D}) = 12.6$ objects/ha. After ordering the bootstrap estimates, the tenth smallest value ($j = (B + 1)\alpha = 10$) was found to be $\hat{D}_{(10)} = 53.6$ and the tenth largest value was $\hat{D}_{(390)} = 100.7$, giving an approximate 95% confidence interval for D of (53.6, 100.7) objects/ha. This compares with $\hat{D} \pm 1.96 \cdot \widehat{\text{se}}(\hat{D}) = (60.8, 100.4)$ha by the more traditional method. Assuming the distribution of $\hat{D}$ is log-normal and using the result of Burnham et al. (1987: 212), we obtain the interval (63.1,

102.9) objects/ha. Note that the lower limit is smaller for the bootstrap method. This is because cosine adjustments to the half-normal fit sometimes generated a fitted detection function with a flatter shoulder than that of the half-normal. If no adjustments to the half-normal fit are allowed, the bootstrap should duplicate the analytic method, except asymptotic normality is not assumed when setting confidence limits. Applying this with $B = 399$ gives $\widehat{se}(\hat{D}) = 10.8$ and an approximate 95% confidence interval for D of (62.5, 102.7), which is shifted slightly to the right of the symmetric analytic interval, reflecting the greater uncertainty in the upper limit, but agrees well with the interval calculated assuming the distribution of $\hat{D}$ is log-normal.

Variances of functions of the fitted density, such as $\hat{\rho}$ or $\hat{r}_{1/2}$, may be estimated using the methods of Section 3.4, or from the above bootstrap method, replacing the bootstrap estimate of density $\hat{D}_{(i)}$ by the appropriate estimate, such as $\hat{\rho}_{(i)}$ or $\hat{r}_{1/2_{(i)}}$. Adopting the analytic approach,

$$\frac{\partial \hat{\rho}}{\partial \hat{\sigma}^2} = 1/\sqrt{(2\hat{\sigma}^2)}$$

so that

$$\widehat{se}(\hat{\rho}) = \sqrt{(2\hat{\sigma}^6)}/\sqrt{\left\{ \sum_{i=1}^{n} (r_i^2 - 2\hat{\sigma}^2)^2 \right\}} = 0.61 \text{ m}$$

and

$$\widehat{se}(\hat{r}_{1/2}) = \widehat{se}(\hat{\rho}) \cdot \sqrt{(\log_e 2)} = 0.51 \text{ m}$$

By comparison, the bootstrap method yields $\widehat{se}(\hat{\rho}) = 1.12$ m, with 95% confidence interval (12.62, 16.84) m, and $\widehat{se}(\hat{r}_{1/2}) = 0.93$ m, with 95% confidence interval (10.51, 14.02) m. If no cosine adjustments are allowed, as above, we get $\widehat{se}(\hat{\rho}) = 0.66$ m, with 95% confidence interval (12.55, 15.00) m, and $\widehat{se}(\hat{r}_{1/2}) = 0.55$ m, with 95% confidence interval (10.45, 12.49) m. These results are in good agreement with the analytic results.

We noted earlier that the AIC value for the preferred analysis of the example data was almost the same as that using the uniform key with a single polynomial adjustment. However, the latter model gave an estimated density of 60.9 objects/ha, with 95% confidence interval [48.2, 77.0]. Thus the true parameter value, $D = 79.6$ objects/ha, is outside the confidence interval. The bootstrap option within DISTANCE was implemented with $B = 200$ replicates, to obtain a variance for $\hat{h}(0)$ that

157

allows for estimation of the number of polynomial terms required. It gave $\widehat{se}\{\hat{h}(0)\} = 0.000783$, compared with the analytic estimate of $\widehat{se}\{\hat{h}(0)\} = 0.000581$, which is conditional on a single term adjustment to the uniform key. Thus the variance is larger as expected, and the revised 95% confidence limit for D is [46.6, 79.3]. The true density is therefore still just outside the interval, probably because the uniform + polynomial model gives a negatively biased estimate of density for this data set. To attempt to improve the variance estimate corresponding to $\hat{D} = 60.9$, a component of variance corresponding to model misspecification bias should be estimated. We do this by generating 199 bootstrap samples, and analysing each resample by the three models of Table 5.1 that gave competitive AIC values, namely uniform + cosine, uniform + polynomial and half-normal + Hermite polynomial. In each resample, the bootstrap estimate of density is taken to be the estimated density under the model with the smallest AIC. Under this rule, the uniform + cosine model was selected in 49 of the 199 replicates, the uniform + polynomial model in 92, and the half-normal + Hermite polynomial model in the remaining 58. The 95% percentile confidence interval was [48.0, 94.5] objects/ha, which is wider than the intervals obtained by assuming that the selected model is the correct model, and comfortably includes the true parameter value, $D = 79.6$.

In the above bootstrap implementations, the sampling unit was taken to be the individual point. This is valid if points are randomly distributed through the study area, and provides a good approximation if points are arranged as a regular grid. To reduce travel time between points, transect lines are sometimes defined, and counts are made at regular points along each line. If the spacing between lines is similar to the distance between neighbouring points on the same line, then the point may still be taken as the sampling unit. However, if separation between lines is large, then the line should be taken as the sampling unit. Thus lines are selected with replacement until the number of lines in the resample is equal to the number in the real sample, or, if the number of points per line is very variable, until the number of points in the resample is as close as possible to the number in the real sample. If a line is selected, the data from all points on that line are included in the resample.

5.8 Estimation when the objects are in clusters

If point transects are used for objects that are sometimes recorded in clusters during the survey period, the recording unit should be the cluster, not the individual object, and analyses should be based on clusters. In this section, various options for the analysis of clusters are considered. If it is assumed that (i) probability of detection is independent of cluster size and (ii) cluster sizes are accurately recorded, or alternatively that they are estimated without bias at all distances, then $E(s)$ may be estimated by the mean size of detected clusters, $\bar{s}$. Estimated cluster density is then

$$\hat{D}_s = \frac{n \cdot \hat{h}(0)}{2\pi k}$$

and estimated object density is

$$\hat{D} = \hat{D}_s \cdot \bar{s} = \frac{n \cdot \hat{h}(0) \cdot \bar{s}}{2\pi k}$$

Note that the formula for cluster density is identical to that for object density when the objects do not occur in clusters. The formula for the variance of $\hat{D}_s$ is also identical to that given for object density in Section 5.7.1. The variance of object density is now estimated by

$$\widehat{\text{var}}(\hat{D}) = \hat{D}^2 \cdot \left\{ \frac{\widehat{\text{var}}(\hat{D}_s)}{\hat{D}_s^2} + \frac{\widehat{\text{var}}(\bar{s})}{\bar{s}^2} \right\} = \hat{D}^2 \cdot \left\{ \frac{\widehat{\text{var}}(n)}{n^2} + \frac{\widehat{\text{var}}[\hat{h}(0)]}{[\hat{h}(0)]^2} + \frac{\widehat{\text{var}}(\bar{s})}{\bar{s}^2} \right\}$$

where

$$\widehat{\text{var}}(\bar{s}) = \sum_{i=1}^{n} (s_i - \bar{s})^2 / \{n(n-1)\}$$

In practice, larger clusters often tend to be more detectable than small clusters at greater distances, so that $E(s)$, and hence D, are overestimated. This is a form of size-biased sampling (Cox 1969; Patil and Ord 1976; Patil and Rao 1978; Rao and Portier 1985). Bias can be negative if the size of a detected cluster at a large distance from the observer tends to be underestimated. If either bias occurs, then the above method should be modified or replaced.

The simplest approach is based on the fact that size bias in detected clusters does not occur within a region around the point for which detection is certain. Hence, $E(s)$ may be estimated by the mean size of clusters detected within distance v of the point, where $g(v)$ is reasonably close to one, say 0.6 or 0.8. In the second method, a cluster of size s_i at distance r_i from the point is replaced by s_i objects, each at distance r_i. Thus, the sampling unit is assumed to be the object rather than the cluster. For the third method, data are stratified by cluster size (Quinn 1979, 1985). The selected model is then fitted independently to the data in each stratum. If size bias is large or cluster size very variable, smaller truncation distances are likely to be required for strata corresponding to small clusters. The final method estimates cluster density D_s conventionally, as does the first. Then, given the r_i, $E(s)$ is estimated by regression modelling of the relationship between s_i and r_i. All four approaches are illustrated in this section using program DISTANCE.

The data used to illustrate the four methods were simulated from a half-normal detection function without truncation, in which the scale parameter σ was a function of cluster size:

$$\{\sigma(s)\}^2 = \sigma_0^2 \left(1 + b \cdot \frac{s - E(s)}{E(s)} \right)$$

where $\sigma_0 = 30$ m, $b = 0.75$ and $E(s) = 1.85$ for the population. (In Chapter 4, $\sigma(s)$ was assumed to be a linear function of s; for point transects, theoretical considerations suggest that it is more appropriate to assume $\{\sigma(s)\}^2$ is a linear function of s.) Cluster sizes s were generated by simulating values from the geometric distribution with rate $E(s) - 1$ and adding one, and a cluster of size s was detected with probability

$$g(r \mid s) = \exp \left[- \frac{r^2}{2\{\sigma(s)\}^2} \right]$$

The expected sample size was $E(n) = 96$, distributed between $k = 60$ points, with $\text{var}(n) = 2.65 \cdot E(n)$. True densities were $D_s = 283$ clusters/km^2 and $D = 1.85 \times 283 = 523$ objects/km^2. The bivariate detection function $g(r, s)$ is monotone non-increasing in r and monotone non-decreasing in s. The detected cluster sizes are not a random sample from the population of cluster sizes; the mean size of detected clusters $\bar{s}$ has expectation $> E(s)$.

A histogram of the untruncated distance data shows a rather long tail (Fig. 5.4). Truncation at 70 m deleted just under 10% of observations (eight from 92), and allowed the data to be modelled more reliably. The same four models were applied as for Section 5.6: uniform + cosine, uniform + polynomial, half-normal + Hermite polynomial and hazard-rate + cosine. All four models fitted the truncated data well. AICs for the four models were 691.8, 693.7, 692.6 and 693.1, which favour the uniform + cosine model. We therefore use it to illustrate methods of analysis of the example data.

The uniform + cosine model for the untruncated data required three cosine terms to adequately fit the right tail of the data (Fig. 5.4). By truncating the data, only a single cosine term is required (Fig. 5.5), and the size bias in the truncated sample of detected clusters is reduced. The fit of the model was good ($\chi_5^2 = 4.51$; $p = 0.48$). The estimated density of clusters was 258.1 clusters/km^2 ($\widehat{se} = 52.2$), compared with the true value of 283. The mean cluster size from the untruncated sample data was 2.293 ($\widehat{se} = 0.165$), which is biased high due to the size-biased sampling. The scatter plot of cluster size against detection distance (Fig. 5.6) shows wide scatter, but a significant correlation ($r = 0.272$). Truncation at $w = 70$ m reduced this correlation to 0.180. Multiplying the density of clusters by the uncorrected estimate of mean cluster size from data truncated at 70 m ($\bar{s} = 2.202$; $\widehat{se}(\bar{s}) = 0.168$), the density of individuals is estimated as 574.6 objects/km^2 with $\widehat{se} = 115.1$ and 95% confidence interval [389.6, 847.5], which comfortably includes the true density of 523 objects/km^2.

5.8.1 Standard method with additional truncation

Observed mean cluster sizes and standard errors for a range of truncation distances are shown in Table 5.4. The detection function $g(r)$ was estimated using a truncation distance of w, while a truncation distance of $v(v \leqslant w)$ was used to estimate mean cluster size. It seems that 70 m may be too large a truncation distance for unbiased estimation of mean cluster size, but an appropriate distance is difficult to determine, because mean cluster size does not stabilize as the truncation distance is reduced. Possible choices for truncation distance v range between 21.5 m, for which $\bar{s} = 1.650$, and 46.9 m, giving $\bar{s} = 2.030$. If strong size bias is

Table 5.4 Observed mean cluster sizes and standard errors for various truncation distances v. Probability of detection at the truncation distance for cluster size estimation, $\hat{g}(v)$, was estimated from a uniform + 1-term cosine model with $w = 70$ m (Fig. 5.5) for $v \leqslant 70$ m, and from a uniform + 3-term cosine model with $w = 120$ m (Fig. 5.4) for $v = 120$ m

Truncation distance, v(m)	n	$\bar{s}$	$\widehat{se}(\bar{s})$	$\hat{g}(v)$
120.0	92	2.293	0.165	0.005
70.0	84	2.202	0.168	0.07
46.9	67	2.030	0.183	0.30
36.8	52	2.135	0.228	0.50
31.9	38	2.079	0.243	0.60
27.0	31	1.806	0.199	0.70
21.5	20	1.650	0.232	0.80
14.9	10	2.000	0.422	0.90

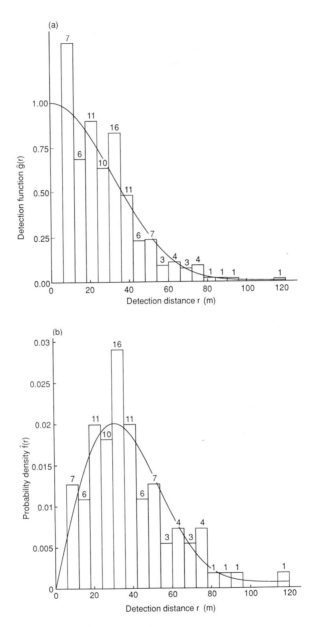

Fig. 5.4. Histograms of the example data using 20 distance categories for the case where cluster size and detection distance are dependent. The fit of a uniform + 3-term cosine detection function to untruncated data is shown in (a), in which frequencies are divided by detection distance, and the corresponding density function is shown in (b).

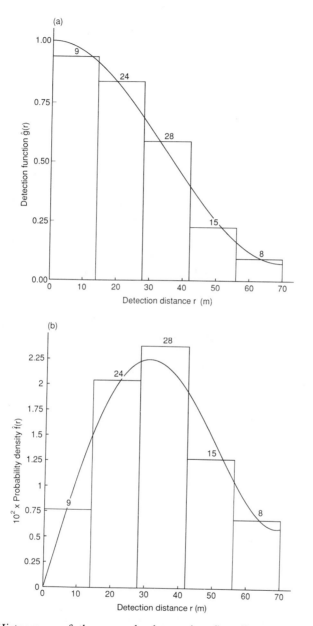

Fig. 5.5. Histograms of the example data using five distance categories and truncation at $w = 70$ m for the case where cluster size and detection distance are dependent. The fit of a uniform + 1-term cosine detection function is shown in (a), in which frequencies are divided by detection distance, and the corresponding density function is shown in (b).

suspected, a reasonable compromise might be $v = 27.0$ m, so that $\bar{s} = 1.806$ with $\widehat{se} = 0.199$. Replacing the estimates $\bar{s} = 2.202$ and $\widehat{se} = 0.168$ by these values, density is estimated as 471.2 individuals/km^2, with $\widehat{se} = 101.5$ and 95% CI [310.4, 715.3]. In view of the difficulty in selecting v, and the sensitivity of the estimate to the choice, another approach seems preferable in this instance.

5.8.2 Replacement of clusters by individuals

If a cluster of size s_i is replaced by s_i objects at the same distance, the assumption that detections are independent is violated, invalidating analytic variance estimates and model selection procedures. Robust methods for variance estimation avoid the first difficulty, but model selection is more problematic. One solution is to select a model taking clusters as the sampling unit, then refit the model (with the same series terms, if any) to the data with object as the sampling unit. Adopting this strategy, a uniform + 1-term cosine model was fitted to the distance data truncated at 70 m, and the following estimates obtained. Number of objects detected, $n = 185$. Estimated density, $\hat{D} = 526.2$ objects/km^2, with analytic $\widehat{se} = 104.6$ and 95% confidence interval [355.3, 779.1]. These estimates are lower than those obtained assuming cluster size is independent of distance, and the point estimate is appreciably closer to the true density of 523 objects/km^2. Average cluster size can be estimated by the ratio of estimated object density (526.2) to estimated cluster density (258.1), giving 2.039.

5.8.3 Stratification

Stratification by cluster size can be an effective way of handling size bias. For the example data, if two strata are defined, one corresponding to individual objects and the other to clusters ($\geqslant$ two objects), sample sizes before truncation are 36 and 56 respectively. If the second stratum is split into clusters of size two and clusters of more than two individuals, the respective sample sizes in the three strata before truncation are 36, 27 and 29. The data were analysed for both choices of stratification.

Results are summarized in Table 5.5. As for the line transect example in the previous chapter, no precision is lost by stratification, despite the small samples from which $f(0)$ was estimated. The estimated densities are lower than that obtained by assuming cluster size is independent of detection distance, as would be expected if size bias is present. Both stratifications yield similar estimated densities, and they bracket the

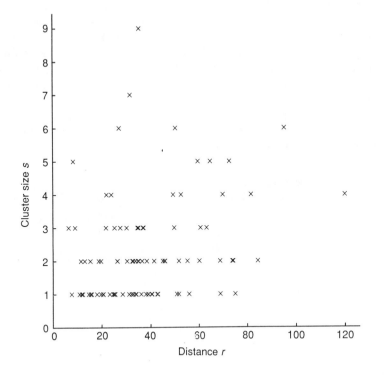

Fig. 5.6. Scatterplot of the relationship between cluster size and detection distance. The correlation coefficient is 0.272 ($w = \infty$).

estimate obtained by the previous method. The true density is 523 objects/km^2, very close to both estimates. Mean cluster size may be estimated by a weighted average of the mean size per stratum, with weights equal to the estimated density of clusters by stratum. Alternatively, $E(s)$ may be estimated as overall $\hat{D}$ from the stratified analysis divided by $\hat{D}_s$ from the unstratified analysis. For two strata, this yields $\hat{E}(s) = 534.3/258.1 = 2.070$, and for three strata, $\hat{E}(s) = 524.0/258.1 = 2.030$. Both estimates are rather higher than the true mean cluster size of 1.85.

5.8.4 Regression estimator

Average cluster size can be estimated from a regression of cluster size on estimated detection probability. This procedure estimates the average cluster size for clusters close to the centreline, where detection is assumed to be certain, and thus size bias is reduced. The loss in precision in correcting for size bias using regression is generally small. The method

165

of regressing $z_i = \log_e(s_i)$ on $\hat{g}(x_i)$ (Section 3.6.3), applied to the example data, yields $\hat{E}(s) = 1.772$ and $\widehat{se}\{\hat{E}(s)\} = \sqrt{[\widehat{var}\{\hat{E}(s)\}]} = 0.125$. The corresponding density estimate is 462.2 individuals/km², with $\widehat{se} = 91.6$, cv = 19.8% and 95% confidence interval [313.8, 680.9]. The estimate $\hat{E}(s)$ is close to the true parameter value of 1.85. The resulting density estimate (462.2) is low relative to the true density (523), although the confidence interval comfortably includes the true value.

Table 5.5 Summary of results for different stratification options. Model was uniform with cosine adjustments; distance data were truncated at $w = 70$ m. True $D = 523$ objects/km²

Cluster size	Sample size after truncation	Efective search radius (m)	$\hat{D}$	$\widehat{se}(\hat{D})$	95% CI for D
All	84	41.3	574.6	115.1	(389.6, 847.5)
1	35	39.2	120.8	24.7	
2–9	49	43.9	413.5	100.5	
All			534.3	103.5	(366.8, 778.3)
1	35	39.2	120.8	24.7	
2	24	41.6	147.1	42.2	
3–9	25	46.0	256.1	80.9	
All			524.0	94.5	(369.0, 744.1)

5.9 Assumptions

The assumptions of point transect sampling are discussed in Section 2.1. There has been considerable confusion on whether objects must be assumed to be randomly distributed, both in the literature and among biologists. If objects are distributed stochastically independently from each other, but with variable rate depending on location, then the assumption that points rather than objects are randomly located suffices unless the rate shows extreme variation over short distances (of the order of a typical detection distance). If the rate can change appreciably in a short distance or if the presence of one object greatly increases the likelihood that another object is nearby (thus violating the assumption that detections are independent events), then given random placement of points, reliable estimation may still be possible provided robust variance estimation methods are used and provided that the results of goodness of fit and likelihood ratio tests (which will tend to give spurious significances) are viewed with suspicion. The more serious the departure from random, independent detections, the larger the sample size required to yield reliable analyses. Robust empirical or resampling methods should always be used for estimating the variance of sample size, as described in Section 5.7, to guard against the effects of clustered detections. The most extreme departures from a random distribution of objects are when the objects occur in well-defined clusters. In such cases, the above problems are avoided by taking the cluster rather than the object to be the sampling unit. Strict random placement of points can be modified. For example, stratification of the study area allows sampling intensity to vary between strata, or a regular grid of points may be randomly superimposed on the area. Use of a regular grid allows the biologist to control the distance between points.

Surveys should be designed to minimize departures from the assumption that probability of detection at the point is unity ($g(0) = 1$). For example, the assumption is likely to be more reasonable for songbirds if the recording time at each point is long (giving each bird time to be detected) or if surveys are carried out in early morning, when detectability may be an order of magnitude higher (Robbins 1981; Skirvin 1981). We do not concur with the argument that early morning should be avoided when carrying out point transects. The reasoning behind it is that bird detectability varies rapidly during the first hour or two of daylight. Although detectability may vary less later in the day, it will also be lower, and densities of some species may be appreciably underestimated. Whenever possible, survey work should be carried out when detectability is greatest, and survey design should allow for variation in detectability. Models that are robust to variable detectability (pooling robust) should be used to analyse the data.

Time of season also determines whether it is reasonable to assume that probability of detection at the point is unity. For multiple species studies, it may be necessary to carry out surveys more than once, say early and late in the season. For any given species, the data collected closest to the time that it is most detectable can then be used. For many songbirds, it may be practical to survey only territorial males.

For point transect sampling, we consider that it is necessary to assume that the detection function has a shoulder because we believe that reliable estimation is not possible if it fails, although small departures from the strict mathematical requirement that $g'(0) = 0$ need not be serious. Unlike line transect data, only a very small proportion of point transect distances is close to zero, because the area covered close to the point is small. Thus, there is a case for designing surveys to ensure that $g(r) = 1$ out to some predetermined distance. If there is an area about the point for which detection is perfect, then different point transect models will tend to give more consistent estimation. When $g'(0) = 0$ but $g''(0) < 0$, the stronger criterion of an area of perfect detection fails. Methods based on squaring detection distances (Burnham *et al.* 1980: 195) and the method due to Ramsey and Scott (1979, 1981b) may then perform poorly. Even when the criterion is satisfied, but the distance up to which detection is certain is close to zero, such methods can be poor.

The mathematical theory assumes that random movement of objects does not occur. In line transect sampling, random movement prior to detection can be tolerated provided average speed of objects is appreciably less than (i.e. up to about one-third of) the speed of the observer (Hiby 1986). The problem is more serious for point transects, for which the observer is stationary. Bias occurs because probability of detection is a non-increasing function of distance from the point, so that objects moving at random are more likely to be detected when closer to the point, leading to overestimation of object density. As noted above, the assumption that $g(0) = 1$ is more plausible if recording time at each point is large, but bias arising from random object movement increases with time at the point; thus recording time at each point is a compromise, and is typically five to ten minutes for songbird surveys.

Response to the observer may take the form of movement towards or away from the observer, or of a change in the probability of detection of the object. Movement towards the observer has a similar effect on the data as random movement, and leads to overestimation of density (Fig. 5.7). Movement away from the observer tends to give rise to underestimation (Fig. 5.7), as does a decrease in detectability close to the point, if this is sufficient to violate the assumption that $g(0) = 1$. An increase in detectability, as when birds 'scold' the intruder, is generally helpful. However, if birds also move in response to the observer, or if females are seldom detected except very close to the point, the detection function might be difficult to model satisfactorily. The effects of response to the observer have been considered by Wildman and Ramsey (1985), Bibby and Buckland (1987) and Roeder et al. (1987).

Bibby and Buckland considered two 'fleeing' models. In the first, each object was assumed to maintain a minimum distance (its 'disturbance radius' r_d) between itself and the observer. The radius was allowed to vary from object to object, and was assumed to follow a negative exponential distribution. The detection function was assumed to be half-normal. If the data were to be analysed using a binomial half-normal model (Section 6.2.1) with the division between near and far sightings set at $c_1 = 30$ m (Chapter 6), and if 50% of detections would fall within c_1 in the absence of evasive behaviour, Bibby and Buckland calculated that the bias in $\hat{D}$ (evaluated by numeric integration) would be -9% when the mean disturbance radius was 10 m, -20% for 15 m, -30% for 20 m and -55% for 40 m. In this case, bias might be deemed 'acceptable' ($< 10\%$ in magnitude) if the mean disturbance radius was

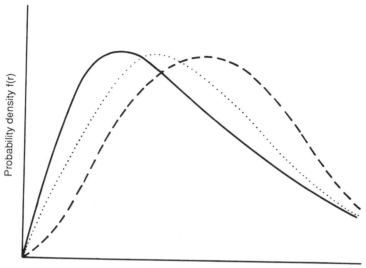

Fig. 5.7. Plots of the real probability density function ($\cdots\cdots$), the apparent function when there is movement away from the observer ($---$), and the apparent function when there is movement towards the observer (———). Estimated density of birds is proportional to the slope of the curve at zero distance from the point, so that density is overestimated when there is movement towards the observer and underestimated when movement is away from the observer.

of the order of one-third the median detection distance or less. In their second fleeing model, many objects close to a sample point become undetectable, because they either leave at the approach of the observer, moving beyond the range of detection, or take to cover, remaining silent until the observer has departed. The probability that an object at distance r is undetectable was modelled as half-normal. In otherwise identical circumstances to the first model, bias in $\hat{D}$ was found to be -24% when the point at which 50% of objects become undetectable was 10 m, -44% for 15 m, -61% for 20 m and -88% for 40 m. They concluded that species for which the second model applied were unsuitable for surveying by the point transect method, but considered that the first model, for which bias was less severe, would apply to most species of woodland songbird that show evasive behaviour.

Roeder *et al.* (1987) also considered two models for disturbance, in which the probability of disturbance was exponentially distributed, being one at distance zero. They then simulated data in which a 'disturbed'

169

object either moved exactly 10 m away from the observer (model 1) or hid (model 2). Their conclusions, based on analyses using the method due to Ramsey and Scott (1979, 1981b), the Fourier series method on squared distances and an order statistic method, were consistent with those reported above. In cases where there is an area of perfect detectability well beyond any effects arising from evasive behaviour, Wildman and Ramsey (1985) showed that their method is still valid under a model in which objects move away from the observer, and can be modified if objects close to the observer are known to hide.

The term 'doughnut' or 'donut' refers to a paucity of observations close to the point, and is generally attributed to object response of one of the above types to the observer. Wildman and Ramsey (1985) used data on the omao or Hawaiian thrush (*Phaeornis obscurus*) as a good example of this. In some instances, a poor choice of model can lead to erroneous identification of a doughnut; the empirical distribution function of detection distances is useful for assessing whether a doughnut really exists.

Distances are assumed to be measured without error (or to be assigned to defined distance intervals without error), but the assumption is less problematic than for line transects in two respects. First, only the observer-to-object distance is required for modelling. This is often easier to measure or estimate than the perpendicular distance of the object from a transect, especially if a detected object is not visible or audible, or has moved, by the time the observer reaches the closest point on the transect to it, or if densities are high, so that the observer may need to keep track of several detections simultaneously. Second, to reduce the problems inherent in estimating perpendicular distances for line transect sampling, sighting distances and sighting angles are often recorded. Effort is often concentrated ahead of the observer, so that measurement errors in the angles often give rise to recorded angles, and hence calculated perpendicular distances, of zero. Such data are notoriously difficult to model. Point transect data do not exhibit this problem; small observer-to-object distances are seldom recorded as zero, and few small distances occur, as the area surveyed close to a point is small.

In songbird point transect surveys on Arapaho National Wildlife Refuge, locations of detected birds were marked, and were later measured to the nearest decimetre (Knopf *et al.* 1988). Such accuracy is not usually possible; for example, up to 90% of detections are purely aural in woodland habitats (Reynolds *et al.* 1980; Scott *et al.* 1981; Bibby *et al.* 1985), so that the location of the bird must be estimated. Consistent bias in distance estimation should be avoided. If distances are overestimated by 10%, densities are underestimated by $100(1 - 1/1.1^2) = 17\%$; if they are underestimated by 10%, densities are overestimated by $100(1/0.9^2 - 1) = 23\%$. Bias in line transect density estimates would be smaller (9% and 11% respectively). **Provided** distance estimation is unbiased on average, measurement errors must be large to be problematic. Permanent markers at known distances are a valuable aid to obtaining unbiased estimates, and good range finders are effective over typical songbird detection distances, at least when the habitat is sufficiently open to use them. Scott *et al.* (1981) suggest that range finders are accurate to ± 1% within 30 m, and to ± 5% between 100 m and 300 m, whereas trained observers are accurate to ± 10–15% for distances to birds that can be seen. In our experience, range finders are often less accurate than ± 5% at distances close to 300 m.

If most objects are located aurally, then the assumption that an object is not counted more than once from the same point may be problematic. For example, a bird may call at one location, move unseen to another location, and again call. It is seldom problematic if the same bird is recorded from different points, unless it is following the observer.

5.10 Summary

Relative to line transects, relatively few distances are recorded close to zero distance in point transect surveys. Thus estimation of the central parameter ($h(0)$ for point transects and $f(0)$ for line transects) is more difficult, and model selection more critical. This was seen for the first example, where estimation was satisfactory if the correct model was selected, but if the uniform + polynomial model was selected, underestimation occurred, even though the model selection criteria indicated that the model was good. One of the contributory factors to this result was that the true detection function was the half-normal, which does not have an area of perfect detection around the point, even though it has a shoulder. Expressing this mathematically, $g''(0) \neq 0$, even though $g'(0) = 0$. If field methods are adopted that ensure an area of perfect detectability, estimation is more reliable, and different models will tend to give very similar estimates of density. The hazard-rate key is best able to fit data that show a large area over which detection is perfect, because the hazard-rate detection function can fit a wide, flat shoulder. It performed relatively badly on the example data sets largely because it tended to fit a flat shoulder to the simulated data, which were generated from the half-normal, which possesses a rounded shoulder.

To estimate densities reliably from point transect sampling, design and field methods should be carefully determined, following the guidelines of Chapter 7, and the data should be checked for recording and transcription errors. Histograms of the distance data are a useful aid for gaining an understanding of any features or anomalies, and give an indication of how much truncation is likely to be required. Several potential models should be considered, and model selection criteria applied to choose between them. Special software is essential if efficient, reliable analysis is to be carried out. Variance estimation methods should be chosen for their robust properties; the model that gives the most precise estimate is **not** the best model if either the estimate is seriously biased or the variance estimate ignores significant components of the true variance. A strategy for data analysis is outlined in Section 2.5.

Systematic error in estimated distances must be avoided. Observer training is essential if data quality is not to be compromised (Chapter 7). If more than one observer collects the data, analyses should be attempted that stratify by observer, to detect observer differences. The importance of this is illustrated in Section 8.7. It may prove beneficial to stratify analysis by other factors, such as species, location, habitat, month, year, or any factor that has a substantial impact on detection probabilities. Hypothesis testing may be used to determine which factors affect detection, thus reducing the amount of stratification and increasing parsimony. If the factor is ordinal or a continuous variable, it might enter the analysis as a covariate, so that its effect on detectability is modelled.

If objects occur in clusters, the location of the centre of each cluster and the number of objects in the cluster should be recorded. If clusters occur but are not well-defined, the observer should record each individual object, and its location, and use robust variance estimation methods. It is also useful to indicate which detected objects were considered to belong to the same cluster, so that a comparative analysis can be carried out by cluster. The location of the cluster can then be determined by the analyst, by calculating the geometric centre of the recorded locations.

The checklist of stages in line transect analyses given at the end of Chapter 4 may also be used for point transect analyses. In that checklist, replace 'line' by 'point', 'perpendicular distance' by 'detection distance', and 'Section 4.*' by 'Section 5.*' Also, the rule of thumb for selecting a truncation point w for detection distances is that $\hat{g}(w) \doteq 0.10$, or less satisfactorily, that roughly 10% of observations are truncated (Section 5.3).

6

Extensions and related work

6.1 Introduction

In this chapter, we consider extensions to the theory described in Chapter 3, and we describe distance sampling methods that are closely related to line and point transect sampling. We also examine models that do not fit into the key + adjustment formulation of earlier chapters. The material on these other models is not exhaustive, but is biased towards recent work, and models that may see future use and further methodological development. Most of the older models not described here are discussed in Burnham *et al.* (1980). One of the purposes of this chapter is to stimulate further research by raising some of the issues that are not satisfactorily handled by existing theory.

6.2 Other models

6.2.1 Binomial models

Binomial models are a special case of multinomial models, the theory for which is given in Section 3.4. We examine them briefly, since closed form estimators are available for some underlying models for the detection function; these are sometimes used as indices of abundance, to assess change in abundance with habitat (Section 8.9) or over time.

Line and point transect methods sometimes provide a quick and inexpensive alternative to census methods for generating population abundance indices of songbirds. In areas of thick cover, the observer may rely heavily on aural detection, with perhaps fewer than 10% of detected birds visible. Difficulty in both locating the bird and moving through vegetation make measurement of each detection distance impractical, and the disturbance would also cause many birds to move or change their behaviour. Bibby *et al.* (1985) stated:

> Recording the distance at which each bird was detected would have been desirable but was not practicable when so many were heard and not seen. Overcoming this difficulty might have risked swamping the observer's acuity for other birds when an average of about nine birds was recorded at each five-minute session. A single decision as to whether or not each bird was within 30 m when first detected was easier to achieve in the field and sufficient to permit estimates of density.

Sometimes, therefore, birds are simply recorded according to whether they are within or beyond a specified distance c_1. To help classify those birds close to the dividing distance, permanent markers may be positioned on trees or bushes at distance c_1. Only single-parameter models may be fitted to such data, and it is not possible to test the goodness of fit of any proposed model. The data may be analysed using the multinomial method for grouped data. Because there are only two groups (with the second cutpoint $c_2 = \infty$), the sampling distribution is binomial. As for the models of Chapter 3, numeric methods will be required in general, but below we consider the half-normal binomial model for point transects (Buckland 1987a), for which analytic estimates are available.

$$\text{Define } g(r) = \exp\{-(r/\sigma)^2\}, \ 0 \leqslant r < \infty$$

Then

$$v = \pi\sigma^2$$

and

$$f(r) = 2r \cdot \exp\{-(r/\sigma)^2\}/\sigma^2$$

Given the binomial likelihood, simple algebra yields the maximum likelihood estimate

$$\hat{h}(0) = 2 \cdot \log_e(n/n_2)/c_1^2$$

where n is the number of birds detected and n_2 is the number beyond distance c_1.

Thus,

$$\hat{D} = \frac{n \cdot \hat{h}(0)}{2\pi k} = \frac{n \cdot \log_e(n/n_2)}{c_1^2 \pi k}$$

where k is the number of plots sampled. Suppose for the example of Fig. 5.1 that distances had been recorded simply as to whether they were within or beyond 15 m. Then $c_1 = 15\,\mathrm{m}$, $k = 30$, $n = 144$ and $n_2 = 46$, yielding $\hat{D} = 0.00775\,\mathrm{birds/m^2}$, or 77.5 birds/ha.

The asymptotic variance of $\hat{h}(0)$ is

$$\widehat{\mathrm{var}}\{\hat{h}(0)\} = \frac{4(1/n_2 - 1/n)}{c_1^4}$$

so that the variance of $\hat{D}$ may be estimated using the methods of Chapter 5. For the example, we obtain $\widehat{\mathrm{se}}(\hat{D}) = 10.7$ birds/ha, compared with 11.1 birds/ha when exact distances are analysed.

Two measures of detectability are $r_{1/2}$, the point at which the probability of detection is one half, and ρ, the effective radius of detection. Further algebra yields

$$\hat{r}_{1/2} = \sqrt{\left\{\frac{2 \cdot \log_e 2}{\hat{h}(0)}\right\}}, \quad \text{with} \quad \widehat{\mathrm{var}}(\hat{r}_{1/2}) = \frac{2 \cdot \log_e 2 \cdot (1/n_2 - 1/n)}{c_1^4 \cdot \{\hat{h}(0)\}^3}$$

while

$$\hat{\rho} = \sqrt{\left\{\frac{2}{\hat{h}(0)}\right\}}, \quad \text{with} \quad \widehat{\mathrm{var}}(\hat{\rho}) = \frac{2 \cdot (1/n_2 - 1/n)}{c_1^4 \cdot \{\hat{h}(0)\}^3}$$

Thus for the example we have $\hat{r}_{1/2} = 11.7\,\mathrm{m}$, with $\widehat{\mathrm{se}}(\hat{r}_{1/2}) = 0.6\,\mathrm{m}$, and $\hat{\rho} = 14.0\,\mathrm{m}$, with $\widehat{\mathrm{se}}(\hat{\rho}) = 0.7\,\mathrm{m}$.

The efficiency of this binomial point transect model for estimating density relative to the half-normal model applied to ungrouped detection distances (Ramsey and Scott 1979; Buckland 1987a) is typically around 65%–80% (Buckland 1987a). This loss is relatively small; more serious is that robust models with more than a single parameter cannot be used on binomial data, and there is no information from which to test whether the form of the half-normal model is reasonable. However, bias from fitting an inappropriate model may be consistent between years, so that the method can be useful for providing an index of relative abundance over time at low cost.

Buckland (1987a) also derived analytic results for a linear binomial model for point transects, and found that density estimates for a variety of species are similar under the two models (Section 8.9). In practice, a detection function is unlikely to be approximately linear, so we give just the half-normal model here. The linear model had been considered earlier by Järvinen (1978), but only partially developed.

The choice of c_1 requires some comment. The value that minimizes the variance of $\hat{h}(0)$, of $\hat{r}_{1/2}$ and of $\hat{\rho}$ is $c_1 = 1.78/\sqrt{h(0)}$, which implies that roughly 80% of detections should lie within c_1. Buckland (1987a) finds that estimation is more robust when around 50% lie within c_1. Further, for simultaneously monitoring several species, an average value of 80% across all species may mean that few or no birds of quiet or unobtrusive species are detected beyond c_1. A practical advantage in selecting a smaller value for c_1 than the optimum is that the observer will be more easily able to determine whether a bird is within or beyond c_1.

As a safeguard, two cutpoints, c_1 and c_2, with $0 < c_1 < c_2 < \infty$, might be used, so that the sampling distribution is trinomial. The data could be analysed using the results for the general multinomial distribution in Section 3.4, but detection functions with at most two parameters could be used. Another option would be to use the above binomial model, first using cutpoint c_1, then using cutpoint c_2. If the two density estimates differed appreciably, this might be an indicator that the model is not robust. Otherwise the two estimates might be averaged. For surveys of several species, the first cutpoint might be used for quieter, more unobtrusive species, and the second for louder, more obvious species.

Järvinen and Väisänen (1975) developed three binomial models for line transects, in which the detection function was assumed to be linear, negative exponential or half-normal. The last of these is the most plausible, but a closed form estimator is not available for it. Program DISTANCE allows the user to implement this model using numeric methods. Otherwise, its limitations and advantages are very similar to those of the binomial half-normal model for point transects, described above.

Although the goodness of fit of a binomial model cannot be tested, the homogeneity of the binomial data can. Suppose each line or point transect is assigned to one of R groups, which might, for example, be R geographic regions or woods. Then an $R \times 2$ contingency table analysis may be carried out, where the frequencies are n_{ij}, $i = 1, \ldots, R$, $j = 1, 2$. Then n_{i1} is the total count within distance c_1 for group i, and n_{i2} is the total count beyond c_1. If a significant test statistic is obtained, then there is evidence that the detection function varies between groups. This method extends in the obvious way to multinomial models.

A variety of now outdated line transect methods is given in Burnham *et al.* (1980). In particular, a non-parametric binomial method once thought to have promise is the Cox method, derived by Eberhardt (1978a). We no longer recommend this method. However, as a matter of intellectual curiosity, we derived the analogue of the Cox method for point transect sampling. A linear detection function is assumed, so the model is very similar in concept to the linear binomial point transect model of Buckland (1987a). The difference is that the Cox method assumes that data are truncated at a distance for which the linear detection function is non-zero, whereas the method of Buckland, in common with the linear line transect model of Järvinen and Väisänen (1975), assumes data are untruncated in the field; an estimate of the point at which probability of detection becomes zero is provided by the model.

Let the distance data be grouped with the first two cutpoints being c_1 and c_2. Let the corresponding counts in these two intervals be n_1 and n_2 with total $n = n_1 + n_2$. Let k be the number of points sampled. The Cox estimator is derived by assuming that $g(r) = 1 + b \cdot r$ is an adequate model over the range $0 \leqslant r \leqslant c_2$. (It would be better to assume a quadratic form, $g(r) = 1 + b \cdot r^2$, but we use Eberhardt's formulation.) Of course the parameter b is negative. Based on just the counts in these first two intervals, we can get an estimate of b and hence an estimator of density, D. We do not provide the algebraic derivation here. The result is

$$\hat{D} = \frac{n}{k\pi c_2^2}\left[\frac{(c_2/c_1)^3 \cdot (n_1/n) - 1}{(c_2/c_1) - 1}\right]$$

Alternatively,

$$\hat{D} = \frac{n \cdot \hat{h}(0)}{2\pi k}$$

from which one can infer $\hat{h}(0)$. In the simple case of the two intervals having equal widths, Δ (i.e. $c_1 = \Delta$ and $c_2 = 2\Delta$), the result reduces to

$$\hat{D} = \frac{7n_1 - n_2}{4k\pi\Delta^2}$$

In Burnham *et al.* (1980: 169), the general Cox case was given for line transects. The $\hat{f}(0)$ in their publication is not in the same form as used here for $\hat{h}(0)$. For comparison we provide the results below.

177

Cox estimator for point transects:

$$\hat{h}(0) = \frac{2}{c_2^2}\left[\frac{(c_2/c_1)^3 \cdot (n_1/n) - 1}{(c_2/c_1) - 1}\right]$$

Cox estimator for line transects:

$$\hat{f}(0) = \frac{1}{c_2}\left[\frac{(c_2/c_1)^2 \cdot (n_1/n) - 1}{(c_2/c_1) - 1}\right]$$

For this same context of two cutpoints and truncation at c_2, if we assume the detection function has the generalized form $g(y) = 1 + b \cdot y^p$ for $0 \leqslant y \leqslant c_2$ and where p is a known integer $\geqslant 1$, then relevant results for point and line transects are

$$\hat{h}(0) = \frac{2}{c_2^2}\left[\frac{(c_2/c_1)^{2+p} \cdot (n_1/n) - 1}{(c_2/c_1)^p - 1}\right]$$

and

$$\hat{f}(0) = \frac{1}{c_2}\left[\frac{(c_2/c_1)^{1+p} \cdot (n_1/n) - 1}{(c_2/c_1)^p - 1}\right]$$

Corresponding theoretical sampling variances are

$$\text{var}\{\hat{h}(0)\} = \frac{4}{c_2^4}\left[\frac{(c_2/c_1)^{2+p}}{(c_2/c_1)^p - 1}\right]^2 \cdot \frac{p_1(1 - p_1)}{n}$$

and

$$\text{var}\{\hat{f}(0)\} = \frac{1}{c_2^2}\left[\frac{(c_2/c_1)^{1+p}}{(c_2/c_1)^p - 1}\right]^2 \cdot \frac{p_1(1 - p_1)}{n}$$

where $p_1 = E(n_1)/n$, and is estimated by $\hat{p}_1 = n_1/n$.

6.2.2 Empirical estimators

Emlen (1971, 1977) developed a non-mathematical approach for line transect analysis of songbird data. He assumed that a characteristic proportion of birds of any species will be detected in the surveyed area $2wL$. He called this proportion the coefficient of detectability. This corresponds to the parameter P_a defined in Section 3.1. The method typically uses data from only two to four distance categories, and an estimator of the product $E(n) \cdot f(0)$ is determined from a smoothed frequency histogram of perpendicular distances. The density estimate from this model is probably best considered as a rough index of relative abundance, and is not recommended (Burnham *et al.* 1980: 164).

Ramsey and Scott (1979, 1981b) developed a point transect model similar to Emlen's (1971, 1977) line transect model. Suppose cutpoints are defined at distances $c_0 = 0, c, 2c, \ldots, kc$, so that the truncation point $w = kc$. Let $A(0, i)$ be the area of the circle of radius ic, and let $A(i, j)$ be the area of the annulus with inner radius ic and outer radius jc. Thus, $A(i, j) = \pi c^2 (j^2 - i^2)$. Let $n(i, j)$ be the number of birds counted within the annulus. Then the corresponding density $D(i, j)$ may be estimated by

$$\hat{D}(i, j) = \frac{n(i, j)}{A(i, j)}$$

The value i is chosen to be the smallest value such that the likelihood of differing densities within the areas $A(0, i)$ and $A(i, j)$ is at least four times the likelihood of equal densities for all $j > i$. That is, i is the smallest value that satisfies

$$[\hat{D}(0, i)]^{n(0, i)} \cdot [\hat{D}(i, j)]^{n(i, j)} \geq 4 \cdot [\hat{D}(0, j)]^{n(0, j)} \quad \text{for all} \quad j > i$$

Having calculated i,

$$\hat{D} = \frac{n(0, i)}{A(0, i)}$$

with

$$\widehat{\mathrm{var}}(\hat{D}) = \frac{\hat{D}}{A(0, i)}$$

This variance estimate is only valid if birds are randomly distributed throughout the study area, so is likely to be poor, but an empirical estimate may be obtained, for example using the bootstrap. The density estimate is only valid if there is an area of perfect detectability, assumed to extend out to distance ic.

Wildman and Ramsey (1985) developed the 'CumD' estimator, which is similar to the above, but estimates the distance out to which detection is certain without the need to group the data. The estimator is defined for both line and point transects, and observations are transformed to detection 'areas', defined as $a = 2Lx$ for line transects and $a = \pi r^2$ for point transects. These areas are ordered from smallest to largest, giving $a_1 \leq a_2 \leq \cdots \leq a_n$, with empirical distribution function $F_n(a)$. Let $j(0) = a_{j(0)} = 0$, and let $j(1)$ be the largest integer such that

$$d_1 = \frac{j(1)}{a_{j(1)}} = \max\left\{\frac{j}{a_j} \mid j \geq \sqrt{n}\right\}$$

Then for $m = 2, 3, \ldots$ let $j(m)$ be the largest integer such that

$$d_m = \frac{j(m) - j(m-1)}{a_{j(m)} - a_{j(m-1)}} = \max\left\{\frac{j - j(m-1)}{a_j - a_{j(m-1)}} \mid j > j(m-1)\right\}$$

Straight lines linking the points $[a_{j(m)}, j(m)/n]$, $m = 0, 1, 2, \ldots$, form a convex envelope over $F_n(a)$, which is the isotonic regression estimate of $F(a)$ (Barlow et al. 1972) and yields an estimated detection function that is a non-increasing function of distance from the point. The slopes d_m are average estimates of density of detections within annuli of increasing distance from the point. These equate to estimates of object density if all objects within a given annulus are detected. Likelihood ratio tests of the equality of density between the first region and the next $m - 1, m = 2, 3, \ldots$, are used to provide a stopping rule. The smallest value of m, m^* say, is chosen such that the null hypothesis is rejected, and all objects within the corresponding radius a_{j^*}, where $j^* = j(m^*)$, are assumed to have been detected, yielding an estimate of density.

This innovative and intuitively appealing approach is computationally inexpensive and easily programmed. However, probability of detection should be at or close to unity for some distance from the point for the method to yield good estimates; because of the paucity of sightings close to the point, this distance must be appreciably greater for point transects than for line transects for comparable performance. Also, estimation of a distance up to which all objects are detected through hypothesis testing causes the method to underestimate density when sample size is small, as there are too few data for the tests to have much power. Bias in the method is therefore a strong function of sample size, at least for small samples. Investigation of how large sample size should be for the method to perform well would be useful. Wildman and Ramsey (1985) show that it must be very large if the true detection function, expressed as a function of area, is negative exponential. For point transects, this corresponds to the half-normal detection function, when expressed as a function of distance, and bias is still of the order of 23% for a sample size of 10 000.

6.2.3 Estimators based on shape restrictions

Johnson and Routledge (1985) developed a non-parametric line transect estimator based on shape restrictions for which they found 'a general improvement in efficiency over existing estimators.' The method has not seen wide use, but the recent release of software TRANSAN (Routledge and Fyfe 1992) makes it more accessible, and we encourage further evaluation.

The density function $f(x)$ is constrained to be non-negative and to integrate to unity. In addition, Johnson and Routledge added the monotonicity constraint that $f(x)$ must be monotonic non-increasing and the shape constraint that $f(x)$ must have a concave shoulder, followed by a convex tail, separated by a single point of inflection. The range of concavity must be determined, or guessed, by the user, and it is suggested that the percentage of detections that fall within the point of inflection might be as high as 90% or below 50%.

The parameter $f(0)$ is estimated by grouping the distance data, and using the frequencies in a histogram estimator of $f(x)$. Let h_i represent the height of the ith histogram bar, $i = 1, \ldots, u$. Find the adjusted heights, $\tilde{h}_i$, by minimizing

$$\sum_{i=1}^{u} (h_i - \tilde{h}_i)^2$$

subject to the imposed constraints. Then $f(0)$ is estimated by $\hat{f}(0) = \tilde{h}_1$.

Johnson and Routledge used a bootstrap approach to quantify precision, but one that is more sophisticated than the general purpose bootstrap described in Section 3.7.4. For the single-parameter case, a guess can be made of say the lower $100(1 - 2\alpha)\%$ confidence limit for $f(0)$, and bootstrap resamples generated. The parameter $f(0)$ is estimated from each resample, and if the proportion exceeding the estimate from the true data is greater than α, then the current guess of the limit is estimated to be too large, and is reduced. The process is repeated until the true limit is located with adequate precision. Johnson and Routledge suggested that a more efficient search procedure might be developed. A general algorithm for evaluating confidence limits in this way for single parameter problems, which updates the estimated limit after each resample, was given by Buckland and Garthwaite (1990). Its optimal properties were noted by Garthwaite and Buckland (1992).

Johnson and Routledge based their conclusions on estimator performance on a simulation study, in which the Fourier series and half-normal models were compared with the shape restriction estimator. If the procedures recommended in earlier chapters were implemented, more severe data truncation would have been carried out for some of their simulations, and the Fourier series and half-normal models would have been rejected as inappropriate in some. However, the shape restriction method proved to be robust to choice of truncation point and to the true underlying detection function, and therefore merits further investigation and development.

6.2.4 Kernel estimators

(a) *Line transect sampling* There are several methods for fitting probability densities using kernels. They are based on the concept of replacing a point (a detection distance here) by a distribution, centred on that point. This is done for all observations, and the distributions are summed, to provide the estimated density function. Buckland (1992a) compared the kernel estimator of Silverman (1982) with the Hermite polynomial model for fitting the deer data from survey 11 of Robinette *et al.* (1974). To force the algorithm to fit a symmetric density, differentiable at zero (and hence possessing a shoulder), each distance x was replaced by two, x and $-x$. The optimum window width for a normal distribution with standard deviation σ, i.e. $h = 1.06\sigma n^{-0.2}$, yielded a comparable but less smooth fit to the data than the Hermite polynomial model. The kernel method is far less computer-intensive than the methods recommended here, but the kernel estimate of $f(0)$ is highly sensitive to the choice of window width (Buckland 1992a). Further, the kernel method does not readily yield a variance for $\hat{f}(0)$, although the bootstrap may be used, either as described in Section 3.7.4 or using the more sophisticated approach of Garthwaite and Buckland (1992), noted in Section 6.2.3. A final disadvantage of the kernel method is that covariates cannot be incorporated, thus ruling out the methods of Section 3.8.2. One advantage of the kernel method is that observations have only a local effect on the fitted density. For parametric or semi-parametric methods, if the model fails to fit the tail of the distribution well, its fit at zero distance may be adversely affected.

Brunk (1978) developed a kernel method based on orthogonal series, an approach which is a close parallel to the key + adjustment formulation, especially if the adjustment terms are orthogonal to the key, as for the Hermite polynomial model. Buckland (1992a) found that Brunk's method gave unstable estimation relative to the Hermite polynomial model when the data were simulated from a markedly non-normal distribution.

(b) *Point transect sampling* Quang (1990b) proposed a method based on kernel techniques. As in his line transect developments, he assumed that perfect detectability occurs somewhere, but not necessarily at zero distance; that is, that $g(r) = 1$ for some value of $r \geqslant 0$. Given $g(0) = 1$, it was noted in Chapter 3 that density estimation could be reduced to estimation of $h(0) = \lim_{r \to 0} f(r)/r$. Under Quang's formulation, this generalizes, so that the maximum value of the ratio $h(r) = f(r)/r$ over all r must be estimated. He used the kernel method with a normal kernel (Silverman 1986) to estimate $f(r)$ and hence $h(r)$.

The concept of maximizing $h(r)$ is also applicable to series-type models. Suppose a model is selected whose plotted detection function increases with r over a part of its range, thus indicating that objects close to the point are evading detection, either by fleeing or by remaining silent. Then $\hat{h}(0)$, which is the estimated slope of the density at zero, $\hat{f}'(0)$, may be replaced by the maximum value of $\hat{h}(r) = \hat{f}'(r)$ in the point transect equation for estimated density (Section 3.7.1).

6.2.5 Discrete hazard-rate models

The hazard-rate development of Chapter 3 assumed that either the sighting cue or the probability density of flushing time is continuous. Often this is not the case. For example whales that travel singly or in small groups may surface at regular intervals, with periods varying from a few seconds to over an hour, depending on species and behaviour, when the animals cannot be detected.

(a) *Line transect sampling* Schweder (1977) formulated both continuous and discrete sighting models for line transect sampling, although he did not use these to develop specific forms for the detection function.

Let $q(z, x) = \text{pr}\{\text{seeing the object for the first time} \mid \text{sighting cue at } (z, x)\}$

where z and x are defined in Fig. 1.5. Then if the ith detection is recorded as (t_i, z_i, x_i), where t_i is the time of the ith detection, the set of detections comprises a stochastic point process on time and space. The first-time sighting probability depends on the speed s of the observer so that

$$q(z, x \mid s) = Q(z, x) \cdot E\left\{\prod_{i > 1} [1 - Q(z_i, x_i)]\right\}$$

where $Q(z, x)$ is the conditional probability of sighting a cue at (z, x) given that the object is not previously seen; $Q(z, x)$ is thus the discrete hazard. Assuming that detections behind the observer cannot occur, then

$$g(x|s) = \int_0^\infty u(z, x)q(z, x|s)dz$$

where $u(z, x)$ is the probability that a sighting cue occurs at (z, x), unconditional on whether it is detected; $u(z, x)$ is a function of both object density and cue rate.

More details were given by Schweder (1977, 1990), who used the approach to build understanding of the detection process in line transect sampling surveys. In a subsequent paper (Schweder *et al.* 1991), three specific models for the discrete hazard were proposed, and the corresponding detection function for the hazard they found to be most appropriate for North Atlantic minke whale data is:

$$g(x) = 1 - \exp[-\exp\{a' + b' \cdot x + c' \cdot \log_e(x)\}] \qquad (6.1)$$

If we impose the constraints $b' \leq 0$ and $c' < 0$, this may be considered a more general form of the hazard-rate model of Equation 3.7, derived assuming a continuous sighting hazard:

$$g(x) = 1 - \exp[-(x/\sigma)^{-b}]$$

When $b' = 0$, Equation 6.1 reduces to Equation 3.7 with $a' = b \cdot \log_e(\sigma)$ and $c' = -b$. Thus a possible strategy for analysis is to use Equation 3.7 (the standard hazard-rate model) unless a likelihood ratio test indicates that the fit of Equation 6.1 is superior. Both models are examples of a complementary log-log model (Schweder 1990).

(b) *Point transect sampling*   Point transects are commonly used to estimate songbird densities. In many studies, most cues are aural, and therefore occur at discrete points in time. Ramsey *et al.* (1979) defined both an 'audio-detectability function' $g_A(r)$ and a 'visual-detectability function' $g_V(r)$. (Both are also functions of T, time spent at the point seeking detections.)

Let $p(r, t) = \text{pr}\{$ object at distance r is not detected within time $t\}$

Then

$$g_A(r) = 1 - p(r, T) = 1 - \sum_{j=1}^{\infty} [1 - \gamma(r)]^j pr(j)$$

where j is the number of aural cues the object makes in time T, $pr(j)$ is the probability distribution of j, and $\gamma(r)$ is the probability that a single aural cue at distance r is detected. This assumes that the probability of detection of an aural cue is independent of time, the number of cues is independent of distance from the observer and the chance of detecting the jth cue, having missed the first $j - 1$, is equal to the chance of detecting the first cue. Hence the audio-detectability function is of the form

$$g_A(r) = 1 - \psi[1 - \gamma(r)]$$

where

$$\psi(s) = \sum_{j=0}^{\infty} s^j pr(j)$$

is the probability generating function of j.

The visual detectability function, $g_V(r)$, is modelled in a continuous framework, and yields Equation 3.8: $g(r) = 1 - \exp[- k(r)T]$. Ramsey *et al.* (1979) then combined these results to give

$$g(r) = 1 - \psi(1 - \gamma(r)) \cdot \exp[- k(r)T]$$

A detectability function may be derived by specifying (1) a distribution for the number of calls, (2) a function describing the observer's ability to detect a single call, and (3) the function $k(r)$ of visual detection intensities. Ramsey *et al.* considered possibilities for these, and plotted resulting composite detection functions. One of their plots shows an audio detection function in which detection at the point is close to 0.6 but falls off slowly with distance and a visual function where detection is perfect at the point, but falls off sharply. The composite detection function is markedly 'spiked' and would be difficult to model satisfactorily. This circumstance could arise for songbird species in which females are generally silent and retiring, and can be avoided by estimating the male population alone, or singing birds alone, if adults cannot easily be sexed. If data are adequate, the female population could be estimated in a separate analysis.

6.3 Modelling variation in encounter rate and cluster size

6.3.1 On the meaning of var(n)

We have concentrated our modelling and data analysis considerations on the detection function $g(y)$ and, where applicable, the mean cluster size $E(s)$. For both of these data components, we emphasize estimation of parameters as a way to extract and represent the structural (i.e. predictable) information in the data and thus make inferences about population abundance. However, it is also necessary to estimate the sampling variance of $\hat{D}$, var($\hat{D}$), which involves at least one additional set of parameters, namely var(n), which may vary over strata and time. The variance of the counts, var(n), is intended to summarize the non-structural or residual component in the counts. By definition, var(n) = $E[n - E(n)]^2$, thus we must consider whether the **expected** encounter rate is constant for each line or point. Expected encounter rate could vary over a sampled area, in which case there is structural information in the counts. We can go further with this idea by considering the information in the actual spatial positions (x–y coordinates) of detections in the sampled area, given the known distribution of effort (i.e. the locations of lines or points). This entails fitting a relative density distribution model to the large-scale spatial structure of object density as revealed through the spatial variation of encounter rate.

This added level of analysis would require fitting $E[n(x, y)]$, the expected encounter rate as a function of spatial position. Then var(n) is estimated from the residuals about this fitted model for encounter rate, denoted by $\hat{E}(n_i/l_i)$ in the case of line transect sampling. (To obtain results for point transects, replace l_i by 1 and L by k throughout this section.) If we have k replicate lines in a stratum, then var(n) should be estimated as

$$\widehat{\text{var}}(n) = L \frac{\sum_{i=1}^{k} l_i \cdot \left[\frac{n_i}{l_i} - \hat{E}\left(\frac{n_i}{l_i}\right) \right]^2}{k - 1} \tag{6.2}$$

If we assume there is no variation in encounter rate among complete lines within a stratum or time period, then we have

$$E\left(\frac{n_i}{l_i}\right) = \mu \tag{6.3}$$

which is constant for all lines. Proper design can ensure that this assumption is reasonable. Relevant design features are stratification, orientation of lines parallel to density gradients (i.e. perpendicular to density contours) and equal, or appropriate, line lengths. When Equation 6.3 is true, the appropriate estimator of μ is $\hat{\mu} = n/L$, and Equation 6.2 gives, within a stratum,

$$\widehat{var}(n) = L \frac{\sum_{i=1}^{k} l_i \cdot \left[\frac{n_i}{l_i} - \frac{n}{L} \right]^2}{k - 1} \tag{6.4}$$

which we have already recommended.

The critical point is that if Equation 6.3 fails to the extent that there is substantial variation in per line encounter rate, with

$$\mu_i = E\left(\frac{n_i}{l_i} \right)$$

then $\widehat{var}(n)$ from Equation 6.4 is not appropriate as it includes both stochastic (residual) variation and the structural variation among $\mu_1, \ldots, \mu_k$. This latter variation does not belong in var(n), as it represents large-scale variation in the true object density over the study area. This variation is of interest in its own right, but it is difficult to model and estimate. We will return to this point in Section 6.3.3.

A final comment on the meaning of var(n) is in order; as stated above, var(n) is meant to measure the residual (stochastic) variation in detection counts, n. There are two components to this stochastic variation. First, there is always the small-scale, hence virtually unexplainable, spatial variation in locations of objects. Thus even if detection probability was one over the strips or plots in which counts are made, there would be a substantial component to var(n) due to the small-scale sampling variation in the number of objects in the sampled area, a. Second, when detection probability is not one, then there is the further stochastic variation in the counts due to the particular detections made given the number of objects in area a. Although possible, it is neither necessary nor useful to partition var(n) into these two components.

6.3.2 Pooled estimation of a common dispersion parameter b

Often we can assume that the expected encounter rate μ_i is constant for each replicate line or point within a stratum and time of sampling; thus we assume Equation 6.3. As noted above, this equation will hold under proper design of the study. The subject of this section is efficient estimation of var(n); this is a concern when sample sizes per stratum are low.

The basic idea of efficient estimation of var(n), once the μ_i are appropriately modelled, is that we can model the structure of var(n) over strata and/or time. The idea of modelling variances is not new to statistics (e.g. Carroll and Ruppert 1988) and the practice is becoming increasingly common. As a starting point to any such modelling, we recommend the representation

$$\text{var}(n) = b \cdot E(n) \tag{6.5}$$

and then modelling the dispersion parameter b. However, the only case we consider here is for when b may be constant over strata and/or time. This is a common situation in our opinion.

Assume the data are stratified spatially and/or temporally into V distinct data sets, indexed by $v = 1, \ldots, V$. Within each data set, assume some replication exists, with line lengths

$$l_{vj}, j = 1, \ldots, k_v, v = 1, \ldots, V$$

and corresponding counts n_{vj}. Nominally, we must now estimate V separate count variances var(n_v), $v = 1, \ldots, V$. The problem is there may be sparse data for each estimate, due to little replication within data sets (small k_v) or few detections (small n_v, perhaps under ten). Experience, and some theory, suggests that the dispersion parameter b in Equation 6.5 will be quite stable and can be modelled, thereby reducing the number of dispersion parameters that must be estimated. If objects have a homogeneous Poisson spatial distribution by data set, then $b = 1$. This is not a reasonable assumption; we should expect $b > 1$, but still probably in the range 1 to say 4.

An accepted principle in data analysis is that we should fit the data by a plausible but parsimonious model, i.e. a model that fits the data well with few parameters (e.g. McCullagh and Nelder 1989). This principle applies to dispersion parameters as well as to structural parameters such as $f(0)$, $h(0)$, $E(s)$ and $E(n)$. Below we provide formulae for estimating a common dispersion parameter across all V data sets.

Separate estimates for each data set are found by applying Equations 6.4 and 6.5:

$$\widehat{\text{var}}(n_v) = L_v \frac{\sum_{i=1}^{k_v} l_{vi} \left[\dfrac{n_{vi}}{l_{vi}} - \dfrac{n_v}{L_v} \right]^2}{k_v - 1}$$

where $\quad L_v = \sum l_{vi}$

and

$$\hat{b}_v = \frac{\widehat{\text{var}}(n_v)}{n_v}$$

with $k_v - 1$ degrees of freedom. If $b_v = b$ for all V data sets, then under a quasi-likelihood approach (McCullagh and Nelder 1989), the estimator of b is

$$\hat{b} = \frac{\sum_{v=1}^{V} (k_v - 1)\hat{b}_v}{\sum_{v=1}^{V} (k_v - 1)} \tag{6.6}$$

which has $\sum_{v=1}^{V} (k_v - 1)$ degrees of freedom.

The estimator $\hat{b}$ should have good properties because it is based on a general theory. There is, however, an obvious alternative moment estimator:

$$\hat{b} = \frac{\sum_{v=1}^{V} \widehat{\text{var}}(n_v)}{\sum_{v=1}^{V} n_v} = \frac{\sum_{v=1}^{V} n_v \hat{b}_v}{\sum_{v=1}^{V} n_v} \tag{6.7}$$

We performed a limited simulation comparison of these two methods (i.e. Equations 6.6 and 6.7), and failed to distinguish one as inferior, in terms of bias or precision.

The case of a common dispersion parameter and a pooled estimator is analogous to the analysis of variance assumption of homogeneity of error variance and use of the corresponding pooled estimator of error mean square. When b is so estimated, the squared coefficient of variation of $n = n_1 + \ldots + n_v$ is estimated by

$$[\text{cv}(n)]^2 = \frac{\hat{b}}{n}$$

and for any one stratum,

$$[cv(n_v)]^2 = \frac{\hat{b}}{n_v}$$

These squared coefficients of variation are then used in formulae for sampling variances of $\hat{D}$ and $\hat{D}_v$; in particular, we can get the best possible variance estimator of an individual stratum density estimate based on

$$[cv(\hat{D}_v)]^2 = \frac{\hat{b}}{n_v} + [cv\{\hat{f}(0)\}]^2 + [cv\{\hat{E}(s)\}]^2$$

Assuming constant $f(0)$ and $E(s)$ over the V strata, which the above formula implicity does, the limiting factor on the precision of $\hat{D}_v$ is just the sample size n_v.

Further comparative investigations of the two estimators of b would be useful. Use of such a pooled estimator is most compelling when sample sizes, n_v, are small, in which case weighting by sample size seems, intuitively, to be important. Yet in Equation 6.6, the weights ignore actual sample sizes. Perhaps when per survey sample sizes, n_v, are smaller than the number of replicate lines or points, k_v, Equation 6.7 would be better.

Further thoughts on this matter arise by considering an average density estimate over temporally repeated surveys on the same area. Then an average density over the repeated surveys is estimated as

$$\hat{\hat{D}} = \frac{n \cdot \hat{f}(0)}{2L} \hat{E}(s)$$

where

$$L = L_1 + \ldots + L_v$$

The sampling variance on $\hat{\hat{D}}$ is provided by the usual formula but with $[cv(n)]^2$ computed as

$$[cv(n)]^2 = \frac{\sum_{v=1}^{V} \widehat{var}(n_v)}{n^2} = \frac{\hat{b}}{n}$$

where $\hat{b}$ is estimated from Equation 6.7, thus supporting use of Equation 6.7 for temporal stratification. However, this argument does not apply

190

to spatial stratification. (In computing this $\hat{D}$, one should consider whether true density varies by time; if it does, then either $\bar{D}$ may not be relevant, or its variance should include the among D_v variation, which the above $cv(n)$ does not do.)

(a) *An example* In 1989 and 1990, Ebasco Environmental, under contract to the US Minerals Management Service, conducted 13 consecutive aerial line transect surveys for marine mammals offshore of the states of Oregon and Washington, USA (Green *et al.* 1992). The same set of 32 parallel transects (i.e. $k_v = 32$ for all v) was flown each survey during an 18-month period. A given survey took about a week; surveys were a month or more apart in time. Two species of dolphin were of particular interest and generated enough detections to allow density estimation: Risso's dolphin (*Grampus griseus*) and Pacific white-sided dolphin (*Lagenorhynchus obliquedens*). Most of the detections occurred during the eight spring and summer surveys (two in each season in both 1988 and 1989). Table 6.1 presents summary results for $\widehat{var}(n)$ and $\hat{b}$ from these eight surveys.

Table 6.1 Encounter data and dispersion parameter estimates from the study reported on in Green *et al.* (1992) for Pacific white-sided and Risso's dolphins for the spring and summer surveys (indexed by v) in 1988 and 1989 (surveys 5, 6, 7 and 13 were in autumn and winter, and spring survey 9 targeted grey whales). $k_v = 32$ for each survey

Pacific white-sided dolphin				Risso's dolphin			
v	n_v	$\widehat{var}(n_v)$	$\hat{b}_v$	v	n_v	$\widehat{var}(n_v)$	$\hat{b}_v$
1	2	3.28	1.64	1	6	10.31	1.72
2	14	58.46	4.18	2	11	37.33	3.39
3	6	21.29	3.55	3	7	10.51	1.50
4	5	3.81	0.76	4	10	25.30	2.53
8	3	5.49	1.83	8	6	7.05	1.18
10	5	33.47	6.69	10	20	66.71	3.34
11	2	0.93	0.47	11	5	21.46	4.29
12	3	2.23	0.74	12	1	0.72	0.72
Totals	40	128.96	19.86		66	179.39	18.67

Table 6.1 shows that individual survey estimates of the dispersion parameter are quite variable, ranging from 0.47 to 6.69 for Pacific white-sided dolphin and from 0.72 to 4.29 for Risso's dolphin. However, the corresponding sample sizes of total per survey counts are small, ranging from 1 to 20, and averaging, per survey, 5 and 8.25 for Pacific

white-sided and Risso's dolphin, respectively. Because of the small sample sizes for these data, most of the variation in the eight estimates of b for each species is surely sampling variation and not variation in true dispersion over time. We believe it is appropriate and desirable to compute and use a single $\hat{b}$ for each species in this case. Equation 6.7 yields $\hat{b} = 128.96/40 = 3.22$ and $179.39/66 = 2.72$ for Pacific white-sided and Risso's dolphins, respectively. Both estimates are close to the 'default' value $b = 3$ suggested by Burnham et al. (1980: 35–6) for when no estimates are available (such as in the initial planning of a study).

(b) *Basis for the theory* The derivation of $\hat{\mu}_v$, $\hat{b}_v$ and $\hat{b}$ (from Equation 6.6) can be carried out in a quasi-likelihood framework (McCullagh and Nelder 1989: 323–8). The starting point is the model $n_{vi} = \mu_v \cdot l_{vi} + \varepsilon_{vi}$ with $E(\varepsilon_{vi}) = 0$ and $\operatorname{var}(\varepsilon_{vi}) = b_v \cdot E(n_{vi}) = b_v \cdot \mu_v \cdot l_{vi}$ for $i = 1, \ldots, k_v$ and $v = 1, \ldots, V$. A special case of the model is to use $b_v = b$ for all v. Given independence over all i and v, then the optimal estimator of μ_v is obtained as the solution to the equation

$$\sum_{i=1}^{k_v} - l_{vi} \cdot \left[\frac{n_{vi} - \mu_v \cdot l_{vi}}{b_v \cdot \mu_v \cdot l_{vi}} \right] = 0$$

The solution is

$$\hat{\mu}_v = \frac{n_v}{L_v}$$

and this is true regardless of whether or not the dispersion parameter b varies by stratum. Direct application of quasi-likelihood theory gives the estimate of separate dispersion parameters as

$$\hat{b}_v = \frac{1}{k_v - 1} \cdot \sum_{i=1}^{k_v} \frac{(n_{vi} - \hat{\mu}_v \cdot l_{vi})^2}{\hat{\mu}_v \cdot l_{vi}}$$

$$= L_v \cdot \frac{\sum_{i=1}^{k_v} l_{vi} \left[\dfrac{n_{vi}}{l_{vi}} - \dfrac{n_v}{L_v} \right]^2}{(k_v - 1) \cdot n_v}$$

Applying the same theory to the special case of a constant dispersion parameter produces the result

192

$$\hat{b} = \frac{1}{\sum\limits_{v=1}^{V}(k_v - 1)} \cdot \left[\sum\limits_{v=1}^{V} \sum\limits_{i=1}^{k_v} \frac{(n_{vi} - \hat{\mu}_v \cdot l_{vi})^2}{\hat{\mu}_v \cdot l_{vi}} \right]$$

This is equivalent to

$$\hat{b} = \frac{\sum\limits_{v=1}^{V} \left(\dfrac{L_v}{n_v}\right) \sum\limits_{i=1}^{k_v} l_{vi} \left(\dfrac{n_{vi}}{l_{vi}} - \dfrac{n_v}{L_v}\right)^2}{\sum\limits_{v=1}^{V}(k_v - 1)}$$

$$= \frac{\sum\limits_{v=1}^{V} \dfrac{1}{n_v} \cdot (k_v - 1) \cdot \widehat{\mathrm{var}}(n_v)}{\sum\limits_{v=1}^{V}(k_v - 1)}$$

$$= \frac{\sum\limits_{v=1}^{V}(k_v - 1) \cdot \hat{b}_v}{\sum\limits_{v=1}^{V}(k_v - 1)}$$

In principle there are ways to test $H_0 : b_v = b$ for all v. However, the motivation for getting a pooled estimate of an assumed constant dispersion parameter is strongest with sparse data, in which case the tests we are aware of are not reliable. Testing $H_0 : b_v = b$ in this distance sampling context with sparse data is an area in need of research.

6.3.3 Modelling spatial variation in encounter rate

Total numbers, or density, in an area can be reliably estimated even if there are predictable trends in density over the area (though care must then be taken with the spatial allocation of lines or points). An example of predictable trends would be a consistent year-to-year density gradient with distance from coastline in some marine mammals, or a biologically significant association of, for example, some dolphin species with measurable oceanographic features such as surface temperature. The distance sampling methods presented in this book can be embedded in point process sampling theory to allow density surfaces to be fitted to the spatial coordinates of detection locations and even to relate such surfaces to measurable covariates. The information for this modelling is contained in the spatial locations of detections as represented in an x–y coordinate system and in any covariates measured at those locations and elsewhere over the study area.

When important trends in density exist within strata, the main effect on theory presented here is with respect to estimation of var(n) using Equation 6.4. Improved estimation of var(n) requires modelling encounter rate to give reliable estimation of

$$\mu_{vi} = E\left(\frac{n_{vi}}{l_{vi}}\right)$$

and then use of Equation 6.2 by stratum. With carefully designed studies, Equation 6.4 will be reliable. However, we think there is substantial benefit to be gained by the addition of biological information and understanding, made possible by modelling the density of the population over the study area. Thus, while the estimation of var(n) is our motivation for mentioning point process modelling of encounter rate, the benefits to be gained go beyond improved sampling variance estimators.

Some basic theory for embedding distance sampling in a point process model of the population over the study area has been developed by Schweder (1974, 1977) and Burdick (1979). We give here our own view of how one can conceptualize this modelling of encounter rate; some simplifications are made below compared to a completely general theory. Again, we consider line transect sampling; theory simplifies for point transect sampling because points may be treated as dimensionless.

Let $D(x, y)$ represent the intensity function for a point process model of objects over area A. (One can think of $D(x, y) \cdot dx \cdot dy$ as the expected local density about point x, y.) The density parameter D that we have focused on in this book can be defined by

$$\frac{1}{A} \int \int_A D(x, y)dx\, dy = D$$

Note that D is really an average density over the study area; also, technically the above double integral gives $E(N/A)$, whereas $D = N/A$. The symbol A is given a dual role here, both as the scalar size of the area and as a symbol for the set of points defining the study area. In a point process model, the expected number of points in any area $a \subset A$ is

$$E(N|a) = \int \int_a D(x, y)dx\, dy$$

194

For one realization of the process, the probability of finding one or more points in the area a is $1 - \exp[- E(N|a)]$.

The relevant surveyed areas for line transects are the sample strips of length l_i and width $2w$, thus changing our symbolism some, we can write

$$E(n_i|l_i) = \int \int_{2wl_i} D(x, y)\, p(x, y)\, dx\, dy$$

The above still denotes an integral over an area. Now we have added $p(x, y)$, which is the conditional probability of actually detecting an object at coordinates (x, y), given an object is there. To simplify the formulation of the problem, we translate the local coordinate system for the above double integral so that the x-coordinate is the transect centreline, and the y-coordinate is perpendicular to that line. (Given a straight line, this is a linear translocation-rotation coordinate transformation.) Thus

$$E(n_i|l_i) = \int_0^{l_i} \int_{-w}^{w} D(x, y)\, p(x, y)\, dy\, dx$$

The next step is critical. In the above coordinate representation, for a given y, variations of $p(x, y)$ in x (i.e. along the line of travel) either do not exist or are irrelevant (in the absence of covariates to explain such spatial variations). Therefore, we can now replace $p(x, y)$ by the detection function $g(y)$. In practice, the scale will always be such that l_i is at least an order of magnitude larger than w. Hence we can safely model the local intensity at any point x in this strip over which integration occurs as the intensity which applies on the transect centreline; hence intensity is assumed to be independent of y. That is, in the strip of area $2wl_i$, we assume the model $D(x) = D(x, y)$. Under these simplifying assumptions, the above double integral becomes

$$E(n_i|l_i) = \int_0^{l_i} \int_{-w}^{w} D(x)\, g(y)\, dy\, dx$$

$$= \left[\int_{-w}^{w} g(y)\, dy \right] \cdot \left[\int_0^{l_i} D(x)\, dx \right]$$

195

$$= 2wl_i \left[\frac{1}{w} \int_{-w}^{w} g(y) \, dy \right] \cdot \left[\frac{1}{l_i} \int_{0}^{l_i} D(x) \, dx \right]$$

$$= 2wl_i \cdot P_a \cdot \bar{D}_i$$

where $\bar{D}_i$ is the average density along the ith transect. Summing over transects, we get

$$E(n \mid L) = \bar{D} \cdot 2wL \cdot P_a$$

where

$$\bar{D} = \frac{\sum\limits_{i=1}^{k} 2wl_i \cdot \bar{D}_i}{2wL}$$

is the average density along the sample of lines used.

It is the above $\bar{D}$ that line transect methods actually estimate (Burdick 1979). Either the design of the line placement must produce $\bar{D} \equiv D$ (random line placement has the purpose of achieving $E(\bar{D}) = D$; expectation is with respect to randomization of line placement), or we must model $D(x, y)$, fit the model from the sample data of spatial coordinates of detected objects, and compute the overall $\hat{D}(= \hat{N}/A)$ from

$$\frac{1}{A} \int \int_{A} \hat{D}(x, y) \, dx \, dy = \hat{D} \tag{6.8}$$

Also, from $\hat{D}(x, y)$ and $\hat{P}_a$ we can then get

$$\hat{E}(n_i \mid l_i) = \hat{P}_a \int \int_{2wl_i} \hat{D}(x, y) \, dx \, dy$$

$$= 2wl_i \cdot P_a \cdot \hat{\bar{D}}_i$$

If $\bar{D}_i = \bar{D}$ for all $i = 1, \ldots, k$, then per line encounter rates are constant and we have

$$\hat{E}\left(\frac{n_i}{l_i}\right) = \frac{n}{L}$$

196

and Equation 6.4 is valid.

The additional information from distance sampling, which is relevant to $D(x, y)$, is in the set of coordinates of detection locations, (x_{ij}, y_{ij}), $j = 1, \ldots, n_i$, $i = 1, \ldots, k$, and any covariates taken at these locations. To extract this spatial information, these points are treated, conceptually, as a sample of size n from the probability density function defined by

$$d(x, y) = \frac{D(x, y)}{2wL \cdot \bar{D}}$$

over the disjoint areas represented by the k lines traversed. Using some form of parametric model for $d(x, y)$, one fits the model to these (x, y) data by standard statistical methods, thereby getting $\hat{d}(x, y)$. This is not a trivial undertaking, but it is possible. The normalization of $d(x, y)$ to integrate to one over the sample area of transects is necessary for identifiability reasons in the fitting of $\hat{d}(x, y)$. Using $\bar{D}(\equiv \hat{D})$ obtained from standard line transect analyses, one obtains the desired rescaling:

$$\hat{D}(x, y) = 2wL \cdot \bar{\hat{D}} \cdot \hat{d}(x, y)$$

If $D(x, y)$, hence $d(x, y)$, is taken as constant over the entire study area, then $d(x, y)$ integrates over the study area to $A/(2wL)$, and from Equation 6.8, we get $\hat{D} = \bar{D}$. However, if $D(x, y)$ varies substantially over the study area and lines are poorly placed, this approach used in conjunction with Equation 6.8 could give a much less biased estimate of $D = N/A$.

We suggest simplifying the process of relating the (x, y) detection location data to $D(x, y)$ by projecting each location perpendicularly onto the line and using that point as the recorded detection location. At the scale (much larger than w) over which important variations occur in density, this redefined detection location is acceptable. Detection locations now become distances along the lines, and the problem is effectively reduced to one dimension and numerical scaling-integrations become one-dimensional line integrals. Now all the locational data (x_{ij}, y_{ij}), $j = 1, \ldots, n_i$, $i = 1, \ldots, k$, fall on a (disjoint) 'line' in the study area. Thus the pdf $t(x, y)$, which is really one-dimensional, of a detection location is

$$t(x, y) = \frac{D(x, y)}{\gamma}$$

where the normalizing γ is the line integral

$$\gamma = \sum_{i=1}^{k} \int_{l_i} D(x, y) \, dx \, dy$$

Note that γ is not identifiable from the (x, y) location data alone. Any parametric model for $D(x, y | \underline{\theta})$ now generates the parametric likelihood

$$\mathscr{L}(\underline{\theta}) = \prod_{i=1}^{k} \prod_{j=1}^{n_i} t(x_{ij}, y_{ij} | \underline{\theta})$$

Standard numerical MLE and model selection methods can be applied; at each iteration, γ must be recomputed by numerical line integration. Once the MLE $\hat{\underline{\theta}}$ is found, then γ is estimated as $L \cdot \hat{D}$ from the usual line transect estimation of D and we can get

$$\hat{D}(x, y) = t(x, y | \hat{\underline{\theta}}) \cdot L \cdot \hat{D}$$

(A similar, but by no means identical, development of basic theory is possible for point transects.) More sophistication can be added if the parameters $\underline{\theta}$ affecting density are modelled as functions of covariates recorded at the locations of detections (and available for a grid of points over A).

In many data sets, even when $E(s)$ is also estimated, we see that the contribution of $\widehat{\mathrm{var}}(n)$ to $\widehat{\mathrm{var}}(\hat{D})$ is large, often greater than 50%, and sometimes in excess of 70%. We suspect that Equation 6.3 often fails and a more detailed analysis of the data to estimate varying encounter rates would be useful. The scientifically critical part of this procedure is what to use as a model for $D(x, y)$, or equivalently, $t(x, y)$. The technically difficult part is computing the integrals that are needed, and fitting $t(x, y)$ (or $d(x, y)$, but that is unnecessarily harder). These integrals and the fitting will generally be done by numerical methods. Additional sampling variance is introduced by $\hat{D}(x, y)$, so there must be a worthwhile reduction in the bias of $\widehat{\mathrm{var}}(n)$ and/or bias of $\hat{D}$, and the detection of important spatial trends in density, to justify this additional analysis. General software and methods for these analyses need to be developed. We expect to see this subject area implemented for practical use in the next five to ten years. This spatial modelling in terms of x–y coordinates is a necessary first step to the incorporation of spatially varying covariates that affect object density.

6.3.4 *Modelling variation in cluster size*

The modelling strategies for encounter rate outlined above may also be applied to cluster size. Spatial and temporal variation in mean cluster size is common, and as with encounter rate, this structural variation should if possible be modelled. A simple method of achieving this is to stratify in space and time before estimating mean cluster size. When sample sizes within strata are small, a common dispersion parameter $c = \text{var}(s)/E(s)$ might be assumed. Suppose the stratification yields V data sets. If $\hat{E}(s_v) = \bar{s}_v$, $v = 1, \ldots, V$, then the variance of $\hat{E}(s_v)$ is estimated by the sample variance of observed group sizes, $\widehat{\text{var}}(s_v)$, divided by n_v. The dispersion parameter is estimated by

$$\hat{c} = \frac{\sum_v \left\{ (n_v - 1) \cdot \frac{\widehat{\text{var}}(s_v)}{\hat{E}(s_v)} \right\}}{\sum_v (n_v - 1)}$$

Then

$$\widetilde{\text{var}}(\bar{s}_v) = \frac{\widehat{\text{var}}(s_v)}{n_v} = \frac{\hat{c} \cdot \bar{s}_v}{n_v}$$

If size bias in the sample of detected clusters is suspected, then $E(s_v)$ and $\text{var}(s_v)$ should be estimated by one of the methods outlined in Section 3.6 before application of the above equation for $\hat{c}$. That method yields estimators $\hat{E}(s_v)$ and $\widehat{\text{var}}[\hat{E}(s_v)] = d_v \cdot \widehat{\text{var}}(s_v)$ for some value d_v, from which the variance of $\hat{E}(s_v)$ is estimated by

$$\widetilde{\text{var}}[\hat{E}(s_v)] = \hat{c} \cdot \hat{E}(s_v) \cdot d_v$$

Modelling the spatial variation in cluster size may be seen as an alternative to dividing an area into geographic strata. The latter is an attempt to create sub-areas in which spatial variation in cluster size is small, whereas the former allows mean cluster size to vary as a continuous function through the area. Similarly, temporal variation in cluster size may be modelled. Having fitted a surface for mean cluster size, using perhaps generalized linear or generalized additive modelling techniques, mean cluster size can be estimated for the study area as a whole, or for any part of it, and if temporal variation is also modelled, mean cluster size can also be estimated at different points in time.

6.3.5 *Discussion*

The spatial models described in general terms in Section 6.3.3 and alluded to above are also applicable to the parameters $f(0)$ (line transects), $h(0)$ (point transects) and g_0. Similarly, estimation of these parameters by individual strata, combined with the assumption of a common dispersion parameter, are options available to the analyst. However, these parameters are unlikely to vary spatially to the same degree that encounter rate and mean cluster size do. Additionally, estimation of $f(0)$, $h(0)$ or g_0 is bias-prone when samples are small, whereas estimation of encounter rate and mean cluster size are less problematic. For these two reasons, the case for spatial modelling of the detection process, or for estimating $f(0)$, $h(0)$ or g_0 separately within strata, is less compelling than for encounter rate and mean cluster size.

In principle, it is possible to model the density surface, allowing for spatial and temporal variation in individual parameters, together with the effects of environmental conditions on parameters, effects of cluster size or observer on probability of detection, and so on. In practice, considerable software development would be necessary, and if the principle of parsimony was ignored, implementation of such general models would be prevented by numerical difficulties. Section 6.8 on a full likelihood approach lays out the philosophy and structure around which more general modelling could be developed. A simpler, if less comprehensive, strategy is to fit a spatial and, where relevant, temporal model for each parameter in turn. By fitting these models independently, with inclusion of covariates such as environmental factors where required, a spatial surface can be estimated for each of the parameters encounter rate, mean cluster size, $f(0)$ or $h(0)$, and, where relevant, g_0. Density can then be estimated at any point in the study area (and at any time in the study period, if temporal variation is modelled) by combining the estimates from each surface at that point. Abundance can then be estimated for any selected part or the whole of the study area by evaluating the density estimate at a grid of points, and implementing numerical integration. Variances may be estimated using resampling methods, to avoid the assumption that the surfaces for the different parameters are independently estimated.

6.4 Estimation of the probability of detection on the line or point

For both line and point transects it is usual to assume that the probability of detection on the centreline or at the point is unity; that is,

$g_0 = 1$. In practice the assumption is often violated. For example, whales that travel in small groups or that dive synchronously may pass directly under a survey vessel without being detected, or birds in the canopy of high forest directly above an observer may be unrecorded unless they call or sing. As shown in Section 3.1, it is easy to include the component g_0 in the general formula for line and point transects; far more difficult is to obtain a valid estimate of g_0. Most of the methodological development for estimating g_0 has arisen out of the need to estimate the size of cetacean stocks from line transect surveys, so that the effects of commercial or aboriginal takes on the stocks may be assessed. A summary of the evolution of ideas, mostly within the Scientific Committee of the International Whaling Commission, follows.

(a) *Line transect sampling* Early attempts to estimate g_0 were based on the models of Doi (1971, 1974). These were exceptionally detailed models, incorporating the effects of whale dive times and blow times, whale aggregation, response to the vessel, vessel speed, observer height above sea-level, physiological discrimination of the observers, number of observers on duty, binocular specification and angular velocity of eye scanning. The models gave rise to estimates of g_0 with very high estimated precision, but different model assumptions led to estimates that differed appreciably from each other (Best and Butterworth 1980; Doi *et al.* 1982, 1983). In other words, by making many detailed assumptions, the estimator has high precision but at the expense of high bias, and the validity of the approach is questionable.

Butterworth (1982a), using the approach of Koopman (1956), developed a continuous hazard-rate model very similar to that described in Section 3.2, and used it to examine mathematical conditions under which $g_0 < 1$. Butterworth *et al.* (1982) used this formulation to derive a formula for g_0 that was a function of vessel speed:

$$g_0(v) = 1 - \exp(- \alpha/v)$$

where v is vessel speed and α depends on the form of the hazard function. They argued that, if the whales were stationary and vessel speed zero, there would be infinitely many chances to detect whales on the centreline, so that $g_0(0) = 1$. If the true hazard is such that $g_0(v) < 1$ for $v > 0$, and if the specific hazard is independent of vessel speed, then the ratio of estimated whale density assuming $g_0 = 1$ at two different speeds will estimate the ratio of g_0 values at those speeds:

$$\frac{\hat{D}_{v_1}}{\hat{D}_{v_2}} = \frac{\hat{g}_0(v_1)}{\hat{g}_0(v_2)} = \frac{1 - \exp(- \alpha/v_1)}{1 - \exp(- \alpha/v_2)}$$

Solving this equation for α and substituting in the above equation for $g_0(v)$ with $v = v_2$ say yields an estimate of $g_0(v_2)$. Surveys of minke whales in the Antarctic are generally carried out on board vessels travelling at 12 knots.

Butterworth *et al.* (1982) reported on variable speed experiments in which $v_1 = 6$ or 7 knots and $v_2 = 12$ knots. They obtained estimates of $g_0(12)$ close to 0.7, but precision was poor, and estimates did not differ significantly from 1.0. Despite using a hazard-rate formulation, they assumed that the detection function was negative exponential, and Cooke (1985) criticized this; if the form is negative exponential at one speed, then it can be shown mathematically under the above model that the form cannot be negative exponential at the other speed. Cooke (1985) proposed a method based on the ratio of sightings rate at the two speeds. However, he noted that expected precision of estimates from this approach is low even when all the assumptions of the method are satisfied, and listed other reservations about the approach. Hiby (1986) also showed that random whale movement at a speed of 3 knots generated large bias in the sighting rate of a vessel travelling at 6 knots. This bias could be corrected for if the true average speed of movement was known, but he questioned whether it could be reliably estimated. Butterworth (1986) applied Cooke's approach, using four different methods of confidence interval estimation, to two data sets, both separately and combined. He found that most of the 95% intervals for $g_0(12)$ spanned the entire range [0,1], and in every case, the upper limit exceeded 1.0. In one case the lower limit also exceeded 1.0. Given the unresolved difficulties, the method has not been used again.

Zahl (1989) continued development of methods to analyse variable speed data, and put the use of variable numbers of observers on the sighting platform into the same theoretical framework. In common with Schweder (1990), he argued that discrete hazard-rate models are more appropriate than continuous models for whale data. He developed such a model in conjunction with a generalization of Cooke's (1985) method. While precision was improved (Zahl, unpublished), it remained poor, and he did not address the problem of random whale movement.

Butterworth *et al.* (1982) also described a parallel ship experiment. Although designed for examining whether whale movement was affected by the presence of a vessel, they noted that the expected proportion of sightings seen from both platforms in such an experiment can be estimated from the fitted detection function. Their estimates were inconsistent with results from the variable speed experiment, which they attributed to the use of the negative exponential for modelling perpendicular distances. Butterworth *et al.* (1984) extended the method to provide estimates of g_0, assuming a generalized exponential form for the detection function. Buckland (1987c) generalized their results to provide g_0 estimates from parallel ship data using any model for the detection function, and allowing a different detection function for each vessel. He then analysed the data assuming the continuous hazard-rate model of Section 3.2. This resolved some of the inconsistencies observed in parallel ship data, but the observed duplicate sightings distribution departed significantly from the distribution predicted by the method. If some whales exhibit behaviour that makes them particularly visible relative to other whales, then the duplicate sightings proportion may be higher than anticipated at greater distances, and the observed data show such an effect. Two additional problems remained unresolved. The first is identification of duplicate detections (i.e. whether a sighting made by one platform corresponds with one made by the other), especially in areas of high whale density; the second is assessment of whether estimates of g_0 from parallel ship experiments are valid for correction of abundance estimates derived from normal survey data.

Schweder (1990) proposed new parallel ship experiments, in which one vessel is not only to one side of but also behind the other. He showed using a discrete hazard-rate model (Section 6.2.5) that sightings of cues from the two platforms cannot be considered independent. By placing one vessel behind the other, whales that are below the surface when the first vessel passes may be visible to the second vessel. Results from experiments advocated by Schweder, together with further methodological development, are given in Schweder *et al.* (1991), who estimated g_0 for North Atlantic minke whales. They mapped out surfacings as recorded by one observer in terms of relative position to the other, coding duplicate sightings as 1 and those missed by the reference observer as 0. The hazard probability of sighting was estimated from these data, and integrated multiplicatively, using stochastic simulation. The surfacing pattern of minke whales was estimated from monitoring two whales to which a VHF transmitter had been attached. This allowed them to estimate g_0 without the assumption that the observers detect animals independently. Instead, the assumption of independence is transferred to individual surfacings; conditional on an animal surfacing at a given location, the probability of detection of that surfacing is assumed to be independent between platforms. They obtained $\hat{g}_0 = 0.43$, with 95% confidence interval (0.32, 0.54). This interval took account of uncertainty in whether detections were duplicates.

Several authors have noted that, if a negative exponential model is assumed for the detection function and a correction factor e is defined to allow for deviations of the true detection function from this model, then although neither e nor g_0 can be estimated robustly or with good precision, their ratio, called the '*eh*' factor, where $h = 1/g_0$, can

(Butterworth *et al.* 1984; Cooke 1985; Kishino *et al.* 1988). However, the method is still sensitive to the assumption that sightings from the two platforms are independent. Following a comparison of the performance of the negative exponential, exponential power series, hazard-rate, Fourier series and Hermite polynomial models (Buckland 1987b), the Scientific Committee of the International Whaling Commission adopted the hazard-rate model in preference to the negative exponential model.

The variable speed and parallel ship methods both require special experiments. These take the vessels away from survey work, and g_0 during the experiments may not be typical of g_0 values during normal survey mode; for example experiments are carried out in areas of high whale density, so that sample size is adequate. The field procedure most widely used currently is the 'independent observer' method; an additional observation platform is used (for example a second crow's nest vertically below the main one), and observer(s) search independently of the observer(s) on the primary platform. Observers on one platform are not advised of sightings made from the other. Few resources beyond those needed for normal survey mode are required, and the method is therefore often incorporated into normal survey mode.

The methods and problems of analysis are similar to those for a parallel ship experiment with inter-ship distance set to zero, and there is again a strong case for using discrete hazard-rate models in any future methodological development. Identification of duplicate cues is simpler than for two ships, since the bridge can more easily coordinate information from the two platforms, and matches are more easily made when both detections are observed from almost the same position. Exact recording of times of cues aids the identification of duplicate sightings, especially when the same cue is seen from both platforms; if feasible, a whale detected from one platform should be located say from the bridge and followed so that if it is later detected from the other platform, it may be more easily identified as a duplicate.

Two further methods of examining independent observer data have arisen from the Southern Hemisphere minke whale subcommittee of the International Whaling Commission. The first uses one platform to confirm the positions of a sample of schools, and then plots the proportion of these sightings detected by the other platform by distance from the centreline, which should provide an empirical fit to the detection function (with $g(0) \leqslant 1$) for the second platform. Butterworth and Borchers (1988) describe this approach, and apply it assuming a negative exponential detection function (their 'DNE' method). The second method does not require that duplicate sightings are identified, and is discussed in Hiby and Hammond (1989). It uses information on pairs of sightings from between and within platforms that are definitely not duplicates to prorate sightings that may be duplicates, without the necessity to identify whether any specific pair of sightings is a duplicate. D.L. Borchers (personal communication) has noted a theoretical shortcoming of this approach, and recommends that it not be used.

One of the most troublesome aspects of estimating g_0 is that different sources of heterogeneity can give rise to substantial bias. Bias in $\hat{g}_0$ might be positive or negative, depending on the type of heterogeneity, and how it is modelled. Observer heterogeneity can arise through different sighting efficiencies for different observers, and through variable sighting efficiency of a single observer through time. Platform heterogeneity is similar in nature. The same observer may have different sighting efficiencies from different platforms, and the relative efficiency of different platforms may vary with environmental or other factors. Environmental heterogeneity affects the efficiency of both the observer and the platform, and environmental variables such as sea state and temperature are likely to affect behaviour of the whales. Individual animals will in any case exhibit heterogeneous behaviour, which leads to too many duplicate detections from double-counting methods, and hence to negatively biased estimates of abundance. Because of the confounding between the various sources of heterogeneity, it is usually not possible to model heterogeneity adequately even when carefully designed experiments are carried out to estimate g_0.

Traditional methods of handling heterogeneity include stratification and covariance analysis, and both are potentially useful for reducing the effects of heterogeneity on estimates of g_0. Generalized linear modelling provides a framework for implementing both approaches. For example, stratification by observer can be achieved by introducing one parameter per observer, and sea state (Beaufort scale) may be incorporated as a regression variable (covariate). The method can be taken further. Suppose parameters are defined for each observer and each platform. Then interaction terms between observer and platform may be introduced. This method is more reliable if each observer is on duty for an equal time on each platform, according to an appropriate design. If observer performance is thought to vary with time, the time that the observer has been on duty (or function(s) of that time) may be entered as a covariate. However, problems arise in practice because different covariates may be highly correlated with each other, and because there may be considerable confounding between stratification factors. Further, the quality of data from which g_0 might be estimated is often poor, so that a realistic model may have more parameters than the data can support, and unquantifiable bias may arise either through practical difficulties in data collection or through inappropriate model specification.

The methods developed for handling heterogeneity in closed population mark–recapture by Huggins (1989, 1991) and Alho (1990) are relevant whenever data from independent platforms are used to estimate g_0. Indeed, probability of detection as a function of distance, estimated by $\hat{g}(x)$, is itself a covariate in this framework. The method is illustrated in a slightly different context in Section 6.12; adding $\hat{g}(x)$ as a covariate allows it to be applied here.

We consider each source of heterogeneity, and use simple examples to illustrate the effects on estimation. For these examples, it is supposed that each detection function is flat and equal to g_0 out to some distance d, and that g_0 is estimated by the Petersen (1896) two-sample mark–recapture estimate. Although this approach is simplistic, it serves to illustrate concepts. Thus we have

$$\hat{g}_{i0} = n_i/\hat{N} \quad \text{for platform} \quad i, i = 1, 2 \tag{6.9}$$

$$\hat{g}_0 = 1 - \prod_{i=1}^{2} [1 - \hat{g}_{i0}] = \hat{g}_{10} + \hat{g}_{20} - \hat{g}_{10}\hat{g}_{20} \tag{6.10}$$

for both platforms combined where n_i = number of detections from platform i and

$$\hat{N} = \frac{n_1 n_2}{n_{12}}, \text{ with } n_{12} = \text{number of detections made from both platforms.}$$

Suppose that there is a single observer on each of two platforms, one of whom sees all the whales on the centreline ($g_{10} = 1.0$) while the other sees only one half ($g_{20} = 0.5$). Suppose further that in the first half of the experiment, 50 whales (or whale schools) pass within distance d of the vessel, and the observers see 50 and 25 of these whales respectively. For the second half of the survey, they switch platforms, and again 50 whales pass, of which the first observer sees 50 and the second observer 25. Then if g_0 is estimated by platform using Equation 6.9, $n_1 = n_2 = 75$ and $n_{12} = 50$, so that $\hat{g}_{10} = \hat{g}_{20} = 2/3$, and $\hat{g}_0 = 8/9$. In fact, $g_0 = 1$, so that abundance is overestimated by $100 \times (9 - 8)/8 = 12.5\%$. In this case, the problem may be solved by applying Equation 6.9 to observers instead of platforms, giving $n_1 = 100$ and $n_2 = n_{12} = 50$, so that $\hat{g}_{10} = 1.0$, $\hat{g}_{20} = 0.5$ and $\hat{g}_0 = 1.0$. However, the above argument may now be reversed; if for a given observer g_0 is less for one platform than the other, bias arises for exactly the same reason. A solution is to estimate g_0 separately for each observer on each platform. This allows for an observer effect, a platform effect and an interaction between them. Data may be too sparse to support such an approach; Equation 6.9 and generalizations of it are unstable for small values of n_{12}. Another solution is to assume there is no interaction effect, so that g_0 is assumed to be an additive function of an observer and a platform effect. This solution may prove satisfactory when the time spent by each observer on each platform is subject to a randomized experimental design. However, it assumes that g_0 for a single observer or platform remains constant throughout the survey.

Platform heterogeneity arising from heterogeneity in environmental conditions would be problematic if for example sighting conditions were better from one platform in some conditions and from the other in different conditions. Thus reflected sunlight may cause one vessel in a parallel ships survey to miss many whales that pass between the vessels and are detected by the other vessel. This will lead to negatively biased estimates of g_0 and positively biased estimates of abundance, since each ship will be affected in this way during different periods of the survey; from a theoretical point of view, it is identical to the problem of the above example, for which platform efficiency changes when the observers swap platforms. However, it is more difficult to design a survey to achieve balance for environmental effects, and appropriate parameterization is problematic.

Observer efficiency may vary over time due to factors such as mood, health, comfort, time on duty, etc. Both observer and platform efficiency will vary with environmental conditions. Bias from these sources will tend to be less; at any given time, implications are similar to the cases considered above, but over time, bias changes. If no observer or platform is consistently more efficient than another, average bias from this source may tend to be small. Additionally, environmental conditions might be introduced into the model as covariates, so that g_0 is related to environmental conditions by a regression model, which might be a generalized linear or non-linear model. Again, data inadequacies may severely constrain the model options.

The fourth class of heterogeneity is heterogeneous behaviour of animals. If some whales are more easily detected than others, then double-counting methods suffer the same bias as two-sample mark–recapture experiments on populations with heterogeneous trappability. Obvious whales are likely to be seen by both observers, whereas unobtrusive whales may be missed by both. The number of whales is therefore negatively biased and $\hat{g}_0$ is positively biased. Suppose that of 160 whales passing within distance d of the vessel, 80 are certain to be seen from each of two platforms and $g_0 = 0.25$ for both platforms for the remaining 80 whales. Assuming independence between platforms, the expected numbers of whales detected are $n_1 = n_2 = 100$, $n_{12} = 85$. Equations 6.9 and 6.10 yield $g_{10} = g_{20} = 0.85$ and $g_0 = 0.9775$. In fact, $g_{10} = g_{20} = 0.625$ and $g_0 = 0.71875$, so that abundance is underestimated by 26.5%. This problem might be partially solved by stratifying detections by animal behaviour, or by type of cue.

Both heterogeneity in whale behaviour and cues that occur only at discrete points in time generate positive bias in the g_0 estimate if the effects are not allowed for, whereas heterogeneity in observer ability and in ease of detection from the respective platforms may lead to negative bias. Schweder's (1990) methods, referenced above, remove the bias that arises from heterogeneity in whale behaviour due to differential diving. If sufficient data are available, the effects of observer heterogeneity might be removed by adding a separate parameter for each observer to the model from which g_0 is estimated, for example using generalized linear models (Gunnlaugsson and Sigurjónsson 1990), although it must still be assumed that a single observer is consistent in ability relative to other observers both within and between watch periods. If there is more than one observer on each platform, another option is to select teams of observers so that each team is of comparable ability. Data analysis should then include a test of whether it is reasonable to assume that each team was equally efficient at detecting whales. It may also be necessary to introduce platform-specific parameters, or at least to test whether such parameterization is required.

Methodological development to solve these difficult problems is continuing, and an innovative paper by Hiby and Lovell (unpublished) proposes a sophisticated approach which allows for response by the whale to the vessel and does not assume stochastic independence between the platforms. The approach uses data on duplicate cues rather than duplicate animals, and can be used in conjunction with either line transect sampling or cue counting methodology. However, it does require surfacing rate to be estimated, and whether the effects on the sightings data of both $g_0 < 1$ and response to the vessel are simultaneously quantifiable has yet to be assessed.

We provide below theory from Buckland (1987c), which allows estimation of g_0 given independent detections from two platforms a distance d apart; $d > 0$ corresponds to parallel ship surveys, where the ships are separated by a distance d, and $d = 0$ corresponds to independent observer platforms on the same vessel or aircraft. The theory for the latter case is also given in Hiby and Hammond (1989). A continuous sighting cue is assumed. The method is subject to potentially serious biases, especially if time periods between cues are not short (say over five minutes for shipboard surveys). Also given is a method described by Buckland and Turnock (1992), which is an extension of ideas utilized by the DNE method (Butterworth and Borchers 1988) mentioned above. The latter approach is more robust to the effects of animal heterogeneity and of animal movement.

Suppose the two platforms are labelled A and B, with B a distance d to the right of A (Fig. 6.1). If both platforms are on the same vessel, then $d = 0$. Suppose a cluster of whales is detected at perpendicular

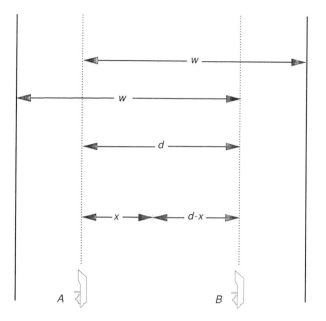

Fig. 6.1. Illustration of parallel platform method. The platforms are a distance d apart, so that a detection at distance x from the centreline of platform A is a distance $d - x$ from the centreline of platform B. If the detected object is to the left of A's centreline, x is negative. A truncation distance of w from the centreline of the farther platform is used.

distance x from A, where x is negative if the detection is to the left of A, and positive if it is to the right. Let the probability that A detects the cluster be $g_A(|x|)$. In terms of notation used elsewhere in this book, $g_A(|x|)$ corresponds to $g(x) \cdot g_0$; thus $g_A(0)$ is not assumed to be unity, but is the value of g_0 for platform A. Further, let the probability that B detects the cluster be $g_B(|d - x|)$, assumed to be independent of $g_A(|x|)$. Then the probability that the cluster is detected by both platforms is $g_A(|x|) \cdot g_B(|d - x|)$.

Now suppose that detected clusters within a distance w of each platform are analysed, where $d/2 < w < \infty$. Define

$$\mu_A = \int_{d-w}^{w} g_A(|x|) \cdot dx$$

$$\mu_B = \int_{d-w}^{w} g_B(|d - x|) \cdot dx$$

209

and

$$\mu_{AB} = \int_{d-w}^{w} g_A(|x|) \cdot g_B(|d-x|) \cdot dx$$

Further, let

$$v_A = \frac{\mu_A}{g_A(0)}, \quad v_B = \frac{\mu_B}{g_B(0)} \quad \text{and} \quad v_{AB} = \frac{\mu_{AB}}{g_A(0) \cdot g_B(0)}$$

Standard line transect analysis of the data from each platform, or of the pooled data from both platforms if the detection function can be assumed to be the same for both, yields estimates of $g(|x|)/g(0)$ for each platform. Hence v_A, v_B and v_{AB} are estimated using numerical or analytic integration. If n_A clusters are detected within the strip of width $2w - d$ (Fig. 6.1) from platform A, n_B from B and n_{AB} from both, then n_{AB}/n_B estimates μ_{AB}/μ_B.

Hence
$$\hat{g}_A(0) = \frac{\hat{v}_B \cdot n_{AB}}{\hat{v}_{AB} \cdot n_B}$$

Similarly,
$$\hat{g}_B(0) = \frac{\hat{v}_A \cdot n_{AB}}{\hat{v}_{AB} \cdot n_A}$$

Variances for these estimates may be found for example using the bootstrap.

If w is chosen such that $g(|x|)$ is more or less constant for $0 \leqslant x \leqslant w$, then $v_A = v_B = v_{AB} = 2w - d$, so that

$$\hat{g}_A(0) = \frac{n_{AB}}{n_B} \quad \text{and} \quad \hat{g}_B(0) = \frac{n_{AB}}{n_A}$$

which may be obtained directly from the two-sample mark–recapture estimate of Petersen (1896).

The following method, due to Buckland and Turnock (1992), is more robust. Suppose an observer or team operates normal line transect sampling techniques from a primary platform (platform A), and an independent observer or team searches a wider area from a secondary platform (platform B), ahead of the normal area of detectability for the primary platform, making no attempt to detect most animals close to the line. The method does not require the assumption that probability of detection on the centreline is unity. If detections from the secondary platform are made before animals move in response to the presence of the platforms, the method is unbiased when responsive movement occurs before detection from the primary platform. Further, if sighting distances and angles of secondary detections are measured without error, the method is unbiased when sighting distances or angles recorded by the primary platform are subject to bias or error. For example, if the primary platform is a ship and the secondary platform is a helicopter hovering above a detection, ship's radar may be used.

Data from the primary platform are used to estimate the encounter rate (number of detections per unit distance), while data from the secondary platform allow the effective width of search from the primary platform to be estimated. The secondary platform may be thought of as confirming the position of a sample of animals, and the proportions of these detected by the primary platform allow estimation of the detection function, without the necessity to assume $g_0 = 1.0$.

The probability of detection of an animal from one platform should be independent of whether the animal is detected by the other. Detections made by observers on the secondary platform should be at least as far ahead of the primary platform as the maximum distance at which animals are likely to move in response to the presence of the observation platforms. Any secondary detections that occur at shorter distances should be truncated before analysis. Secondary observers should also search out as far as the greatest distance perpendicular to the transect line from which animals would be able to move into the normal detectable range of the primary platform. For animals that are only visible at regular points in time, such as whales with a regular dive cycle, the normal detectability range of the secondary observers should exceed the distance travelled by the primary observers during the course of a single, complete cycle. Conceptually, duplicate detections are expected to be sighted from the secondary platform first, but if the above conditions are met, the analysis may include duplicates first sighted from the primary platform.

Of those animals that pass within the normal detectability range of the secondary observers, the proportion actually detected need not be high, although if few duplicate detections occur, precision will be poor. If the secondary platform cannot be in operation throughout the survey, it should operate during representative, preferably random, subsets of the survey effort.

Secondary observers need not detect all animals on the centreline of either platform. However, the perpendicular distance of each secondary detection from the centreline of the primary platform must be recorded. It is also necessary to determine whether any animal detected by the secondary platform is also detected by the primary platform. Thus, any animal detected by one platform should be monitored by that platform, or by a third platform, until either the other platform detects the animal or it passes beyond the area searched by the other platform. If animals occur in groups, group size should be recorded by both platforms, either to help identify duplicate detections or, if duplicates are reliably identified and animal groups are well-defined, to validate group size estimates.

Suppose for the moment that both platforms operate throughout the survey. The sightings data from the primary platform are analysed independently of the data from the secondary platform, to yield a conventional line transect density estimate $\hat{D}_A$, calculated assuming no movement and $g_0 = 1$. If animals occur in groups, we define these to be group densities (number of animal groups per unit area), rather than animal densities. The estimates may be biased either because of movement in response to the observation platforms or because probability of detection on the centreline is less than unity. However, for the subset of duplicate detections, the position of the animals prior to any response to the platforms is known. A detection function may therefore be fitted to the distances, as recorded by the secondary platform, of duplicate detections from the centreline of the primary platform. In addition, a detection function, relative to the centreline of the primary platform, is fitted to all secondary detections. An asymptotically unbiased density estimate, $\hat{D}_u$, is calculated as follows.

Let $g_A(x') =$ probability that an animal detected from the secondary platform at perpendicular distance x' from the centreline of the primary platform is subsequently detected from the primary platform, with $g_A(0) \leqslant 1.0$

$w =$ truncation distance for perpendicular distances x'

$f_A(x') =$ probability density of perpendicular distances, prior to responsive movement, of animals subsequently detected by the primary platform

$$= g_A(x')/\mu, \quad \text{with} \quad \mu = \int_0^w g_A(x')\, dx'$$

$n_B =$ number of secondary detections

$n_{AB} =$ number of detections made from both platforms (duplicate detections)

$n_A =$ number of primary detections

$f_B(x') =$ probability density of perpendicular distances from the primary platform centreline for secondary detections

212

$f_{AB}(x') =$ probability density of perpendicular distances from the primary platform centreline for duplicate detections, as recorded by the secondary platform

$f(x) =$ probability density of perpendicular distances x recorded from the primary platform

$L =$ length of transect line

Then the conventional (biased) estimate of density is

$$\hat{D}_A = \frac{n_A \cdot \hat{f}(0)}{2L}$$

and the asymptotically unbiased estimate is given by

$$\hat{D}_u = \frac{n_A \cdot \hat{f}_A(0)}{2L \cdot \hat{g}_A(0)}$$

where

$$\hat{f}_A(0) = \frac{\hat{g}_A(0)}{\int_0^w \hat{g}_A(x') \, dx'}$$

and

$$\hat{g}_A(x') = \frac{n_{AB} \cdot \hat{f}_{AB}(x')}{n_B \cdot \hat{f}_B(x')}$$

The probability densities $f_{AB}(x')$ and $f_B(x')$ are estimated by standard line transect methods. The critical assumptions of the method are as follows.

1. No animals beyond the range of detectability of the secondary platform are able to move into the range of detectability from the primary platform.
2. It is always possible to determine whether an animal detected by the secondary platform is also detected by the primary platform.
3. Given that an animal passes the secondary platform at perpendicular distance x', its probability of detection from the primary platform is independent of whether it was detected by the secondary platform.

213

4. Perpendicular distances of animals detected by the secondary platform from the centreline of the primary platform are measured exactly, or at least without bias.

If the secondary platform is not in continuous operation, the above procedure is carried out on data collected while both platforms were in operation and a correction factor is calculated as

$$c = \frac{\hat{D}_u}{\hat{D}_A}$$

Density for the entire survey area is estimated by $c \cdot \hat{D}$, where $\hat{D}$ is estimated from the sightings data from the primary platform for the full survey, calculated assuming $g_0 = 1$. To estimate the variance of $\hat{D}_u$ or of c analytically, the correlation between the estimated densities $\hat{f}_B(x')$ and $\hat{f}_{AB}(x')$ and between n_{AB} and both n_B and n_A must be accounted for, and a robust method should be used to estimate the variance in sample sizes. Variance can be estimated more simply and more robustly by applying a resampling method. For example, bootstrap variances may be obtained by resampling from the sightings and effort data from both platforms by day or by search leg (Section 3.7.4).

In the presence of random or responsive movement, $\hat{g}_A(0)$ is not a valid estimate of g_0, since animals close to the centreline when detected by the secondary platform may have moved away from it before detection by the primary platform, and similarly, some away from the centreline may approach it. Thus $\hat{g}_A(0)$ is biased low for g_0 (Buckland and Turnock, 1992). In this case, the method provides a single correction for both sources of bias; stronger assumptions are required to separate the two components of the correction.

The above methods were applied to Dall's porpoise data. The fitted densities $\hat{f}_B(y)$ and $\hat{f}_{AB}(y)$, estimated assuming a hazard-rate model, are shown in Fig. 6.2. The estimate of $g_A(0)$ was 0.597, and the multiplicative correction factor was 0.130 ($\widehat{se} = 0.050$; 95% percentile confidence interval [0.075, 0.262]). Thus the corrected density estimate is less than one seventh the uncorrected estimate. For these data, the combination of strong attraction of porpoise towards the ship and the close approach most porpoise were able to make before detection by the shipboard observers led to an estimate of porpoise density that was an order of magnitude too high.

The approach outlined above is relatively insensitive to observer, platform and environment heterogeneity, provided the secondary platform is in operation continuously, or at least during a representative sample of time periods in the survey. Animal heterogeneity is more

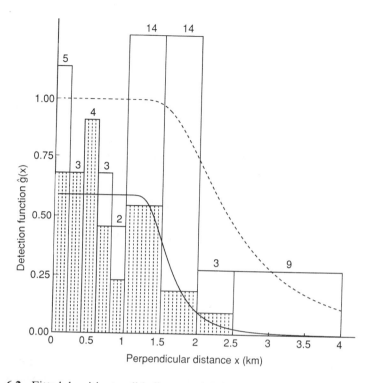

Fig. 6.2. Fitted densities to all helicopter sightings (upper curve) and to duplicate sightings (lower curve), Dall's porpoise, 1984. The hatching indicates the number of duplicate detections in each perpendicular distance interval (as recorded by the helicopter), and the open bars correspond to detections made by the helicopter alone.

problematic, but if the method of detection from the secondary platform is such that the probability of detection from the primary platform is independent of, or only weakly dependent on, whether an animal was detected by the secondary platform, then estimation should be reliable. It may prove necessary to stratify by behaviour of the animal or, if animals occur in groups, by group size to satisfy this requirement.

The field methods of our example, for which a helicopter searches ahead of a survey ship, illustrate one application of the method. If it is not practical to use a helicopter, but normal survey mode is to search with the naked eye, then a simpler solution might prove effective. Suppose an independent observer platform is available on the same vessel or vehicle as the primary platform. Instruct the independent observer to scan through binoculars, searching at distances beyond those typically scanned by the other observers. It does not matter that the restricted field of view may cause the observer to miss many animals close to the centreline. He or she should follow any detection until either the other observers detect it or it has passed abeam. If his or her average detection distance is substantially greater than that of the other observers, the method might prove satisfactory for the case of diving species, provided the dive cycle is sufficiently short that each animal group surfaces within detectability range at least once. He or she should strive to detect animals beyond the maximum distance they are likely to respond significantly to the vessel, even at the expense of reducing the overall number of detections made. He or she should concentrate effort ahead of the vessel, because the above method corrects for non-uniform effort, and effort searching abeam is wasted, since the other observers are unlikely to detect any animals at large perpendicular distances. If normal searching mode is through hand-held binoculars, the secondary observer could use powerful, tripod-mounted binoculars. By ensuring that the secondary observer searches beyond the normal detectability range from the primary platform, bias from heterogeneity between animals is reduced, especially if the mode of searching from the secondary platform is very different from that from the primary, as would be the case if the secondary platform is a helicopter and the primary platform is a ship or is ground-based.

In practice, it may prove difficult to operate a secondary platform, especially in poor sighting conditions. Even if sufficient detections can be made at distances beyond the range over which animals respond to the observer, it may not be possible to track detected animals, to determine whether they are also detected by the primary platform. In designing line transect surveys, priority should be given to ensuring that g_0 is as close to unity as possible and that detections are made prior to responsive movement. Only if g_0 might be appreciably less than unity or if substantial responsive movement prior to detection is suspected should the methods outlined above be considered.

The methods of Huggins (1989, 1991) and Alho (1990), developed for mark–recapture models, provide a flexible framework for estimation of g_0, which may prove superior to the above methods. As noted earlier, their application is illustrated in a similar context in Section 6.12. By first fitting a detection function to the pooled perpendicular distance data, the estimated probability of detection for an object at distance x, $\hat{g}(x)$, can be included as a covariate. The method used in Section 6.12 then yields an estimate of the probability of detection unconditional on x, and hence of g_0. We encourage further development of this approach.

(b) *Point transect sampling* Estimation of g_0 has seldom been considered for point transect sampling, although in their hazard-rate for-

216

mulation, Ramsey *et al.* (1979) note that g_0 need not be unity. In particular, if only aural cues are recorded, then birds that do not call or sing will not be detected, irrespective of their position. The cue counting method for estimating whale numbers (described later) is very similar theoretically to point transects, and g_0 estimation is more important in this context. Problems are similar to the line transect case, and are well described by Hiby and Ward (unpublished), who propose a model that allows for the discrete nature of cues and yields estimates of g_0.

6.5 On the concept of detection search effort

The detection of objects in distance sampling requires some type of active search effort. This will often be visual, so that observers must have some visual search pattern. Koopman (1980) discusses ideas on the search and detection process. We suggest that it is useful to consider some concept of search effort, and we pursue this suggestion here for line transects. (Detections are often by aural cues in point transects, in which case it is not clear to us how to model search effort other than as time spent at the point.)

Conceptually, searching effort has its own distribution about the centreline for line transects. Can we separate this concept of search effort from some concept of 'innate' detectability? To a limited (but useful) extent, we think the answer is 'yes'. Let $e(x)$ be relative searching effort at distance x, and let E be total absolute effort over all perpendicular distances. Then the perpendicular distance distribution of total effort is $E(x) = E \cdot e(x)$. Total absolute effort, E, is conceptual because we do not precisely know what constitutes total effort, given that there are subjective aspects to the detection process; we do not know how to measure E on a meaningful scale. However, relative effort at distance $x \cdot dx$ could be the relative time spent searching at perpendicular distance $x \cdot dx$. This measure of $e(x)$ is sensible and could be measured, in principle. Usually, we require that total effort E is sufficient to ensure $g(0) = 1$. Therefore, we will use the norm $e(0) = 1$ to scale $e(x)$. We maintain, and assume, that $e(x)$ should be non-increasing in perpendicular distance x.

We consider here some useful heuristic thinking, while recognizing that this is not the best mathematical approach. Use of a hazard-rate approach is coherent, but is not required for the points we wish to make. For the detection function, write

$$g(x) = d(x) \cdot e(x) \qquad (6.11)$$

where $d(x)$ is some innate, or standard, detection function, as for example for some optimal effort, $e_0(x)$. By assumption, both $d(0)$ and $e(0)$ are unity and both functions are non-increasing in x. Based on Equation 6.11, $g'(x) = d'(x) \cdot e(x) + d(x) \cdot e'(x)$, so that $g'(0) = d'(0) + e'(0)$. It follows that if both effort and innate detectability have a shoulder, then so does $g(x)$. However, if effort is poorly allocated perpendicular to the line, then we can get $g'(0) < 0$, i.e. no shoulder in the distance data, even when $d'(0) = 0$, which means that a shoulder is innately possible with a proper search effort design.

For line transect surveys in which there is visual searching for objects, especially aerial and ship surveys, histograms of the detection distances all too commonly have the shape of Fig. 6.3. It seems unlikely that the innate detectability would drop off this sharply; it is more likely that the data reflect an inadequate distribution of search effort or another field problem, such as heaping at zero distance or attraction of animals to the vessel before detection. We focus on effort here.

In order to pursue this idea mathematically, we need to be able to conceptualize innate detectability, $d(x)$. Although we may want to think of $d(x)$ as detectability under some optimal searching pattern $e_0(x)$, it is not possible to define an actual detection function, $g(x)$, free of some implicit or explicit underlying detection effort distribution. For $x \leqslant w$,

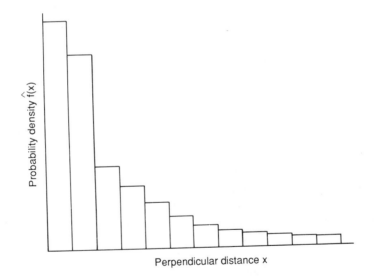

Fig. 6.3. Line transect data exhibiting a shape that is encountered too often; the idealized histogram estimator of the density function $f(x)$ suggests a narrow shoulder followed by an abrupt drop in detection probability, and a long, heavy tail.

we might allow $e(x)$ to be distributed as uniform $(0, w)$, but this becomes unreasonable for large w, and impossible as $w \to \infty$. Still, for small to moderate w, we could define $e_0(x)$ to be uniform. Then for a survey with this effort function, $g(x)$ would reflect the innate detectability of the object at perpendicular distances $x \leqslant w$.

Our intuition that detectability should not fall off sharply with increasing distance should be applied to $d(x)$. In most line transect sampling with which we have experience, the assumptions that $d(0) = 1$ and that $d(x)$ has a shoulder, i.e. $d'(0) = 0$, seem reasonable to us. (Many marine mammal surveys are exceptions to the first assumption (Section 6.4), and potentially to the second.) When the observed data appear not to exhibit a shoulder, we should bear in mind that the data really came from the detection function

$$g(x) = d(x) \cdot e(x)$$

and hence the probability density function of the observed perpendicular distance data is

$$f(x) = \frac{d(x) \cdot e(x)}{\displaystyle\int_0^w d(x) \cdot e(x)\, dx}$$

If $g(x)$ is as shown in Fig. 6.4, the data may primarily reflect effort $e(x)$, not innate detectability $d(x)$. For any data set, we would like to know the general nature of the effort distribution $e(x)$ to assess our faith in the assumptions that $g(x)$ has a shoulder and satisfies $g(0) = 1$.

Desirable patterns for search effort should be addressed at the design stage, and observers should be trained to follow them; Fig. 6.4 suggests that the search pattern was poor. We suspect that in aerial and ship surveys, there are often two distinct search modes occurring simultaneously: (1) intense scanning of the area near the centreline for much of the time, and (2) occasional scans at greater distances and over large areas, with more lateral effort. This may occur because one observer 'guards' the centreline, searching with the naked eye, while another scans a wider area with binoculars, or a single observer may search mostly with the naked eye, with occasional scans using binoculars. Data then come from the composite probability density function, as indicated in Fig. 6.5. In this case, most choices of histogram interval will obscure the shoulder.

For the innate detectability, $d(x)$, a shoulder should exist. Assume that an object on the centreline is moved just off the line. In an aerial transect

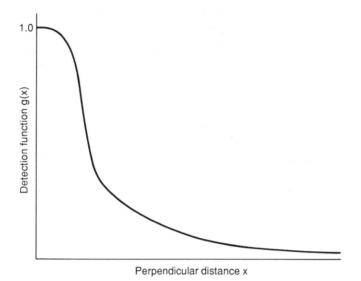

Fig. 6.4. An undesirable relative effort function $e(x)$ can give rise to the detection function shown here, and hence to data that exhibit the features of Fig. 6.3. Relative effort should be expended to ensure that the detection function has a wider shoulder relative to the tail than is shown here.

survey (Fig. 6.6), the maximum angle of declination to the object would change from 90° to perhaps 89° or 88°. Assuming the observer's view is not obstructed, the perceived properties of the object and detection cues will barely change. There is continuity operating, so that $g(x)$ will be almost the same at $x = 0$ as at a small increment from zero. Given continuity, we maintain that it is not reasonable for $d(x)$ to be spiked (i.e. $d'(x) < 0$).

We turn our attention now to considering what an optimal $e_0(x)$ might be. We consider an aerial survey (Fig. 6.6), although the theoretical approach applies more generally. In Fig. 6.6, the angle of declination ψ is also the angle of incidence of vision, with $\pi/2 \geqslant \psi \geqslant 0$. If objects are assumed to be essentially flat and detection probability is proportional to the perceived area of an object, then the same object when moved further away shows less area and so is less detectable. The best you could achieve in this case is an innate detectability $d(x)$ proportional to

$$\cos(\theta) = \cos\left[\tan^{-1}\left(\frac{x}{h}\right)\right]$$

220

This only allows for the loss in perceived object area due to the oblique view of the object as ψ decreases. Using heuristic arguments, if we add the effect of perpendicular distance off the centreline and generalize the result, we get the **form**

$$d(x) = [1 - e^{-\lambda O}] \cdot \frac{\cos\left[\tan^{-1}\left(\frac{x}{\sigma}\right)\right]}{1 + \left(\frac{x}{\sigma}\right)^2}, \quad 0 < x \text{ and for some scale factor } \sigma,$$

as a plausible innate detection function. Here, O is the true area of the object. We would want λO such that $d(0) \doteq 1$, in which case the tail behaviour of $d(x)$ (i.e. as x gets large) is

$$d(x) \doteq \frac{1}{1 + \left(\frac{x}{\sigma}\right)}$$

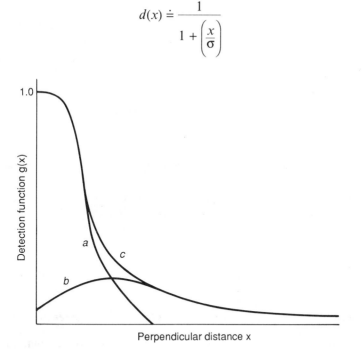

Fig. 6.5. The detection function c, which is the same as in Fig. 6.4, can arise from a mixture of curves a and b corresponding perhaps to two observers, one (a) 'guarding' the transect line and the other (b) scanning laterally; such minimal overlap of effort is undesirable.

221

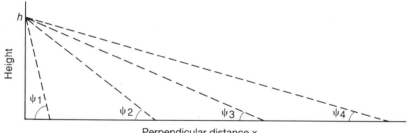

Fig. 6.6. The distance h represents (eye) height of an observer, and detections at various perpendicular distances x, indicated by the dashed lines, occur for angles of declination, ψ. For some types of visual cue, the cue strength depends critically upon ψ.

This is a very slow drop-off in detection probability. In fact, it cannot be used as a basis for general theory because it corresponds to $\int d(x)\,dx \to \infty$. This $d(x)$ does, however, give some sort of plausible upper bound on innate detectability, hence on possible effort. That is, we have reason to expect for any effort $e(x)$, properly scaled on x,

$$e(x) \leq \frac{\cos[\tan^{-1}(x)]}{1 + x^2}$$

Also, note that in this simple situation, innate detectability would have a definite shoulder.

Motivated by an aerial line transect mode of thinking, we could express our effort in terms of a distribution on the angle ψ (Fig. 6.6). To make derivations easier, we focus on u, $0 \leq u \leq 1$, where

$$u = \frac{2}{\pi} \cdot \psi = \frac{2}{\pi} \cdot \tan^{-1}\left(\frac{x}{h}\right)$$

and define $q(u)$ as the pdf of u. For greater generality, we use $u = \frac{2}{\pi} \cdot \tan^{-1}\left(\frac{x}{\sigma}\right)$; now the distribution of effort is proportional to

$$\frac{q[u(x)]}{\dfrac{dx}{du}}$$

222

where dx/du is evaluated at $u = \dfrac{2}{\pi} \cdot \tan^{-1}\left(\dfrac{x}{\sigma}\right)$, giving

$$e(x) \propto \frac{\dfrac{2}{\pi} \cdot q\left[\dfrac{2}{\pi} \cdot \tan^{-1}\left(\dfrac{x}{\sigma}\right)\right]}{\sigma\left[1 + \left(\dfrac{x}{\sigma}\right)^2\right]}, \qquad 0 \leqslant x$$

The proportionality constant is determined by the convention that $e(0) = 1$.

Let the effort be uniformly distributed over ψ, so that $q(u) \equiv 1$ for all u and

$$e(x) = \frac{1}{\left[1 + \left(\dfrac{x}{\sigma}\right)^2\right]}, \qquad 0 \leqslant x$$

If effort is uniform on $\cos(\psi)$, then we spend more time looking away from the centreline, and the result is

$$e(x) = \frac{\cos\left[\tan^{-1}\left(\dfrac{x}{\sigma}\right)\right]}{1 + \left(\dfrac{x}{\sigma}\right)^2}$$

It is interesting that if either ψ or $\cos(\psi)$ is uniform, the tail behaviour of the induced effort distribution is

$$e(x) \rightarrow \frac{1}{\left[1 + \left(\dfrac{x}{\sigma}\right)^2\right]} \rightarrow \left(\dfrac{\sigma}{x}\right)^2$$

Note that for the hazard-rate model of $g(x)$ for large x

$$g(x) \rightarrow \left(\dfrac{\sigma}{x}\right)^b$$

which of course includes the case of $b = 2$. Because effort decreases at large perpendicular distances, we would expect the applicable b to be $\geqslant 2$.

It is also useful to consider total effort, E, and its likely influence on $g(x)$ near $x = 0$. Now Equation 6.11 must be replaced by the more coherent form (justified by a hazard-rate argument)

$$g(x) = 1 - e^{-E \cdot e(x)} \qquad (6.12)$$

with $e(0) = 1$, but where $e(x)$ is not mathematically identical to the $e(x)$ function considered above. Consider what happens at $x = 0$ as a function of total effort, E: $g(0) = 1 - \exp(-E)$. Some values of $g(0)$ as E increases are as follows:

E	$g(0)$
1	0.6321
2	0.8647
4	0.9817
8	0.9997
10	1.0000

It is obvious upon reflection, as the above illustrates, that if effort is inadequate, detection probability on the line can be less than unity even if innate detection probability at $x = 0$ is one. More interesting is what might happen to a shoulder under inadequate detection effort. Analytically from Equation 6.12, we have

$$g'(x) = \{1 - g(x)\} \cdot E \cdot e'(x)$$

and with E finite, we can write

$$g'(0) = (1 - g(0)) \cdot E \cdot e'(0)$$

If total effort is large enough to achieve $g(0) = 1$, we are virtually sure that $g'(0) = 0$, regardless of the shape of the relative effort, $e(x)$ (provided $e(x)$ is not pathologically spiked at $x = 0$, with $e'(0) = -\infty$). Also if $e(x)$ has a shoulder, then $g'(0) = 0$. This could occur with insufficient total effort, E, to ensure $g(0) = 1$, hence the presence of a shoulder in the data is no guarantee that $g(0) = 1$.

The case in which relative effort has no shoulder is interesting. As noted above, it is possible that $e'(0) < 0$ and yet $g'(0) = 0$. As an example, consider the spiked relative effort $e(x) = e^{-x}$ for $E = 15$, so that

$$g(x) = 1 - e^{-15e^{-x}}$$

and

$$g'(x) = (-e^{-15e^{-x}})(15e^{-x})$$

A few values are given below:

x	$g(x)$	$g'(x)$
0.0	1.0000	-0.00001
0.1	1.0000	-0.00002
0.5	1.0000	-0.00102
1.0	0.9960	-0.02215
1.5	0.9648	-0.11778

Even though effort is spiked at $x = 0$, $g(x)$ has a distinct shoulder. However, if total effort is decreased, this shoulder will vanish and $g(0) = 1$ will fail.

The result illustrated above is due to a threshold effect. Once effort is large enough to achieve $g(0) = 1$, more effort cannot push $g(0)$ higher, but it can increase $g(x)$ for values of $x > 0$. We conclude that if there is sufficient total effort expended, then a shoulder is expected to be present even with a spiked relative effort function. The converse is disturbing: if total effort is too little, we can expect $g(0) < 1$, and there may be no shoulder. We emphasize the implications of guarding the centreline; if this is done, then as total effort decreases, more of the relative effort is likely to go near the centreline. This forces $e(x)$ to decrease more quickly, ultimately becoming spiked. The end result might be that we would have $g(0) < 1$, and $g(x)$ might be spiked (no shoulder). The data analysis implications are that if $\hat{g}(x)$ is, or is believed to be, spiked, then there is a basis to suspect that $g(0)$ is less than one. Conversely, if there is a shoulder, then there is a greater chance that $g(0) = 1$.

6.6 Fixed versus random sample size

6.6.1 Introduction

Theory and application of distance sampling has been almost exclusively in terms of fixed line lengths (and a fixed number of replicate lines) or fixed time spent at each point, for a fixed number of points. This approach means that line lengths $l_1, \ldots, l_k$ (and k itself), and of course L, are *a priori* fixed measures of sampling effort; that is, they are known before traversing the transects. It is then sample size n (overall and per line) that is random. Similarly, for point transects, k is fixed, time spent at each point is fixed and number of detections is a random variable. In principle it is possible to do the reverse: fix the total sample size to be achieved and traverse the line(s) until that predetermined n is reached, or count at a point until a predetermined sample size is reached. This sampling scheme results in L, or time at the point, being a random variable.

The purpose of this section is to provide some results comparing the cases of random and fixed n, under simplistic but tractable assumptions, and to comment upon this alternative design. We conclude that the two schemes (fixed L and random n, or fixed n and random L), under some idealized conditions, are not importantly different in their statistical properties. Primarily, field (i.e. applied) considerations dictate the choice between sampling schemes.

A common example contrasting fixed and random effort sampling is provided by the (positive) binomial and negative binomial distributions. For the binomial distribution, sample size is fixed at n and we record the number of successes, $\tilde{y}$, in n independent binary trials. For the negative binomial, we fix the number of successful trials (y) and sample the binary events until y successes occur, so that the number of trials, $\tilde{n}$, is random. (The added notation, '$\sim$', is needed here to indicate which variable is random.) The corresponding probabilities, expectations and variances are given below for the positive and negative binomial cases respectively:

$$\Pr\{\tilde{y} = i | n\} = \binom{n}{i} p^i (1 - p)^{n-i} \qquad \begin{aligned} E(\tilde{y}) &= np \\ \mathrm{var}(\tilde{y}) &= np(1 - p) \end{aligned}$$

$$\Pr\{\tilde{n} = i + y | y\} = \binom{i + y - 1}{y - 1} p^y (1 - p)^i \qquad \begin{aligned} E(\tilde{n}p) &= y \\ \mathrm{var}(\tilde{n}p) &= E(\tilde{n}p)(1 - p) \end{aligned}$$

Despite the differences in the two sampling schemes, the sampling variances are essentially the same. In particular, with reference to a fixed n under the direct (binomial) sampling approach, if we could select y for the inverse sampling such that $y = np$, then both sampling methods would have the same sampling variance.

Moreover, the respective MLEs and their variances are, for the positive binomial,

$$\hat{p} = \frac{\tilde{y}}{n}, \quad \text{var}(\hat{p}) = \frac{p(1-p)}{n}$$

and for the negative binomial,

$$\hat{p} = \frac{y}{\tilde{n}}, \quad \text{var}(\hat{p}) = \frac{p(1-p)}{E(\tilde{n})}$$

Thus, again, if we design the inverse sampling so that $E(\tilde{n}) = n$, there is no important large sample difference between the two approaches.

Another example more related to distance sampling is use of randomly placed quadrats versus a sample of random points, with the data being distance to the nearest plant (e.g. Patil *et al.* 1979b). In quadrats, the area is fixed and counts are random. In nearest neighbour sampling, the plant count is fixed but the area sampled is random. Under an appropriate matching of the effort expended under the two schemes, the corresponding density estimates have almost equal large sample sampling variances when plants are randomly distributed (Holgate 1964).

We surmise that this relationship holds for most positive and negative sampling schemes, i.e. there exist pairs of schemes such that the sampling variance of the parameter estimator is almost the same under either approach. In line transects, we have either L as fixed and n as random, or we fix n and traverse a random line length until n detections are made. To be consistent with the usual definitions of direct (positive) and indirect (negative, or inverse) sampling, we label these as below:

Positive case	n fixed	$\tilde{L}$ random
Negative case	$\tilde{n}$ random	L fixed

Comparability of sampling variances requires that comparable effort be used in both schemes; this translates into the pair of relationships $E(\tilde{L}|n) = L$ and $E(\tilde{n}|L) = n$.

6.6.2 Line transect sampling with fixed n and random $\tilde{L}$, under Poisson object distribution

We examine here some properties of $\hat{D}$ under such comparable schemes assuming a homogeneous Poisson distribution of objects, a constant detection function everywhere in the sampled area (spatially invariant $g(x)$), and independent detections. For the (usual) L-fixed case under these assumptions, $\tilde{n}$ has a Poisson distribution with mean $2LD/f(0)$. For $\tilde{L}$ random, we assume a random starting point for the line and we

227

move along it until n detections are made. Thus, there are $n - 1$ random inter-observational segments, of length $\tilde{l}_i$, which add to $\tilde{L}$. The first segment is from the starting point to the point perpendicular to the first detection. In general, the ith segment of length $\tilde{l}_i$ is the distance travelled between points perpendicular to detections $i - 1$ and i, where $i = 0$ is defined to be the starting point. Assuming that the number of objects in any area of size a, including the total area ($a = A$), is Poisson with mean aD, then it can be shown that $\tilde{l}$ is an exponentially distributed random variable with mean $E(\tilde{l}) = f(0)/(2D)$. The pdf of $\tilde{l}$ is

$$f_{\tilde{l}}(\tilde{l}) = \frac{2D}{f(0)} \cdot \exp\left[-\frac{2\tilde{l}D}{f(0)}\right]$$

By the assumptions we have made here, the $\tilde{l}_1, \ldots, \tilde{l}_n$ are independent. Because $\tilde{L}$ is the sum of independent exponential random variables, it is known that

$$\tilde{L} = \sum_{i=1}^{n} \tilde{l}_i$$

is distributed as a gamma $[n, f(0)/(2D)]$ distribution, so it has pdf

$$f_{\tilde{L}}(\tilde{L}) = \frac{\left[\dfrac{2D}{f(0)}\right]^n \cdot \tilde{L}^{n-1} \cdot \exp\left[-\dfrac{2\tilde{L}D}{f(0)}\right]}{(n-1)!}$$

It is also easily established that

$$E(\tilde{L}\,|\,n) = \frac{n \cdot f(0)}{2D}$$

which leads to the estimator

$$\tilde{D} = \frac{n \cdot \hat{f}(0)}{2\tilde{L}} \tag{6.13}$$

$\hat{f}(0)$ is computed conditional on n exactly as in the case of fixed L and random $\tilde{n}$, so $\hat{f}(0)$ is the same estimator in either sampling scheme.

Compare the estimator in Equation 6.13 to that when $\tilde{n}$ is random:

$$\hat{D} = \frac{\tilde{n} \cdot \hat{f}(0)}{2L} \tag{6.14}$$

Under a sampling theory approach, the two estimators have different expressions for small sample bias. For random $\tilde{L}$, from Equation 6.13,

$$E(\tilde{D}) = \frac{n}{2} \cdot E\left[\frac{\hat{f}(0)}{\tilde{L}} \Big| n\right]$$

Given a Poisson distribution of objects and constant $g(x)$, it is reasonable to assume that $\tilde{l}$ and x are independent. Then the above becomes

$$E(\tilde{D}) = \frac{n}{2} \cdot E\left[\frac{1}{\tilde{L}} \Big| n\right] \cdot E[\hat{f}(0)|n]$$

Under the gamma distribution of $\tilde{L}$,

$$E\left[\frac{1}{\tilde{L}} \Big| n\right] = \left[\frac{1}{n-1}\right] \cdot \left[\frac{2D}{f(0)}\right]$$

which yields

$$E(\tilde{D}) = \frac{n}{n-1} \cdot D \cdot \frac{E[\hat{f}(0)|n]}{f(0)}$$

When L is fixed,

$$E(\hat{D}) = D \cdot \frac{E[\hat{f}(0)|n]}{f(0)}$$

so there is little difference between the two sampling schemes for large n in this example. For $\tilde{L}$ random, the bias associated with $1/\tilde{L}$ could be eliminated by using

$$\tilde{D} = \frac{(n-1) \cdot \hat{f}(0)}{2\tilde{L}}$$

This adjustment for bias when n is fixed and $\tilde{L}$ is random seems generally appropriate.

An asymptotic formula for the variance of $\tilde{D}$ in Equation 6.13 is

$$\text{var}(\tilde{D}) = \left[\frac{f(0) \cdot n}{2E(\tilde{L}|n)}\right]^2 \cdot \left[\frac{\text{var}(\tilde{L}|n)}{[E(\tilde{L}|n)]^2}\right] + \left[\frac{n}{2E(\tilde{L}|n)}\right]^2 \cdot \text{var}[\hat{f}(0)|n]$$

$$= D^2 \cdot \left[[\text{cv}(\tilde{L})]^2 + [\text{cv}\{\hat{f}(0|n)\}]^2\right] \tag{6.15}$$

For the Poisson distribution of objects and a spatially invariant $g(x)$, so that $\tilde{l}$ is exponential, we have

$$\text{var}(\tilde{l}) = [E(\tilde{l})]^2 = \left[\frac{f(0)}{2D}\right]^2$$

and

$$\text{var}(\tilde{L}|n) = n \cdot \left[\frac{f(0)}{2D}\right]^2$$

Using these results and Equation 6.15 gives

$$\text{var}(\tilde{D}) = D^2 \cdot \left[\frac{1}{n} + [\text{cv}\{\hat{f}(0|n)\}]^2\right]$$

The variance of $\hat{D}$ in Equation 6.14 is

$$\text{var}(\hat{D}) = D^2\left[\frac{1}{E(\tilde{n}|L)} + [\text{cv}\{\hat{f}(0|\tilde{n})\}]^2\right]$$

Under comparable effort, in which case $E(\tilde{L}|n) = L$ and $E(\tilde{n}|L) = n$, it is clear that for large samples, $\text{var}(\tilde{D}) \doteq \text{var}(\hat{D})$.

The condition under which the two sampling schemes have almost the same variance for estimated density is that the coefficients of variation for $\tilde{n}$ and $\tilde{L}$ are equal:

$$\frac{\text{var}(\tilde{n}|L)}{[E(\tilde{n}|L)]^2} = \frac{\text{var}(\tilde{L}|n)}{[E(\tilde{L}|n)]^2}$$

This relationship holds for the above case.

6.6.3 *Technical comments*

Assuming independent detections, then a general formula for the cumulative distribution function of $\tilde{l}$ is

$$F_{\tilde{l}}(\tilde{l}) = 1 - \sum_{i=0}^{\infty} \left[1 - \frac{1}{w \cdot f(0)}\right]^i \cdot \Pr\{\tilde{N} = i \,|\, a = 2w\tilde{l}\} \tag{6.16}$$

The probability of moving distance $\tilde{l}$ and detecting no objects is $1 - F_{\tilde{l}}(\tilde{l})$. Assume the area examined for detections is of width $\pm w$ about the line. Then the unconditional probability of detecting an object is $1/[w \cdot f(0)]$. The event that there are no detections in distance $\tilde{l}$ happens if there are no objects in the area of size $2w\tilde{l}$, or if there is one object but it is not detected, or there are two objects and both remain undetected, and so forth. The joint probability that there are i objects in the area of size $2w\tilde{l}$ and all are undetected is (under the assumptions made)

$$\left[1 - \frac{1}{w \cdot f(0)}\right]^i \cdot \Pr\{\tilde{N} = i \,|\, a = 2w\tilde{l}\}$$

For $i = 0, 1, 2, \ldots$, these events are mutually exclusive, hence we get Equation 6.16. For the Poisson case,

$$\Pr\{\tilde{N} = i \,|\, a = 2w\tilde{l}\} = \frac{\exp(-2w\tilde{l}D) \cdot (2w\tilde{l}D)^i}{i!}$$

Thus

$$F_{\tilde{l}}(\tilde{l}) = 1 - \sum_{i=0}^{\infty} \left[1 - \frac{1}{w \cdot f(0)}\right]^i \cdot \frac{\exp(-2w\tilde{l}D) \cdot (-2w\tilde{l}D)^i}{i!}$$

$$= 1 - \exp[-2w\tilde{l}D] \cdot \exp\left\{\left[1 - \frac{1}{w \cdot f(0)}\right] \cdot (2w\tilde{l}D)\right\}$$

$$= 1 - \exp[-2\tilde{l}D/f(0)]$$

which is the cdf of an exponential distribution. Notice also that w drops out of $F_{\tilde{l}}(\tilde{l})$ in this example.

Closed form results can also be derived if a negative binomial distribution is assumed for $\Pr\{\tilde{N} = i | a = 2\tilde{l}w\}$ (Burnham *et al.* 1980: 197). However, we have not perceived a simple way to derive results for the random $\tilde{L}$ case without making strong assumptions about $\Pr\{\tilde{N} = i | a = 2w\tilde{l}\}$ and independence of detections.

Finally, we describe how to find $\Pr(\tilde{n} | L)$, as this is needed to compare the two schemes. Let $\tilde{N}$ be the number of objects in the searched strip of area $2wL$. The event $\tilde{n} = i$ arises as the sum of mutually exclusive events: $\tilde{N} = i$ and all i objects are detected, $\tilde{N} = i + 1$ and only i are detected, $\tilde{N} = i + 2$ and only i are detected, and so forth. The probability formula is

$$\Pr(\tilde{n} | L) = \sum_{\tilde{N} = i}^{\infty} \Pr\{\tilde{n} = i | \tilde{N}\} \cdot \Pr\{\tilde{N} | a = 2wL\} \tag{6.17}$$

For example, under the assumptions of a spatially constant detection function and independent detections, $\Pr\{\tilde{n} = i | \tilde{N}\}$ is binomial:

$$\Pr\{\tilde{n} = i | \tilde{N}\} = \binom{\tilde{N}}{i} \cdot \left[\frac{1}{w \cdot f(0)}\right]^{i} \cdot \left[1 - \frac{1}{w \cdot f(0)}\right]^{\tilde{N} - i}$$

For the Poisson case,

$$\Pr\{\tilde{N} | a = 2wL\} = \frac{\exp(-2wLD) \cdot (2wLD)^{\tilde{N}}}{\tilde{N}!}$$

Applying Equation 6.17 with these distributions gives

$$\Pr(\tilde{n} | L) = \frac{\exp[-2LD/f(0)] \cdot [2LD/f(0)]^{i}}{i!}$$

which is a Poisson distribution.

6.6.4 *Discussion*

Having fixed n and random $\tilde{L}$ is often not a practical design in line transect sampling. In particular, when planes or helicopters are used, you cannot set out to fly a random distance; $\tilde{L}$ cannot exceed the fuel capacity of the plane. For most methods of traversing the line(s), the distance to travel must be specified in advance. This also allows an accurate cost estimate for a study, which is generally necessary. Further, in most studies, if effort ceased when the required sample size was reached, the observer would have to return to some point at the study boundary, and it would be wasteful not to seek and record detections for the maximum possible time in the study area.

Representative coverage of the area being sampled is also an important consideration. Lines, or points, are allocated so as to achieve a representative sampling of the area. This is critical to allow valid inference from the sample to the entire area. If $\tilde{L}$ was random, one might finish before completing the *a priori* determined sample of lines, or finish the sample of lines and still need to sample more. With random line length, it is difficult to assure a representative sample over the area of interest; thus there is more danger of substantial bias in $\tilde{D}$ due to unbalanced spatial coverage.

Point transect sampling is potentially more amenable to a fixed n strategy. The fixed n can be set on a per point basis. Then a representative sample of k points can be selected and every point can be visited. The amount of time at each point will vary. An upper bound could be put on time, leading to a mixed strategy: stay at a point until n detections are made, or until the maximum time is reached. There is potential to develop the theory for such a strategy. However, it is still not especially practical to ask a recorder to be aware of when n total detections are made, and then to stop effort. Also, should this n be a total for all species, or for one target species?

We have not here presented any theory for point transects with fixed n and random time, as that theory is more difficult to conceptualize. For the typical application to birds, detections depend on cue generation, which would have their own temporal distribution. This adds another level of complexity to the case of a scheme with fixed n and random time.

Even for a fixed line length scheme, there is information in the interobservational distances as defined here. For example, the $\tilde{l}_i$ may be used to assess the spatial distribution of objects (Burnham *et al.* 1980: 196–8). Under the (unlikely) hypothesis of a Poisson spatial distribution and constant $g(x)$, $\tilde{l}$ is an exponential random variable. There are many tests available for the null hypothesis that a random sample is from an exponential pdf. The distribution of $\tilde{l}$ under other object distributions can be determined by methods presented here. The information contained in the $\tilde{l}_i$ is reduced in practice, because they are likely to be serially correlated. However, if independence can be assumed, the information in the $\tilde{l}_i$ might be used to provide better estimates of the residual variation in the $\tilde{n}$; this subject may be worthy of study. The concept is that

$$\widehat{\text{var}}(\tilde{n}|L) = \left(\frac{n}{L}\right)^2 \cdot \widehat{\text{var}}(\tilde{L}|n)$$

might hold true. Given independence of the interobservational distances,

$$\widehat{\text{var}}(\tilde{L}|n) = \frac{\sum\limits_{i=1}^{n}(\tilde{l}_i - \tilde{L}/n)^2}{n-1}$$

thus giving an alternative estimator of $\text{var}(\tilde{n}|L)$. The above variance estimator for $\tilde{L}$ also provides the basis for an empirical estimator of $\text{var}(\tilde{D})$ for the fixed n scheme if such sampling is practical and valid.

A final important comment is required about the relative merits of random and fixed line lengths. Often, lines are *a priori* of fixed (frequently equal) length, in which case all the fixed length (and random n) theory holds. However, designs or field practice often result in unequal line lengths, for example when lines are placed at random but then cross from one side to the other of an *a priori* defined area, or when bad weather causes effort termination during a ship survey, so that a transect is shorter than intended. Either of these instances gives the appearance of some stochasticity in line length, hence one might consider that the set of lines has a random component that should be accounted for in variances (and perhaps biases) of estimates (Seber 1979). We disagree with this thinking; it is entirely appropriate to condition on achieved line length in these and other cases, provided the stochastic variations in length are unrelated to the density of objects, or if it is not possible to fit a model that relates variation in line length to object density.

It does not follow that random line length theory applies, simply because the survey design or field protocol results in varying line lengths. The theory applies only if there is information about density D in the probability distribution of line lengths. Even in the case of randomly placed lines running across a predefined area, there is no information about object density in the probabilistic distribution of line length by itself. Moreover, in this case the line lengths are known before they are ever traversed, hence there is every reason in theory to consider line length as a fixed ancillary (i.e. it affects the precision of $\hat{D}$ but contains no direct information about D) in all the usual designs and field protocols.

Once we consider line lengths as known and fixed prior to data collection, or after the fact we condition on actual line lengths when appropriate, then some potential statistical methods are not relevant. For example, it is not relevant to apply finite population sampling ratio estimation theory to encounter rate; such ratio estimation theory leads to slightly different formulae for var(n) than we give in this book. The key point here is that line lengths l_i may often differ; this does not make them random in any sense that concerns us, especially if we know the actual line lengths before they are traversed. It is proper to take these line lengths as fixed unless there is a probability distribution on possible line lengths which depends on the density parameter of interest. The latter is only likely to be true under the scheme of fixed n, in which random linear effort continues until n detections are made, or under adaptive sampling schemes, in which sampling effort increases when areas of high density are found.

6.7 Efficient simulation of distance data

6.7.1 The general approach

To produce simulated distance data requires the Monte Carlo generation of sample size n, detection distances $y = x$ or r, and for the clustered case, cluster sizes s. The efficient way to do this is first to generate sample size according to some discrete distribution, $p(n)$, then generate n distances and cluster sizes based on the bivariate sampling distribution of distance and s. The alternative is first to generate spatially distributed clusters, and independently for each cluster, a cluster size s. Then determine for each cluster whether it gets detected according to some detection function, $g(y|s)$. This method is indirect and inefficient. The purpose of this section is twofold: to show how to simulate distance data directly and to outline explicitly how the simulated data of Chapters 4 and 5 were generated.

The following general approach is recommended. First, decide on a detection function; it will be bivariate if cluster size is to vary and there is to be size-biased detection. Otherwise, $g(y)$ depends only on distance. For the clustered case, decide on a probability distribution, $\pi(s)$, of cluster sizes in the yet-to-be-sampled population. Also select a sampling distribution for sample size, $p(n)$, such as Poisson or negative binomial. It is then possible to specify the exact parameterization of $p(n)$. To simulate data for k replicate lines or points, first generate independent sample sizes $n_1, \ldots, n_k$ according to $p(n)$. If objects do not occur in clusters, just generate n_i independent distances, $i = 1, \ldots, k$, according to the probability density function of distances, $f(y)$. This function is determined by, and computed from, $g(y)$, and the context (line or point transects). If cluster size varies but detection is independent of size, then for each generated distance y, produce independently a value of the random variable s according to $\pi(s)$.

The case of size-biased detection requires a two step process of either generating y from its marginal density function, then s from the size-biased distribution $\pi^*(s|y)$, or the reverse (which we recommend): generate an observed cluster size from the marginal size-biased distribution of detections, $\pi^*(s)$, then generate y from $f(y|s)$. The detailed theory for these distributions, given $g(y|s)$ and $\pi(s)$, is in Section 3.6.6.

The distribution of n, $p(n)$, must have, at least implicitly, $E(n)$ as one of its parameters, where

$$E(n) = \frac{2LD_s}{f(0)} \quad \text{(line transects)}$$

$$E(n) = \frac{2\pi k D_s}{h(0)} \quad \text{(point transects)}$$

(π by itself refers to 3.14159...). The density D_s denotes density of clusters. We use $f(0)$ and $h(0)$ in these representations to facilitate the case without truncation ($w = \infty$).

We now summarize quantities that must be specified to simulate distance data. These quantities are interrelated and hence cannot be independently set; in particular, we recommend that either $E(n)$ or sampling effort (L or k) is specified, and the other quantity is computed. Constants to be specified are w and k, and for line transects, line lengths l_i, $i = 1, \ldots, k$, which sum to L. Separately specified parameters are D_s and $E(s)$, and any additional parameters in $p(n)$ other than $E(n)$. Fundamental distributions to specify are $p(n)$ and $\pi(s)$. Finally, there is the detection function, $g(y|s)$, which, in conjunction with $\pi(s)$, determines the sampling distributions of y and s. In general we would need to compute $f(0)$, $h(0)$, $\pi^*(s)$, $f(x|s)$ and $f(x|s)$ numerically. From Section 3.6.6, formulae necessary in simulation of line transect data are

$$f(0|s) = \frac{1}{\displaystyle\int_0^w g(x|s)\,dx} \tag{6.18}$$

$$f(0) = \frac{1}{\displaystyle\sum_{s=1}^{\infty} \frac{\pi(s)}{f(0|s)}} \tag{6.19}$$

$$\pi^*(s) = \left[\frac{f(0)}{f(0|s)}\right]\pi(s) \tag{6.20}$$

and

$$f(x|s) = f(0|s) \cdot g(x|s) \qquad (6.21)$$

Formulae necessary in simulation of point transect data are

$$h(0|s) = \frac{1}{\displaystyle\int_0^w r \cdot g(r|s)\,dr} \qquad (6.22)$$

$$h(0) = \frac{1}{\displaystyle\sum_{s=1}^{\infty} \frac{\pi(s)}{h(0|s)}} \qquad (6.23)$$

$$\pi^*(s) = \left[\frac{h(0)}{h(0|s)}\right]\pi(s) \qquad (6.24)$$

and

$$f(r|s) = h(0|s) \cdot r \cdot g(r|s) \qquad (6.25)$$

If the distribution of n is assumed to be Poisson, then

$$p(n) = \frac{e^{-E(n)}[E(n)]^n}{n!} \qquad (6.26)$$

A useful parameterization of the negative binomial is

$$p(n) = \frac{\Gamma(\theta + n)}{\Gamma(\theta)\,\Gamma(n+1)}\,(1-\tau)^n \cdot \tau^\theta, \quad 0 < \theta,\ 0 < \tau < 1,\ 0 \leqslant n \qquad (6.27)$$

which has

$$E(n) = \theta \cdot \frac{1-\tau}{\tau}$$

and

$$\mathrm{var}(n) = \frac{E(n)}{\tau}$$

237

In point transects all with a fixed observation time, τ and θ can be the same over different points (within a stratum). For line transects, the l_i usually vary, and we recommend keeping τ constant while letting θ vary by line length, because this gives coherent results: $n_1 + \ldots + n_k$ is then distributed as negative binomial with parameters τ and $\theta_1 + \ldots + \theta_k$. Thus, we can arbitrarily vary the line lengths and still have sample sizes (individually and totals) as negative binomial random variables. Under this strategy, $1/\tau$ is the variance inflation factor relative to a Poisson (random) spatial distribution of objects. For the case of $\tau = 1$, use the Poisson distribution for Monte Carlo generation of sample sizes.

Consider the line transect case with one long line (i.e. ignore replicate lines) of length L and objects not clustered. We would first specify $g(x)$, then compute $f(0)$, by numerical integration if need be. It is important to keep straight the units of measurement in a simulation, because with real data, detection distances and line length are often in different units, such as metres and kilometres. Also, the units of $f(0)$ are the reciprocal of the distance units used for x. Say we get $f(0) = 10$, with units on x taken as kilometres. Then effective strip width is 0.1 km or 100 m.

In this hypothetical example, next we specify $D_s = 2$ clusters/km^2 and $E(n) = 70$. Now we determine L from $E(n) = 2LD_s/f(0)$: $70 = 2 \cdot 2 \cdot L/10$, or $L = 175$km. If we want n to be Poisson, then we generate it from a Poisson with mean 70. Given n, generate $x_1, \ldots, x_n$ from the pdf $f(x) = f(0) \cdot g(x)$.

We illustrate the approach in more detail for the simulation generation of examples in Chapters 4 and 5.

6.7.2 The simulated line transect data of Chapter 4

The half-normal bivariate detection function may be taken as

$$g(x \mid s) = \exp\left[-\frac{1}{2}\left(\frac{x}{\sigma(s)}\right)^2\right]$$

where we model the scale parameter, σ, as a function of cluster size (c.f. Quinn 1979; Drummer and McDonald 1987; Ramsey et al. 1987; Otto and Pollock 1990). In particular, the form $\sigma(s) = \sigma \cdot s^\alpha$ has been much used. We think this is a reasonable model for data analysis; however, for simplicity of theory, we used a linear form for line transects:

$$\sigma(s) = \sigma_0\left[1 + b \cdot \frac{s - E(s)}{E(s)}\right] \tag{6.28}$$

subject to the constraint

$$b \leq \frac{E(s)}{E(s) - 1}$$

and assuming $0 \leq b$, although to a limited extent, negative values of b are mathematically possible. This form is also suggested by us because it puts the problem into the framework of generalized linear models (McCullagh and Nelder 1989). The case $b = 0$ corresponds to detection probability independent of cluster size. For $w = \infty$, $f(0\,|\,s)$ and $f(x\,|\,s)$ are closed form:

$$f(0\,|\,s) = \left[\sqrt{\frac{2}{\pi}}\right] \cdot \frac{1}{\sigma(s)}$$

$$f(x\,|\,s) = \left[\sqrt{\frac{2}{\pi}}\right] \cdot \frac{1}{\sigma(s)} \cdot \exp\left[-\frac{1}{2}\left(\frac{x}{\sigma(s)}\right)^2\right] \tag{6.29}$$

Applying Equation 6.19, we have

$$f(0) = \left[\sqrt{\frac{2}{\pi}}\right] \cdot \frac{1}{\displaystyle\sum_{s=1}^{\infty} \pi(s) \cdot \sigma(s)} = \left[\sqrt{\frac{2}{\pi}}\right] \cdot \frac{1}{E\{\sigma(s)\}}$$

The form of $\sigma(s)$ in Equation 6.28 is convenient because we can explicitly evaluate its expectation with respect to $\pi(s)$; in fact, $E\{\sigma(s)\} = \sigma_0$ for any value of the parameter b. Thus, for any extent of size bias under this model,

$$f(0) = \left[\sqrt{\frac{2}{\pi}}\right] \cdot \frac{1}{\sigma_0}$$

From Equation 6.20,

$$\pi^*(s) = \left[\frac{f(0)}{f(0\,|\,s)}\right]\pi(s) = \frac{\sigma(s)}{\sigma_0}\,\pi(s)$$

$$= \left[1 + b \cdot \frac{s - E(s)}{E(s)}\right]\pi(s) \qquad s = 1, 2, \ldots \tag{6.30}$$

The expected value of s in the sample of detected clusters is

$$E^*(s) = \sum s \cdot \pi^*(s) = E(s) + b \cdot \frac{\text{var}(s)}{E(s)}$$

A simple choice for $\pi(s)$ is to let $s - 1$ have a Poisson distribution with mean $E(s) - 1$:

$$\pi(s) = \frac{e^{-[E(s) - 1]} \cdot [E(s) - 1]^{s-1}}{(s - 1)!} \qquad s = 1, 2, \ldots$$

We used this $\pi(s)$ in the Chapter 4 examples and then created a table (in the computer) of the values of $\pi^*(s)$, from Equation 6.30, for specified $E(s)$, b and σ_0. Then a value of a detected cluster size was generated by standard Monte Carlo methods. Given a value of s, then x was generated according to the distribution of Equation 6.29. This was done by generating a χ^2 variate on 1 df and calculating $x = \sigma(s) \cdot \sqrt{\chi^2}$.

The distribution of sample detections in Chapter 4 was negative binomial, set up with $\tau = 0.4$, so that the variance inflation factor is 2.5. The choice of 12 replicate lines was arbitrary. Other choices made: $\sigma_0 = 10\text{m}$ and $E(n) = 96$. It was then convenient to use $L = 48\text{km}$ and keep the encounter rate constant over replicate lines (whose lengths varied). These choices and decisions produce, by design, the result

$$E(n_i) = \theta_i \frac{1 - \tau}{\tau} = 2l_i \qquad i = 1, \ldots, 12$$

which implies $\theta_i = 64 l_i / 48$.

In metres, $f(0) = 0.0798$, hence with a conversion factor to units of per km^2, density of clusters is

$$D = \frac{96 \times 0.0798}{2 \times 48} \times 1000 = 79.8 \text{ clusters/km}^2$$

The simulation was set up in such a way that density of individuals is $79.8 \cdot E(s) = 239.4$ regardless of the value of b; b only determines the degree of size bias. Values of b used were 0 and 1 (with the same set of $n_1, \ldots, n_{12}$).

6.7.3 The simulated size-biased data of Chapter 5

The generation of the simulated data for point transects with size bias used the same half-normal bivariate detection function as for the line transect case:

$$g(r|s) = \exp\left[-\frac{1}{2}\left(\frac{r}{\sigma(s)}\right)^2\right]$$

However, the relevant formulae are now Equations 6.22–6.25. In particular, we have for $w = \infty$,

$$h(0|s) = \frac{1}{[\sigma(s)]^2}$$

and

$$f(r|s) = \frac{r}{[\sigma(s)]^2} \cdot \exp\left[-\frac{1}{2}\left(\frac{r}{\sigma(s)}\right)^2\right] \qquad (6.31)$$

Using Equation 6.23 with the above expression for $h(0|s)$, we have

$$h(0) = \frac{1}{\sum\limits_{s=1}^{\infty} \pi(s) \cdot \{\sigma(s)\}^2} = \frac{1}{E[\{\sigma(s)\}^2]}$$

Thus we choose to parameterize $\sigma(s)$ as

$$\{\sigma(s)\}^2 = \sigma_0^2 \cdot \left[1 + b \cdot \frac{s - E(s)}{E(s)}\right]$$

subject to the constraint

$$b \leqslant \frac{E(s)}{E(s) - 1}$$

This model gives

$$h(0) = \frac{1}{\sigma_0^2}$$

and

$$h(0|s) = \frac{1}{\sigma_0^2 \cdot \left[1 + b \cdot \dfrac{s - E(s)}{E(s)}\right]}$$

Hence, from Equation 6.24,

$$\pi^*(s) = \left[\frac{h(0)}{h(0|s)} \right] \pi(s) = \frac{\sigma(s)}{\sigma_0} \cdot \pi(s)$$

$$= \left[1 + b \cdot \frac{s - E(s)}{E(s)} \right] \pi(s) \qquad s = 1, 2, \ldots$$

which is the identical $\pi^*(s)$ in the line transect case in Equation 6.30.

For the Chapter 5 size-biased example data, we choose $\pi(s)$ to be the geometric distribution,

$$\pi(s) = (1 - \beta)^{s-1} \cdot \beta \qquad 0 < \beta < 1 \qquad s = 1, 2, \ldots$$

For this distribution, $E(s) = 1/\beta$. From the above expression for $\pi^*(s)$, we have for this example

$$\pi^*(s) = \left[1 + b \cdot \frac{s - E(s)}{E(s)} \right] \cdot (1 - \beta)^{s-1} \cdot \beta \qquad s = 1, 2, \ldots \qquad (6.32)$$

Note that $b = 1$ corresponds to considerable size bias and gives the simple form

$$\pi^*(s) = s \cdot (1 - \beta)^{s-1} \cdot \beta^2 \qquad s = 1, 2, \ldots$$

For the example in Section 5.8, we used $E(s) = 1.85$, hence $\beta = 0.54054$, and $b = 0.75$. These values serve to specify $\pi^*(s)$ completely. Also, we set $\sigma_0 = 30\,\text{m}$, which, together with $b = 0.75$, serves to specify $g(r|s)$, $h(0|s)$ and $f(r|s)$. The latter is given by Equation 6.31; that density function has cumulative distribution function

$$f(r|s) = 1 - \exp\left[-\frac{1}{2}\left(\frac{r}{\sigma(s)} \right)^2 \right]$$

Consequently, for this example, a random r, given an s drawn from Equation 6.32, was produced as

$$r = -2 \cdot \sigma(s) \cdot \log_e(1 - u)$$

where u is a uniform pseudo-random variable on the interval 0 to 1.

The variation in counts, n, was generated from the negative binomial distribution with variance inflation (dispersion) factor set at 2.65, so that $\tau = 0.37736$. The encounter rate per point was set at 1.6 and then k was set at 60 points to give, overall, $E(n) = 96$. In terms of Equation 6.27 and associated results, this means that on a per point basis (which is how the random counts are generated), we specified

$$E(n_i) = \theta \, \frac{1 - \tau}{\tau} = 1.6$$

and

$$\text{var}(n_i) = \frac{E(n_i)}{\tau} = 2.65 \, E(n_i) \qquad i = 1, \ldots, k$$

Hence, τ is as above and $\theta = 0.52416$. As a consequence of the choice of model and parameters in this example, the density of clusters is

$$\frac{96 \times [1/30]^2}{2\pi \times 60} \times 1\,000\,000 = 283 \;\; \text{clusters/km}^2$$

and density of individuals is $1.85 \times 2.83 = 523$ objects/km^2.

6.7.4 Discussion

We have recommended simulating the pair (y, s) by generating s from its marginal distribution $\pi^*(s)$, and then y from the conditional distribution, $f(y|s)$. The algebra for this was straightforward in the above two examples. Consider, however, the reverse process for the point transect example above: generate r from $f(r)$ then s from $\pi^*(s|r)$. The relevant formulae are

$$f(r) = \frac{1}{\sigma_0^2} \left[\sum_{s=1}^{\infty} (1 - \beta)^{s-1} \cdot \beta \cdot r \cdot \exp\left\{ -\frac{1}{2} \left(\frac{r}{\sigma(s)} \right)^2 \right\} \right]$$

and

$$\pi^*(s|r) = \frac{(1 - \beta)^{s-1} \cdot \exp\left[-\frac{1}{2} \left(\frac{r}{\sigma(s)} \right)^2 \right]}{\displaystyle\sum_{s=1}^{\infty} (1 - \beta)^{s-1} \cdot \exp\left[-\frac{1}{2} \left(\frac{r}{\sigma(s)} \right)^2 \right]}$$

243

Use of these formulae would entail much more computing than the use of $\pi^*(s)$ and $f(r|s)$. Heuristically, this is because there is only a finite (and small, usually) number of possible values for s, whereas infinitely many values of r can occur. Therefore, with each new r, one must recompute the entire distribution $\pi^*(s|r)$ before s can be generated.

In some real applications, cluster sizes potentially range from one to thousands, for example dolphin surveys on some species. To simulate the essence of such applications, it is not necessary for s to vary over this set of values. A set of a hundred (or fewer) values should suffice (e.g. s taking values 1–10, 15, 20, 30, 50, 75, 100–900 by 100, 1000–5000 by 500). Keeping the range set of s small will greatly speed up simulations.

Closed form expressions underlying simulations will be the exception. Even in the above line transect examples, if we take w as finite, numerical integration must be used to find the necessary quantities given by Equations 6.18 to 6.21. Expect to use numerical integration; fortunately for purposes of simulation, the numerical methods need not be highly sophisticated.

Sometimes we only want to explore statistical properties of estimators of $f(0)$, $h(0)$, $g(x|s)$, $g(r|s)$ and $E(s)$, and not properties of $\hat{D}_s$ and $\hat{D}$. In this event, it is not necessary to generate a random sample size n for each replicate. In fact, it is better to fix n and do, say, 1000 replicates at that n, and repeat the process over a set of values of n.

6.8 Thoughts about a full likelihood approach

6.8.1 Introduction

In principle, analysis of distance data could be based on a full likelihood, $\mathcal{L}(D,\underline{\theta})$, for all data components. The focus is on average density D in the study area; we represent all the other parameters by $\underline{\theta}$. These other parameters appear in the probability components for n, $y = x$ or r, and s. An advantage of having a full likelihood is that it allows the computation of profile likelihood intervals for D. The disadvantages are the need to specify probability models for n and s. We have avoided assuming any probability model for n by using an empirical estimator of $\text{var}(n)$, and getting confidence intervals assuming $\hat{D}$ is log-normally distributed. Similarly, a point estimator and sampling variance of $E(s)$ can be obtained in a regression framework, so no probability model is required for $\pi(s)$ or $\pi^*(s)$ (the distribution of s in the entire population, and in the detected sample, respectively). Probability models, and likelihood inference, have only been used for the distance part of the data, and we use generalized approaches to ensure robust inference for n and s. The purpose of this section is to demonstrate how a full likelihood approach could be used, to make some comparisons of profile likelihood intervals to traditional and log-based intervals, and to comment on other advantages of a full parametric (likelihood) approach.

6.8.2 Full likelihood for line transects: simple examples

(a) *Half-normal* $g(x)$, *Poisson* n Assume that objects are spatially distributed as a homogeneous Poisson process, that the detection function is half-normal, that $w = \infty$, and that objects are single entities (i.e. we take $s = 1$). Then data from replicate lines may be collapsed into just the total count, n, for total line length L, and the perpendicular distances $x_1, \ldots, x_n$. The detection function $g(x)$ and pdf $f(x)$ are

$$g(x) = \exp\left\{-\frac{1}{2}\left(\frac{x}{\sigma}\right)^2\right\}$$

and $\quad f(x) = \left[\sqrt{\frac{2}{\pi\sigma^2}}\right] \cdot \exp\left\{-\frac{1}{2}\left(\frac{x}{\sigma}\right)^2\right\} \qquad 0 \leqslant x < \infty, \, 0 < \sigma$

The probability distribution of n is

$$\Pr(n) = \frac{\exp\{-2LD/f(0)\} \cdot \{2LD/f(0)\}^n}{n!}$$

(Section 6.6). We have

$$f(0) = \sqrt{\frac{2}{\pi\sigma^2}}$$

so there are only two parameters, σ and D, although it is sometimes simpler to leave $f(0)$ in the formulae. The full likelihood is

$$\mathcal{L}(D, \sigma) = \Pr(n) \cdot \left[\prod_{i=1}^{n} f(x_i)\right]$$

$$= \frac{\exp\{-2LD/f(0)\} \cdot \{2LD/f(0)\}^n}{n!} \cdot \prod_{i=1}^{n}\left[\left(\sqrt{\frac{2}{\pi\sigma^2}}\right) \cdot \exp\left\{-\frac{1}{2}\left(\frac{x_i}{\sigma}\right)^2\right\}\right]$$

245

We define

$$T = \sum_{i=1}^{n} x_i^2$$

and simplify $\mathcal{L}(D,\sigma)$ by collapsing terms where possible and by dropping some multiplicative constants, giving

$$\mathcal{L}(D, \sigma) = \exp\{- LD\sigma\sqrt{(2\pi)}\} \cdot \{LD\sigma\sqrt{(2\pi)}\}^n \cdot \left[\frac{1}{\sigma^n} \cdot \exp\left\{-\frac{T}{2\sigma^2}\right\}\right]$$

or simplified as much as possible,

$$\mathcal{L}(D,\sigma) = \exp\left\{-\left(LD\sigma\sqrt{(2\pi)} + \frac{T}{2\sigma^2}\right)\right\} \cdot D^n \qquad (6.33)$$

Note that we ignore multiplicative constants in the likelihood function. The joint MLEs from Equation 6.33 are

$$\hat{\sigma} = \sqrt{\frac{T}{n}}$$

and

$$\hat{D} = \frac{1}{L} \sqrt{\frac{n}{2\pi T}} \equiv \frac{n \cdot \hat{f}(0)}{2L}$$

Standard likelihood theory can be used to derive the theoretical variance of $\hat{D}$, which may be expressed in a variety of ways:

$$\text{var}(\hat{D}) = \frac{1.5 D \cdot f(0)}{2L} = D^2 \cdot \left[\frac{1.5}{E(n)}\right] = D^2 \cdot \left[\frac{1}{E(n)} + \frac{1}{2E(n)}\right]$$

$$= D^2 \left[\{\text{cv}(n)\}^2 + \{\text{cv}(\hat{f}(0))\}^2\right]$$

Thus, these results are all exactly the same as what are derived by the 'hybrid' method of using $E(n) = 2LD/f(0)$, the Poisson variance of n, and a likelihood only for the distance data.

In general, the hybrid approach with empirical estimation of var(n) will be almost fully efficient for $\hat{D}$ and is more robust as no distribution need be assumed for n. One advantage of a full likelihood approach is the possibility of using a profile likelihood interval for D. Such intervals can be expected to perform better than $\hat{D} \pm 1.96 \, \widehat{se}(\hat{D})$ or log-based intervals because the likelihood function encodes information about the sampling distribution of $\hat{D}$, thus allowing for non-normality of $\hat{D}$ (or $\log_e(\hat{D})$). Below, we give some general explanation of profile likelihoods, then derive the profile log-likelihood function for D for the above example.

(b) *Profile likelihood intervals*   Let $\mathscr{L}(D, \underline{\theta})$ be the full likelihood such as that given in Equation 6.33. The profile likelihood is symbolically

$\mathscr{L}(D, \hat{\underline{\theta}}(D)) = \text{maximum value of } \mathscr{L}(D, \underline{\theta}) \text{ over } \underline{\theta} \text{ for any given value of } D$

This is then a function of just the single parameter D. For computing profile likelihood intervals, it is convenient to use the following function:

$$\phi(D) = 2 \{\log_e \mathscr{L}(\hat{D}, \hat{\underline{\theta}}) - \log_e \mathscr{L}(D, \hat{\underline{\theta}}(D))\} \qquad (6.34)$$

In Equation 6.34, $\mathscr{L}(\hat{D}, \hat{\underline{\theta}})$ is equivalent to $\mathscr{L}(\hat{D}, \hat{\underline{\theta}}(\hat{D}))$, where $\hat{D}$ is the MLE of D. The function $\phi(D)$ is a pivotal quantity, asymptotically distributed as a single degree of freedom chi-square, χ_1^2. This approximate distribution of $\phi(D)$ holds better at small sample sizes than the assumed normality of $\hat{D}$ underlying the use of $\hat{D} \pm 1.96 \, \widehat{se}(\hat{D})$. A $100(1 - \alpha)\%$ profile confidence interval for D is given as the set of all values of D such that $\phi(D) \leqslant \chi_1^2(\alpha)$, where $\chi_1^2(\alpha)$ is the $1 - \alpha$ percentile point of the χ_1^2 distribution (3.84 for a 95% confidence interval). We only need the interval endpoints, which are the two solutions to the equation

$$\phi(D) = 2 \{\log_e \mathscr{L}(\hat{D}, \hat{\underline{\theta}}) - \log_e \mathscr{L}(D, \hat{\underline{\theta}}(D))\} = \chi_1^2(\alpha) \qquad (6.35)$$

Barndorff-Neilson (1986) described the theory underlying the method, including ways to improve on the approximation of $\phi(D)$ as a χ_1^2 random variable.

(c) *Profile formulae, half-normal g(x), Poisson n*   Starting with the likelihood in Equation 6.33, we first must find the maximum for σ given any fixed value of D. The steps are summarized below:

$$\log_e \mathscr{L}(D, \sigma) = - LD\sigma\sqrt{(2\pi)} - \frac{T}{2\sigma^2} + n \cdot \log_e(D)$$

and

$$\frac{\partial \log_e \mathcal{L}(D, \sigma)}{\partial \sigma} = - LD\sqrt{(2\pi)} + \frac{T}{\sigma^3} = 0$$

From the above equation,

$$\hat{\sigma}(D) = \left[\frac{T}{LD\sqrt{(2\pi)}}\right]^{1/3}$$

The joint MLEs $\hat{D}$ and $\hat{\sigma}$ are given above, hence finding the expression in Equation 6.34 is now merely algebraic manipulation:

$$\log_e \mathcal{L}(D, \hat{\sigma}(D)) = - \frac{3n}{2}\left[\frac{D}{\hat{D}}\right]^{2/3} + n \cdot \log_e(D)$$

from which

$$\log_e \mathcal{L}(\hat{D}, \hat{\sigma}) = - \frac{3n}{2} + n \cdot \log_e(\hat{D})$$

Reduced to a very simple form, we get for this example

$$\phi(D) = 3n \cdot [(D/\hat{D})^{2/3} - 1 - \log_e \{(D/\hat{D})^{2/3}\}] \qquad (6.36)$$

To get a profile likelihood interval for D, we substitute the values of $\hat{D}$ and n in Equation 6.36, tabulate $\phi(D)$ for a range of D, and pick off the two solutions to Equation 6.35. It can be useful to plot $\phi(D)$, as is shown in another context by Morgan and Freeman (1989).

Below we look at some numerical examples comparing different confidence intervals. However, first we determine Equation 6.34 explicitly for a few more examples. These, and the above, are overly simplistic compared to real data, but only very simple cases lead to analytical, or even partially analytical, solutions for $\phi(D)$.

(d) *Negative exponential g(x), Poisson n* The negative exponential $g(x)$ is not a desirable detection function, but for $w = \infty$ and a Poisson n, we can derive closed form results for this case. Some formulae are

$$g(x) = \exp\left\{-\frac{x}{\lambda}\right\}$$

$$f(x) = \frac{1}{\lambda} \cdot \exp\left\{-\frac{x}{\lambda}\right\} \qquad 0 \leq x < \infty, \ 0 < \lambda$$

and $f(0) = 1/\lambda$. The full likelihood is

$$\mathscr{L}(D, \lambda) = \frac{\exp\{-2LD/f(0)\} \cdot \{2LD/f(0)\}^n}{n!} \cdot \prod_{i=1}^{n}\left[\frac{1}{\lambda} \cdot \exp\left\{-\frac{x_i}{\lambda}\right\}\right]$$

Defining

$$T = \sum_{i=1}^{n} x_i$$

we simplify $\mathscr{L}(D, \lambda)$ to

$$\mathscr{L}(D, \lambda) = \exp\left\{-\left(2LD\lambda + \frac{T}{\lambda}\right)\right\} \cdot D^n$$

The joint MLEs are $\hat{\lambda} = \dfrac{T}{n}$ and $\hat{D} = \dfrac{n^2}{2LT} \equiv \dfrac{n \cdot \hat{f}(0)}{2L}$

From likelihood theory,

$$\text{var}(\hat{D}) = \frac{D}{L\lambda} = D^2 \cdot \left[\frac{2}{E(n)}\right] \equiv D^2 \left[\{\text{cv}(n)\}^2 + \{\text{cv}(\hat{f}(0))\}^2\right]$$

Fixing D in $\mathscr{L}(D, \lambda)$, we find $\hat{\lambda}(D)$ as follows:

$$\log_e \mathscr{L}(D, \lambda) = -2LD\lambda - \frac{T}{\lambda} + n \cdot \log_e(D)$$

and

$$\frac{\partial \log_e \mathscr{L}(D, \lambda)}{\partial \lambda} = -2LD + \frac{T}{\lambda^2} = 0$$

so that

$$\hat{\lambda}(D) = \sqrt{\frac{T}{2LD}}$$

Finally, we derive

$$\phi(D) = 4n \cdot \left[(D/\hat{D})^{1/2} - 1 - \log_e \{ (D/\hat{D})^{1/2} \} \right] \qquad (6.37)$$

(e) *Negative exponential* $g(x)$, *Poisson* n, *Poisson* s To the above example, we add the feature of varying cluster size, but with detection probability independent of cluster size, s. Let s be Poisson with mean κ. The parameter D is the density of individuals, not clusters. In the Poisson model, as given above, for counts of clusters, the density parameter is cluster density, D_s, not D. To parameterize this likelihood component in terms of density of individuals, we must replace D_s by D/κ. The full likelihood for this model is

$$\mathcal{L}(D,\lambda,\kappa) = \frac{\exp\{ -2LD\lambda/\kappa \} \cdot \{ 2LD\lambda/\kappa \}^n}{n!} \cdot \prod_{i=1}^{n} \left[\frac{1}{\lambda} \cdot \exp\left\{ -\frac{x_i}{\lambda} \right\} \cdot \left\{ \frac{\exp(-\kappa) \cdot \kappa^{s_i}}{s_i!} \right\} \right]$$

Using $\bar{s}$ to denote mean cluster size and $\bar{x} = T/n$, this likelihood can be reduced to

$$\mathcal{L}(D, \lambda, \kappa) = \exp\left\{ -\left(\frac{2LD\lambda}{\kappa} + \frac{T}{\lambda} + n\kappa \right) \right\} \cdot D^n \cdot \kappa^{n(\bar{s} - 1)}$$

and

$$\log_e \mathcal{L}(D, \lambda, \kappa) = -\frac{2LD\lambda}{\kappa} - \frac{T}{\lambda} - n\kappa + n \cdot \log_e(D) + n(\bar{s} - 1) \cdot \log_e(\kappa)$$

The hybrid and full likelihood results agree here; in particular,

$$\hat{D} = \frac{n \cdot \bar{s}}{2L\bar{x}}$$

$$\mathrm{var}(\hat{D}) = D^2 \left[\frac{3}{E(n)} \right]$$

To get the profile likelihood, we need $\hat{\lambda}(D)$ and $\hat{\kappa}(D)$. Closed form results do not seem to exist. However, from the two partial derivatives set to zero,

$$\frac{\partial \log_e \mathcal{L}(D, \lambda, \kappa)}{\partial \lambda} \quad \text{and} \quad \frac{\partial \log_e \mathcal{L}(D, \lambda, \kappa)}{\partial \kappa}$$

we can derive the equations

$$\hat{\kappa}(D) = \frac{\bar{x}}{\hat{\lambda}(D)} + \bar{s} - 1 \qquad (6.38)$$

and

$$\hat{\lambda}(D) = \sqrt{\left\{ \frac{n\bar{x}}{2LD} \cdot \left[\frac{\bar{x}}{\hat{\lambda}(D)} + \bar{s} - 1 \right] \right\}} \qquad (6.39)$$

The function $\phi(D)$ can be written as

$$\phi(D) = 6n \cdot \left[\hat{\kappa}(D) - \bar{s} - \log_e \left[\left(\frac{D}{\hat{D}} \right)^{1/3} \right] - (\bar{s} - 1) \cdot \log_e \left[\left(\frac{\hat{\kappa}(D)}{\bar{s}} \right)^{1/3} \right] \right] \qquad (6.40)$$

To compute $\phi(D)$, choose a value of D, solve Equation 6.39 iteratively (easily done as $\hat{\lambda}(D)$ is a stable fixed point), compute $\hat{\kappa}(D)$, then compute Equation 6.40 (also using $\hat{D}$, which is closed form).

6.8.3 Full likelihood for point transects: simple examples

(a) *Negative exponential g(r), Poisson n* Results for point transects can be obtained for a couple of simple cases. Here we assume a Poisson distribution for n, a negative exponential detection function, $g(r) = \exp(-r/\lambda)$, and k randomly placed points. Basic theory then gives $E(n) = 2\pi k D\lambda^2$, and the pdf of detection distance r is

$$f(r) = \frac{r \cdot \exp\left\{ -\frac{r}{\lambda} \right\}}{\lambda^2} \qquad 0 < r, \ 0 < \lambda$$

The full likelihood is

$$\mathcal{L}(D, \lambda) = \frac{\exp\{ -2\pi k D\lambda^2 \} \cdot \{2\pi k D\lambda^2\}^n}{n!} \cdot \prod_{i=1}^{n} \left[\frac{r_i \cdot \exp\left\{ -\frac{r_i}{\lambda} \right\}}{\lambda^2} \right]$$

which simplifies to

$$\mathcal{L}(D,\lambda) = \exp\left\{-\left(2\pi k D\lambda^2 + \frac{T}{\lambda}\right)\right\} \cdot D^n$$

where $\quad T = \sum_{i=1}^{n} r_i$

The log likelihood is thus

$$\log_e \mathcal{L}(D, \lambda) = -2\pi k D\lambda^2 - \frac{T}{\lambda} + n \cdot \log_e(D)$$

Standard likelihood theory now leads to

$$\hat{\lambda} = \frac{T}{2n} = \frac{\bar{r}}{2} \qquad\qquad \text{var}(\hat{\lambda}) = \frac{\lambda^2}{2E(n)}$$

$$\hat{D} = \frac{n}{2\pi k \hat{\lambda}^2} = \frac{n \cdot \hat{h}(0)}{2\pi k} \qquad \text{var}(\hat{D}) = D^2\left[\frac{3}{E(n)}\right]$$

In order to find the profile likelihood, we solve

$$\frac{\partial \log_e \mathcal{L}(D, \lambda)}{\partial \lambda} = -4\pi k D\lambda + \frac{T}{\lambda^2} = 0$$

getting

$$\hat{\lambda}(D) = \left[\frac{T}{4\pi k D}\right]^{1/3}$$

Using the above to form $\log_e \mathcal{L}(D, \hat{\lambda}(D))$ allows us to find the expression for $\log_e \mathcal{L}(\hat{D}, \hat{\lambda})$, from which we construct a simple representation of $\phi(D)$:

$$\phi(D) = 6n \cdot \left[(D/\hat{D})^{1/3} - 1 - \log_e\{(D/\hat{D})^{1/3}\}\right] \tag{6.41}$$

(b) *Half-normal g(r), Poisson n* Instead of a negative exponential $g(r)$, let us assume $g(r)$ is half-normal; other assumptions are as in the above case. Now basic theory gives $E(n) = 2\pi k D\sigma^2$, and the pdf of detection distances is

$$f(r) = \frac{r \cdot \exp\left\{-\frac{1}{2}\left(\frac{r}{\sigma}\right)^2\right\}}{\sigma^2} \qquad 0 < r, \ 0 < \sigma$$

The full likelihood is

$$\mathcal{L}(D, \sigma) = \frac{\exp\{-2\pi k D\sigma^2\} \cdot \{2\pi k D\sigma^2\}^n}{n!} \cdot \prod_{i=1}^{n}\left[\frac{r_i \cdot \exp\left\{-\frac{1}{2}\left(\frac{r_i}{\sigma}\right)^2\right\}}{\sigma^2}\right]$$

which simplifies to

$$\mathcal{L}(D, \sigma) = \exp\left\{-\left(2\pi k D\sigma^2 + \frac{T}{2\sigma^2}\right)\right\} \cdot D^n$$

for T defined as the total, $T = \sum\limits_{i=1}^{n} r_i^2$

Standard likelihood theory now leads to

$$\hat{\sigma}^2 = \frac{T}{2n} \qquad\qquad \mathrm{var}(\hat{\sigma}^2) = \frac{\sigma^4}{E(n)}$$

$$\hat{D} = \frac{n}{2\pi k\hat{\sigma}^2} = \frac{n \cdot \hat{h}(0)}{2\pi k} \qquad\qquad \mathrm{var}(\hat{D}) = D^2 \cdot \left[\frac{2}{E(n)}\right]$$

To find the profile likelihood, we solve

$$\frac{\partial \log_e\mathcal{L}(D, \sigma)}{\partial\sigma} = -4\pi k D\sigma + \frac{T}{\sigma^3} = 0$$

getting

$$\hat{\sigma}(D) = \left[\frac{T}{4\pi k D}\right]^{1/4}$$

Carrying through the algebra and simplifications, we have

$$\phi(D) = 4n \cdot [(D/\hat{D})^{1/2} - 1 - \log_e \{(D/\hat{D})^{1/2}\}] \qquad (6.42)$$

Notice that Equation 6.42 is identical to 6.37; this we expected, because there is a duality in the mathematics between the case of line transects with a negative exponential detection function and point transects with a half-normal detection function, both with n distributed as Poisson and with $w = \infty$.

6.8.4 Some numerical confidence interval comparisons

We used the above results on $\phi(D)$ and $\text{var}(\hat{D})$ to compute a few illustrative numerical examples of profile, log-based and standard confidence intervals (nominal 95% coverage). To facilitate comparisons, what is presented are the ratios, (interval bound)$/\hat{D}$. Thus, the standard method yields relative bounds as $1 \pm 1.96 \, \text{cv}(\hat{D})$, and the log-based relative bounds are $1/C$ and C, where

$$C = \exp[1.96 \sqrt{\log_e\{1 + [\text{cv}(\hat{D})]^2\}}]$$

Some of our results are based on sample sizes that are smaller than would be justified for real data; our intent is to compare the three methods, and the differences are biggest at small n. The actual coverage of the intervals is not known to us; we take the profile likelihood intervals as the standard for comparison. Results are shown in Tables 6.2, 6.3, 6.4 and 6.5. One reason for the comparisons is to provide evidence that the log-based intervals are generally closer to the profile intervals.

Table 6.2 Some relative 95% confidence intervals, $\hat{D}_{lower}/\hat{D}$ and $\hat{D}_{upper}/\hat{D}$, for the profile, log-based and standard method, for line transects with a half-normal detection function, $w = \infty$, and Poisson distributed sample size n. Equation 6.36 is the basis of the profile interval; results are invariant to the true D and σ

n	Profile interval		log-based interval		Standard interval	
5	0.296	2.610	0.366	2.729	− 0.074	2.074
10	0.437	2.014	0.481	2.081	0.241	1.759
20	0.565	1.660	0.590	1.694	0.462	1.537
40	0.673	1.439	0.687	1.457	0.621	1.380
70	0.744	1.321	0.752	1.330	0.713	1.287
100	0.781	1.263	0.787	1.270	0.760	1.240

Table 6.3 Some relative 95% confidence intervals, $\hat{D}_{lower}/\hat{D}$ and $\hat{D}_{upper}/\hat{D}$, for the profile, log-based and standard method, for line transects with a negative exponential detection function, $w = \infty$, and Poisson distributed sample size n. Equation 6.37 is the basis of the profile interval; results are invariant to the true D and λ

n	Profile interval		log-based interval		Standard interval	
5	0.251	3.076	0.321	3.117	− 0.240	2.240
10	0.389	2.264	0.433	2.309	0.124	1.877
20	0.520	1.803	0.546	1.831	0.380	1.620
40	0.635	1.526	0.649	1.542	0.562	1.438
70	0.711	1.380	0.720	1.390	0.669	1.331
100	0.753	1.311	0.759	1.318	0.723	1.277

Table 6.4 Some relative 95% confidence intervals, $\hat{D}_{lower}/\hat{D}$ and $\hat{D}_{upper}/\hat{D}$, for the profile, log-based and standard method, for point transects with a negative exponential detection function, $w = \infty$, and Poisson distributed sample size n. Equation 6.41 is the basis of the profile interval; results are invariant to the true D and λ

n	Profile interval		log-based interval		Standard interval	
5	0.191	4.056	0.261	3.833	− 0.518	2.518
10	0.319	2.754	0.366	2.729	− 0.074	2.074
20	0.453	2.072	0.481	2.081	0.241	1.759
40	0.575	1.684	0.590	1.694	0.463	1.537
70	0.660	1.487	0.669	1.494	0.594	1.406
100	0.708	1.395	0.714	1.401	0.661	1.339

Table 6.5 Some relative 95% confidence intervals, $\hat{D}_{lower}/\hat{D}$ and $\hat{D}_{upper}/\hat{D}$, for the profile, log-based and standard method, for line transects with a negative exponential detection function (parameter λ), $w = \infty$, Poisson distributed sample size n, and cluster size as Poisson, mean κ. Equation 6.40 is the basis of the profile interval; results are invariant to true D and λ, but depend weakly on true κ; $\hat{\kappa} = 3.0$ was used for these results

n	Profile interval		log-based interval		Standard interval	
5	0.230	3.440	0.261	3.833	− 0.518	2.518
10	0.365	2.440	0.366	2.729	− 0.074	2.074
20	0.497	1.900	0.481	2.081	0.241	1.759
40	0.614	1.583	0.590	1.694	0.463	1.537
70	0.693	1.419	0.669	1.494	0.594	1.406
100	0.737	1.341	0.714	1.401	0.661	1.339

Table 6.2 corresponds to the line transect case in which $g(x)$ is half-normal and objects have a Poisson distribution. The relative interval endpoints in that table depend only upon sample size n, so these results are quite general under the assumed model. The invariance property of the ratios $\hat{D}_{lower}/\hat{D}$ and $\hat{D}_{upper}/\hat{D}$ applies also to Table 6.3 (line transect, negative exponential $g(x)$ and Poisson n) and Table 6.4 (point transect, negative exponential $g(x)$ and Poisson n). The log-based interval is slightly to be preferred to the standard method in Table 6.2, and more strongly preferred for the cases of Tables 6.3 and 6.4. The choice in Table 6.5 (line transect, negative exponential $g(x)$, Poisson n and Poisson s) is unclear. Note that the results in Table 6.3 for line transects with a negative exponential $g(x)$ are identical to results for the same values of n for point transects with a half-normal $g(r)$.

These sort of results on confidence intervals would be interesting to compute for other scenarios. We present the above specific formula for $\phi(D)$ to illustrate the ideas; in particular, the negative exponential $g(y)$ is used only because it is very easy to work with.

Table 6.5 reflects a case where the population of objects is clustered. The relative confidence intervals are for density of individuals. This is an interesting case because the log-based and standard relative confidence intervals do not depend upon D, λ or κ (because the relative intervals do not depend upon the specific values of $\bar{x}$ or $\bar{s}$). The relative profile intervals do not depend upon D or λ (thus the results in Table 6.5 are independent of the choice of D and λ), but they do depend weakly upon κ because the specific value of $\bar{s}$ (three in this example) affects even the relative profile intervals. Heuristically, this seems to be because the sample size of number of individual animals detected increases as $\bar{s}$ increases and the likelihood function uses this information. To illustrate this point, we give below relative profile interval endpoints (based on Equation 6.40 and 95% nominal coverage) for a few values of $\bar{s}$ at $n = 20$:

n	$\bar{s}$	Profile interval	
20	1.25	0.465	2.023
20	3.00	0.497	1.900
20	30.00	0.518	1.813
20	300.00	0.520	1.804

There is quite a noticeable effect here of average group size and this is an effect that is not found in either standard or log-based methods. We speculate that in realistic likelihood models, the profile interval would be generally more sensitive to information in the data than simpler confidence interval methods.

(a) *One more example* Consider line transect sampling, in which $g(x)$ is half-normal, clusters have a homogeneous Poisson distribution, and cluster size is a geometric random variable. Further assume that $g_0 < 1$, but that it can be estimated by an independent source of information, from which it is known that, of m clusters 'on' the line, z are detected. We assume that z is distributed as binomial (m, g_0). The counts, n, will be Poisson with mean $E(n) = 2LDg_0/\{\kappa f(0)\}$ where $\kappa = E(s)$ and D is density of individuals. The geometric distribution is used here in the form $\pi(s) = p^{s-1}(1 - p)$, hence $\kappa = 1/(1 - p)$. Also, $f(0) = \frac{1}{\sigma} \cdot \sqrt{(2/\pi)}$, so that we have

$$E(n) = \sqrt{2\pi}\,\sigma LDg_0(1 - p)$$

Maximum likelihood estimators are

$$\hat{\sigma} = \frac{\sum\limits_{i=1}^{n} x_i^2}{n} = \frac{T}{n}$$

$$\hat{D} = \frac{n \cdot \hat{f}(0) \cdot \bar{s}}{2L\hat{g}_0}$$

$$\hat{p} = \frac{\bar{s} - 1}{\bar{s}}$$

and

$$\hat{g}_0 = \frac{z}{m}$$

and the asymptotic estimated var($\hat{D}$) is

$$\widehat{\mathrm{var}}(\hat{D}) = \hat{D}^2 \cdot \left[\frac{1}{n} + \frac{1}{2n} + \frac{\hat{p}}{n} + \frac{1}{m} \cdot \left(\frac{1}{\hat{g}_0} - 1 \right) \right]$$

The full likelihood of the data entering into $\hat{D}$ is given by the products of the likelihoods of the independent data components:

257

$$\mathcal{L}(D, \sigma, p, g_0) = \frac{\exp\{-[\sqrt{2\pi}\sigma L D g_0(1-p)]\} \cdot \{\sqrt{2\pi}\sigma L D g_0(1-p)\}^n}{n!} \times$$

$$\prod_{i=1}^{n}\left[\left(\sqrt{\frac{2}{\pi\sigma^2}}\right) \cdot \exp\left\{-\frac{1}{2}\left(\frac{x_i}{\sigma}\right)^2\right\}\right] \times \prod_{i=1}^{n}[p^{(s_i-1)} \cdot (1-p)] \times \left[\binom{m}{z}(g_0)^z(1-g_0)^{m-z}\right]$$

Dropping constants and otherwise simplifying this likelihood gives

$$\mathcal{L}(D, \sigma, p, g_0) = \left[\exp\left\{-\left(\sqrt{2\pi}\sigma L D g_0(1-p) + \frac{T}{2\sigma^2}\right)\right\}\right] \times$$

$$[D^n \cdot p^{n(\bar{s}-1)}(1-p)^{2n}(g_0)^z(1-g_0)^{m-z}] \qquad (6.43)$$

Closed form expressions for $\hat{\sigma}(D)$, $\hat{p}(D)$ and $\hat{g}_0(D)$ do not seem possible, but $\phi(D)$ can be computed using numerical optimization. A slight 'trick' simplifies the process of getting $\phi(D)$.

By setting the partial derivatives of $\mathcal{L}$ with respect to σ, p and g_0 to zero, and with D arbitrary, we derived the following results:

$$p(\sigma) = \frac{\bar{s}-1}{\bar{s}+1-\dfrac{T}{n\sigma^2}}$$

$$g_0(\sigma) = \frac{n+z-\dfrac{T}{n\sigma^2}}{n+m-\dfrac{T}{n\sigma^2}}$$

and

$$D = \frac{T}{\sqrt{2\pi}\sigma^3 L g_0(\sigma) \cdot (1-p(\sigma))}$$

While we cannot easily select D and compute $\phi(D)$, we can specify values of σ and compute the unique associated $p(\sigma)$ and $g_0(\sigma)$ that apply for D, which is then computed. These are then the values of $\hat{\sigma}(D)$, $\hat{p}(D)$ and $\hat{g}_0(D)$ to use in computing $\phi(D)$ for that computed value of D. All we need do is select a range of σ which generates a range of D, and then we treat $\mathcal{L}(D, \sigma, p, g_0)$ as a function of D, not of σ. The MLEs are known, so the absolute maximum of $\mathcal{L}$ is known, thus normalizing $\mathcal{L}$ to ϕ is easy.

Table 6.6 Some relative 95% confidence intervals, $\hat{D}_{lower}/\hat{D}$ and $\hat{D}_{upper}/\hat{D}$, for the profile, log-based and standard method, for line transects for the likelihood in Equation 6.43. Sample size n is Poisson, x is half-normal, s is geometric, and g_0 is estimated from z ~binomial (m, g_0)

n	Profile interval		log-based interval		Standard interval	
5	0.238	4.009	0.289	3.457	− 0.376	2.376
10	0.365	2.688	0.395	2.535	0.015	1.985
20	0.488	2.054	0.501	1.996	0.287	1.713
40	0.593	1.721	0.595	1.680	0.472	1.528
70	0.663	1.568	0.658	1.521	0.576	1.424
100	0.700	1.503	0.690	1.449	0.626	1.374

Table 6.6 gives a few numerical results for the model considered here. Sample sizes n and m are the dominant factors influencing the Table 6.6 results. In fact, these results do not depend on true D, L or σ. However, they do depend on $E(s)$ and g_0 too strongly to draw broad conclusions here. Inputs to the likelihood used for Table 6.6 were $T/n = 1$ (so MLE $\hat{\sigma} = 1$), $\bar{s} = 10$ ($\hat{p} = 0.9$) and $z = 16$, $m = 20$($\hat{g}_0 = 0.8$). The log-based intervals are closer to the profile intervals than are the standard intervals.

It is also worth noting that if n (i.e. the line transect sampling effort) is increased while m is fixed, the estimate of g_0 is the weak link in the data. Studies that estimate g_0 need to balance the effort for the two data types. It would be best to collect data on g_0 during the actual distance sampling study to achieve both such balance of effort (with respect to n and m) and relevance of $\hat{g}_0$ to the particular study. If in this example for $n = 100$ we also put $z = 80$ and $m = 100$, then the three relative intervals are more similar, especially profile and log-based:

Profile interval		Log-based interval		Standard interval	
0.726	1.377	0.728	1.373	0.681	1.319

In practice, it is unlikely that such a high proportion of detections (80%) could be considered as 'on' the line, necessitating the use of methods that utilize detections off the line (Section 6.4).

(b) *A general comment on precision*   The relative confidence intervals in Tables 6.2–6.6 have been computed in a variety of cases: line and point transects, some with clustered populations, different detection functions, and one case with an adjustment for $g(0) < 1$. A general conclusion is that sample size has the overwhelming effect on relative precision of

$\hat{D}$. Relative confidence intervals are quite wide at $n = 40$, being roughly $\pm 45\%$ of $\hat{D}$. At $n = 70$ and 100, the intervals are roughly $\pm 35\%$ and $\pm 25\%$, respectively. This level of precision is under very idealized conditions that will not hold in practice for real data. With comparable sample sizes, we expect that relative interval widths will exceed the tabulated values. These results and our experience in distance sampling suggest strongly that reliable, precise abundance estimates from distance data require minimum sample sizes around 100. Coefficients of variation of around 20% (i.e. intervals of $\pm 40\%$) are often adequate for management purposes; the results presented here indicate minimum sample sizes of 40–70 in this circumstance.

6.8.5 Discussion

Reliable analysis of distance sampling data is possible without a full likelihood approach. We recommend a robust approach of empirical estimation of var(n), a semiparametric, likelihood-based estimation of $f(0)$ or $h(0)$ from the marginal distance data, and finally, estimation of $E(s)$ conditional on the observed distances y_i, using a regression approach. Other strategies are possible and use of a bivariate model for $g(y, s)$ is closer to a full likelihood approach. The difficult modelling aspect is to specify general probability models for n and s and it is those steps we bypass.

There are, however, reasons to develop a full likelihood approach: (1) intellectual curiosity, (2) efficiency of estimators and tests if the assumed model is correct, (3) availability of well-developed likelihood based theory for profile likelihood intervals and for model selection such as AIC, (4) the convenience of further developing such models by having parameters as functions of covariates (effort, environmental, spatial factors), and (5) as a necessary part of a Bayesian approach to distance sampling. We consider some of these points and difficulties of the approach.

Models for the marginal function $g(y)$ are abundant and choices also exist for bivariate versions, $g(y, s)$ (e.g. Quinn and Gallucci 1980; Drummer 1985, 1990; Drummer and McDonald 1987; Thompson and Ramsey 1987; Otto and Pollock 1990), and for the distribution $\pi(s)$ of cluster size in the sampled population. Any probability model of a discrete random variable on $s = 1, 2, \ldots$ is a candidate for $\pi(s)$, and if s can take on hundreds of values (such as for dolphin schools), continuous models could be used (such as a log-normal distribution for s). Good probability models for n are more problematic.

The Poisson distribution for n is not reasonable. The negative binomial model might be tenable, but in general, a reasonable model for

$\Pr(n)$ may need more than two parameters. The negative binomial is given by

$$\Pr(n) = \frac{\Gamma(\theta + n)}{\Gamma(\theta) \cdot \Gamma(n + 1)} \cdot (1 - \tau)^n \cdot \tau^\theta, \qquad 0 < \theta, 0 < \tau < 1, 0 \leqslant n$$

which has

$$E(n) = \theta \cdot \frac{1 - \tau}{\tau}$$

$$\mathrm{var}(n) = \frac{E(n)}{\tau}$$

$\Pr(n)$ is not used as parameterized, but rather the relationship $E(n) = 2LD/f(0)$ (line transects), or $E(n) = 2\pi k D/h(0)$, must be imposed on the parameters in the distribution. With a multiparameter $\Pr(n)$, such as the negative binomial, there is no obvious unique way to reparameterize $\Pr(n)$. We suggest it will instead be necessary to optimize the log-likelihood function subject, for example, to the constraint $E(n) = 2LD/f(0)$, where $f(0)$ is replaced by its form as a function of the parameters in the detection function $g(x)$. In some cases, it might be meaningful to associate one parameter in $\Pr(n)$ with $E(n)$, such as having τ a free parameter, and setting

$$\theta = \frac{2LD\tau}{f(0) \cdot (1 - \tau)}$$

There are other generalized distributions possible for n, see for example Johnson and Kotz (1969).

Constructing the full likelihood in the general case is complicated, but not fundamentally difficult, if strong assumptions of independence are made. These independence assumptions are often not reasonable, but robust variances can be found by appropriate quasi-likelihood or bootstrap methods. Under independence and k replicate lines, the probability model for the data (from which the likelihood is derived) is symbolically

$$\Pr(n_i, x_{ij}, s_{ij}, j = 1, \ldots, n_i, i = 1, \ldots, k)$$

$$= \prod_{i=1}^{k} \left[\Pr(n_i | l_i) \cdot \left\{ \prod_{j=1}^{n_i} f(x_{ij} | s_{ij}) \cdot \pi^*(s_{ij}) \right\} \right] \tag{6.44}$$

where

$$\pi^*(s) = \frac{f(0)}{f(0|s)} \cdot \pi(s)$$

is the distribution of cluster sizes given the cluster is detected and $\pi(s)$ is the probability distribution in the entire population. It is $\pi(s)$ we suggest modelling. For point transects, we have k points and

$$\Pr(n_i, r_{ij}, s_{ij}, j = 1, \ldots, n_i, i = 1, \ldots, k)$$

$$= \prod_{i=1}^{k} \left[\Pr(n_i | \text{point } i) \cdot \left\{ \prod_{j=1}^{n_i} f(r_{ij} | s_{ij}) \cdot \pi^*(s_{ij}) \right\} \right] \qquad (6.45)$$

where

$$\pi^*(s) = \frac{h(0)}{h(0|s)} \cdot \pi(s)$$

More on the theory of $f(x|s)$, $f(r|s)$ and $\pi^*(s)$ is given in Section 6.7 along with the explanations of $f(0)$, $h(0)$, $f(0|s)$ and $h(0|s)$ and their relationship to $g(\cdot|s)$.

Theory development in capture–recapture is in some ways more advanced than in distance sampling; capture–recapture is also in some ways a simpler statistical problem. The state of the statistical art in capture–recapture for survival estimation is represented by Lebreton *et al.* (1992), in which inference is based on (full) likelihood models, and model selection is based on Akaike's Information Criterion (AIC; Akaike 1985). We have made use of AIC in model selection for the marginal detection function, but only for that model component. Drummer (1991) uses AIC in a bivariate detection function, $g(x, s)$. The full likelihood approach to capture–recapture is very powerful. Also, using such explicit parametric models allows meaningful modelling of embedded parameters as functions of auxiliary information. These approaches could be similarly useful in distance sample and deserve to be explored.

Survival analysis in capture–recapture deals with only two classes of parameters: survival rates and capture rates. A fully parametric approach to distance sampling would deal with D, the parameters in $\Pr(n)$ (say $\underline{\theta}$), in $g(\cdot)$ (say $\underline{\beta}$), and in $\pi(s)$ (say $\underline{\gamma}$). Moreover, if the locational information in spatial coordinates of detected objects is used, then D is in effect expanded into a fourth class of parameters. Thus in its most general form, distance sampling deals with more classes of parameters than capture–recapture and is in that sense a harder problem.

262

If Bayesian methods are to be fully developed for distance sampling, they will require likelihoods as in Equations 6.44 and 6.45 to augment to priors on the parameters. In either case of a full likelihood or a Bayesian approach, there will be a need for numerical optimization and integration methods, possibly on objective functions with a dozen, or many more, parameters. Even 50 or 100 parameters are not too many for numerical optimization methods (MLEs and profile likelihoods), so this scope of problem is numerically feasible now.

6.9 Distance sampling in three dimensions

Conceptually, line transects can be considered as one-dimensional distance sampling, because only distances perpendicular to the line of travel are used, even though objects are distributed in two dimensions. Point transects sample distances in those two dimensions because radial detection distances are taken at any angle in what could be represented as an x–y coordinate system. In principle, distance sampling can be conducted in three dimensions, such as underwater for fish, or in space for asteroids, where objects can be located anywhere in three dimensions relative to the observer. The observer might traverse a 'line', and record detection distance in two dimensions perpendicular to the line of travel, or remain at a point, recording data in three-dimensions within the sphere centred at the point. Given the assumption of random line or point placement with respect to the three dimensional distribution of objects, the mathematical theory is easily developed for the three-dimensional case. In practice, the third dimension may pose a problem: there may only be a thin layer in three dimensions, and in the vertical dimension, objects may exhibit strong density gradients (e.g. birds in a forest canopy, or fish near the sea surface). Operational problems could be difficult; we do not claim this extension to three dimensions has application, but it is interesting to consider.

Assume that we follow a line randomly placed in three dimensions. Now we sample volume, not area, so D = objects/unit volume; line length is still L. Assume we record distances r for all objects detected out to perpendicular distance w. Counting takes place in a cylinder of volume $v = \pi w^2 L$, rather than a strip of area $a = 2wL$. The statistical theory at a fixed slice through the cylinder perpendicular to the line of travel is just like point transect theory. This sort of sampling (i.e. a 'tube transect') is like 'pushing' the point transect sampling point a distance L in three dimensions.

Aside from volume v replacing area a, we need little new notation: n is the sample size of objects detected in the sampled cylinder of radius

w; p_v is the average probability of detecting an object in the cylinder of volume v; $g(r)$ = probability of detecting an object that is at perpendicular distance r, $0 \leqslant r \leqslant w$; D is the true density of objects in the study space.

A basic starting place to develop theory is the formula

$$E(n) = D \cdot v \cdot p_v = D \cdot \pi w^2 \cdot L \cdot p_v$$

The unconditional detection probability is easily written down because objects are, by assumption, uniformly distributed in space within the cylinder. Therefore, the pdf of the radial distance r for a randomly specified object (before the detection process) is

$$u(r) = \frac{2\pi r}{\pi w^2}$$

The unconditional detection probability is $p_v = E[g(r)]$, where expectation is with respect to pdf $u(r)$. This is a weighted average of $g(r)$:

$$p_v = \int_0^w u(r) \cdot g(r) \, dr = \frac{1}{\pi w^2} \int_0^w 2\pi r \cdot g(r) \, dr$$

Notice that this p_v is identical to the unconditional detection probability in point transects.

A direct approach can be used to derive $E(n)$. Let v_ε be a small volume in the cylinder centred at distance r and position l along the line ($0 \leqslant l \leqslant L$). Thus $D \cdot v_\varepsilon$ = the expected number of objects in volume $v_\varepsilon = 2\pi r \, dr \, dl$, and the expected count of these objects is then $g(r) \cdot D \cdot v_\varepsilon$. $E(n)$ can now be expressed as

$$E(n) = \int_0^L \int_0^w g(r) \cdot D \cdot 2\pi r \, dr \, dl = L \cdot D \cdot \int_0^w 2\pi r \cdot g(r) \, dr = L \cdot D \cdot \pi w^2 \cdot p_v$$

An estimator of D is

$$\hat{D} = \frac{n}{\pi w^2 \cdot L \cdot \hat{p}_v} \quad \text{or} \quad \hat{D} = \frac{n}{L \cdot \hat{\mu}_w}$$

where

$$\mu_w = \pi w^2 \cdot p_v = \int_0^w 2\pi r \cdot g(r) \, dr$$

The sample of distances to detected objects is $r_1, \ldots, r_n$. The pdf of distance r to detected objects is

$$f(r) = \frac{2\pi r \cdot g(r)}{\mu_w} = \frac{r \cdot g(r)}{\int_0^w r \cdot g(r) \, dr}$$

This result is identical to that for point transects and can be proven using the same theory. In fact, slight modifications of point transect theory suffice as a complete theory for line transect sampling in three dimensions. In particular,

$$f'(r) = \frac{2\pi \cdot g(r)}{\mu_w} + \frac{2\pi r \cdot g'(r)}{\mu_w}$$

so if $g'(0)$ is finite and $g(0) = 1$, then $f'(0) = 2\pi/\mu_w$. For consistency with point transects, we use

$$h(0) = f'(0) = \frac{2\pi}{\mu_w}$$

and hence we have

$$\hat{D} = \frac{n \cdot \hat{h}(0)}{2\pi L}$$

Compare this with the point transect estimator,

$$\hat{D} = \frac{n \cdot \hat{h}(0)}{2\pi k}$$

The only difference is that L replaces k.

In fact, all the theory for point transects applies to line transect sampling in three dimensions if we replace k by L. Thus, estimation of $h(0)$ or p_v could be done using program DISTANCE and treating the detection distances, r_i, as point transect data. The case of objects as clusters poses no additional problems, giving our general formulation:

$$\hat{D} = \frac{n \cdot \hat{h}(0)}{2\pi L c \hat{g}_0} \cdot \hat{E}(s)$$

where c is the fraction of the circle, around the line, in which detections are recorded ($c = \phi/2\pi$ for some sector angle ϕ in 0 to 2π). For the clustered case, all the theory in Section 3.6.6 for point transects applies to line transects in three dimensions with k replaced by L. We do not know of any data for three-dimensional line transects as described here; however, if any such studies are ever done, we note that a complete theory for their analysis already exists.

Point transect sampling in two dimensions can be extended to three dimensions. (To people who use the term variable circular plots, such extension becomes a variable spherical plot.) Now the detection distances r are embedded in a three-dimensional coordinate system. There is no existing theory for this type of distance sampling, although theory derivation methods used for line and point transects are easily adapted to this new problem, and we present some results here.

In this case, the observer would be at a random point and record detections in a full (or partial) sphere around that point. For a sphere of radius w, the volume enclosed about the point is

$$v = \frac{4}{3}\pi w^3$$

Given truncation of the data collection process at distance w, the expected sample size of detections at k random points is

$$E(n) = k \cdot D \cdot v \cdot p_v$$

To derive p_v, we note that the pdf of radial distance for a randomly selected object in the sphere is

$$u(r) = \frac{4\pi r^2}{4\pi w^3/3}$$

and $p_v = E[g(r)]$ with respect to $u(r)$

$$p_v = \int_0^w u(r) \cdot g(r) \, dr = \frac{1}{4\pi w^3/3} \int_0^w 4\pi r^2 \cdot g(r) \, dr$$

so that

$$E(n) = k \cdot D \cdot \frac{4}{3} \pi w^3 \cdot p_v = k' \cdot D \cdot \int_0^w 4\pi r^2 \cdot g(r) \, dr$$

An alternative derivation is to consider that the volume, v_ε, of space in the shell at distances r to $r + dr$ is $4\pi r^2 dr$ (to a first order approximation, which is all we need as we let $dr \to 0$). Thus,

$$E(n) = k \cdot D \cdot \int_0^w g(r) \cdot v_\varepsilon \, dr = k \cdot D \cdot \int_0^w 4\pi r^2 \cdot g(r) \, dr$$

Now define μ_w as

$$\mu_w = \int_0^w 4\pi r^2 \cdot g(r) \, dr$$

so that

$$\hat{D} = \frac{n}{k \cdot \hat{\mu}_w} \equiv \frac{n}{k \cdot (4\pi w^3 / 3) \cdot \hat{p}_v}$$

The pdf of detection distance r is

$$f(r) = \frac{4\pi r^2 \cdot g(r)}{\mu_w}, \qquad 0 < r < w$$

Taking second derivatives, we get

$$f''(r) = \frac{8\pi \cdot g(r)}{\mu_w} + \frac{16\pi r \cdot g'(r)}{\mu_w} + \frac{4\pi r^2 \cdot g''(r)}{\mu_w}$$

Hence, if $g(0) = 1$ and both $g'(0)$ and $g''(0)$ are finite (preferably zero as then the estimators have better properties), then

$$f''(0) = \frac{8\pi}{\mu_w}$$

For simplicity of notation, we define $d(0) = f''(0)$, so that

$$\hat{D} = \frac{n \cdot \hat{d}(0)}{8\pi k}$$

267

The estimation problem reduces to fitting a pdf $f(r)$, as given above, to the detection distances $r_1, \ldots, r_n$ based on some model for the detection function, $g(r)$. This will lead to $\hat{d}(0)$ and $\widehat{\text{var}}\{\hat{d}(0)\}$ by any of a variety of statistical methods. Because the variance of $\hat{d}(0)$ is conditional on n,

$$\widehat{\text{var}}(\hat{D}) = \hat{D}^2 \left[[\text{cv}(n)]^2 + [\text{cv}\{\hat{d}(0)\}]^2 \right]$$

As with point transects in two dimensions, the theory for three dimensions can be transformed to look like line transect theory in one dimension. The transform is from radial distance r to the volume sampled, $\eta = \frac{4}{3}\pi r^3$, giving the pdf of η as

$$f(\eta) = \frac{g(\xi)}{\mu_w}, \qquad 0 < \eta < v = \frac{4}{3}\pi w^3$$

where

$$\xi = \left(\frac{3v}{4\pi} \right)^{1/3}$$

Then, if $g(0) = 1$, $f(0) = 1/\mu_w$. As for two-dimensional point transects, we do not recommend that analysis be based on such a transformation (c.f. Buckland 1987a).

The case of objects as clusters with size-biased detection can be developed for this three-dimensional point transect sampling using the methods of Section 3.6.6. First, we would have a conditional detection function, $g(r|s)$, and a distribution of cluster sizes in the entire population, $\pi(s)$. The following result holds for each cluster size:

$$D(s) = \frac{E[n(s)] \cdot d(0|s)}{8\pi k} \qquad 1 \leqslant s$$

The density of clusters irrespective of size is

$$D = \frac{E(n) \cdot d(0)}{8\pi k}$$

Thus, 'dividing' the first by the second of these two formulae, we get

$$\frac{D(s)}{D} = \pi(s) = \frac{E[n(s)]}{E(n)} \cdot \frac{d(0|s)}{d(0)} = \pi^*(s) \cdot \frac{d(0|s)}{d(0)}$$

where $\pi^*(s)$ is the distribution of detected cluster sizes. Summing both sides of the above leads to

$$d(0) = \sum \pi^*(s) \cdot d(0|s)$$

whereas rearranging the formula and summing produces

$$d(0) = \frac{1}{\sum \dfrac{\pi(s)}{d(0|s)}}$$

where all summations are over $s = 1, 2, 3, \ldots$.
   Thus

$$\pi(s) = \frac{\pi^*(s) \cdot d(0|s)}{\sum \pi^*(s) \cdot d(0|s)}$$

and

$$E(s) = \frac{\sum s \cdot \pi^*(s) \cdot d(0|s)}{\sum \pi^*(s) \cdot d(0|s)}$$

from which expressions, estimators of $\pi(s)$ and $E(s)$ are evident.
   Straightforward expressions for $d(0)$ and $d(0|s)$ are

$$d(0) = \frac{2}{\displaystyle\int_0^w r^2 \cdot g(r)\, dr}$$

$$d(0|s) = \frac{2}{\displaystyle\int_0^w r^2 \cdot g(r|s)\, dr}$$

Two more formulae are just stated here:

$$g(r) = \sum g(r|s) \cdot \pi(s)$$

$$f(r|s) = \frac{r^2 \cdot g(r|s)}{\displaystyle\int_0^w r^2 \cdot g(r|s)\, dr}$$

269

Also of interest are conditional distributions of cluster size given detection distance, r. These distributions are useful for exploring $E(s|r)$, where now the s is from the size-biased detected sample. The result is

$$\pi^*(s|r) = \frac{g(r|s) \cdot \pi(s)}{\sum g(r|s) \cdot \pi(s)} \equiv \frac{g(r|s) \cdot \pi(s)}{g(r)}$$

This is exactly the same as for either line or point transect results given in Section 3.6.6.

Perhaps some day three-dimensional point transect data will be taken in deep space or oceans.

6.10 Cue counting

Cue counting (Hiby 1982, 1985; Hiby and Hammond 1989) is a method developed for estimating whale numbers that has very similar design considerations as line transect sampling – and in fact is sometimes carried out simultaneously with line transect sampling – yet theoretically is much more closely related to point transects. An observer scans a sector ahead of the viewing platform – usually an airplane or the bow of a ship – and records the distance to each detected cue. The cue is usually defined to be a whale blow. Cues are recorded irrespective of whether the whale was previously detected, and it is not necessary to estimate school (cluster) size. The method yields estimates of cue density, which can only be converted into whale density by estimating the cue or blow rate, ρ, from separate surveys. Cue density is estimated much as bird density is estimated from point transect data. The observer records only radial distances. Perpendicular distances are not needed, and angles only determine whether a cue is within or outside the observation sector. To estimate cue rate ρ, individual whales are followed, and the observed rate is used as an estimate of the cue rate for the whole population. This is the main weakness of the approach, as relatively few whales can be monitored for sufficiently long periods to obtain reasonable cue rate estimates. Further, these whales may not exhibit typical cue rates; for example whales with high cue rates are less likely to be 'lost' before an estimate can be obtained, and whales monitored over a long time period may change their cue rate in response to the vessel.

Suppose cues are recorded out to a distance w. For cue counting, as for point transect sampling, the area of a ring of incremental width δr at distance r from the observer is proportional to r. It follows that $f(r) = 2\pi rg(r)/v$, where $v = 2\pi\int_0^w rg(r)\,dr$. Given that a cue occurs in the sector of area $c \cdot a$, where ϕ is the sector angle and $a = \pi w^2$, so that

$c = \phi/2\pi$, let the probability that it is seen be P_a. Then this probability is $v/(\pi w^2)$. Thus $a \cdot P_a = v$, which holds as $w \to \infty$. Assuming all cues very close to the observer are seen ($g_0 = 1$), Equation 3.1, with $E(s) = 1$, yields the following estimate of cue 'density' per unit time (i.e. the number of cues per unit area per unit time):

$$\hat{D}_c = \frac{2\pi n}{\phi \hat{v} T}$$

where n is the number of cues recorded in time T. The constant T is the total time that the observer is searching (i.e. 'on effort'), and corresponds to the line transect parameter, L. If the cue rate is estimated as $\hat{\rho}$ cues per unit time per animal, then estimated whale density is

$$\hat{D} = \frac{2\pi n}{\phi \hat{v} T \hat{\rho}}$$

As for point transects, $\hat{v} = 2\pi/\hat{h}(0)$, where $h(0) = \lim_{r \to 0} f(r)/r$, so that

$$\hat{D} = \frac{n \cdot \hat{h}(0)}{\phi T \hat{\rho}}$$

The value of $\hat{h}(0)$ may be obtained by modelling the recorded distances to cues, as if they were distances from a point transect survey. DISTANCE has a cue count option to carry out the above analysis (below). Because successive cues from the same whale, or cues from more than one whale in a pod, may be counted, the distances are not independent observations. This does not invalidate the method, but analytic variances should not be used. The bootstrap, applied by taking say cruise legs as the sampling unit, provides valid variance estimation.

Line transect sampling of whale populations is beset with problems of how to estimate g_0, especially for aerial surveys, where a whale may be below the surface while it is in range of the observer, and for species such as sperm whales, which typically dive for around 40 minutes at a time. Cue counting does not require that all whales on the centreline are detected. Instead, it assumes that all cues occurring immediately ahead of the observer are seen. Thus, of those on the centreline, only whales that are at the surface when the vessel passes are assumed to be detected with certainty. In practice, whales may show vessel avoidance, so that the recorded number of cues very close to the vessel is depressed. Because the area surveyed close to the vessel is small, the effect of vessel avoidance might be expected to be small, unless avoidance occurs at

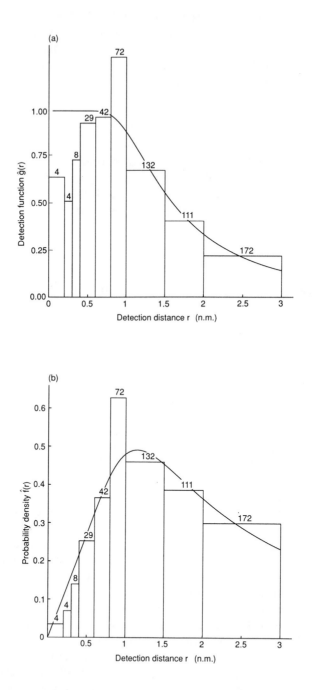

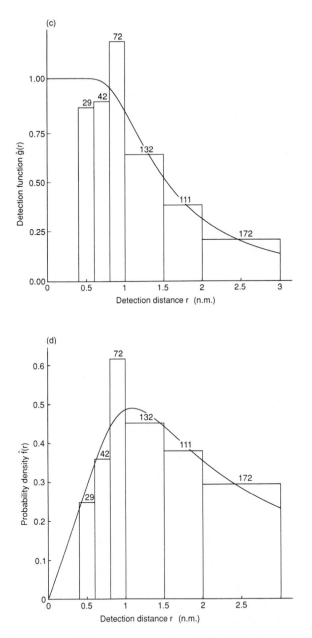

Fig. 6.7. Histograms of the cue count data. Also shown are the fits of the hazard-rate model to the data without left-truncation (a and b) and with left-truncation (c and d). The fitted detection functions are shown in (a) and (c) and the corresponding density functions in (b) and (d).

relatively large distances. If avoidance is suspected, the distance data may be left-truncated. This solution should prove satisfactory provided the effects of vessel avoidance only occur well within the maximum distance for which the probability of detecting a cue is close to unity.

If cues immediately ahead of the vessel might be missed, double-counting methods similar to the independent observer analyses given in Section 6.4 (with $d = 0$) may be used. This has the advantage over those analyses in that it is easier to identify whether a single cue is seen from both platforms, for example by recording exact times of cues, than to identify whether a single animal or animal cluster is seen by both platforms, since the two platforms may see different cues from the same animal.

Cue counting has been used in aerial surveys to estimate fin whale densities near Iceland (Hiby *et al.* 1984) and in shipboard surveys to estimate whale densities in the North Atlantic (Hiby *et al.* 1989) and minke whale densities in the Antarctic (Hiby and Ward 1986a, b; Ward and Hiby 1987). We use here data from Hiby and Ward (1986a) to illustrate the method. Annual surveys of Southern Hemisphere minke whales have been carried out since the 1978–79 season. The first attempt to use cue counting during shipboard surveys occurred on the 1984–85 cruise. Hiby and Ward considered that cues close to the vessel were under-represented, possibly because whales showed vessel avoidance behaviour or because blows close to the vessel were under-recorded by observers. We therefore analysed the data both with no left-truncation and with left-truncation at 0.4 n.m. (nautical mile). The data were right-truncated at 3 n.m. Under the hazard-rate model, frequencies at distances less than 0.4 n.m. are not significantly below expected frequencies, and truncation makes little difference; the only anomaly is the relatively high frequency at 0.8–1.0 n.m. (Fig. 6.7), which may be chance fluctuation, or, more likely, preferential rounding to that distance interval. Hiby and Ward (1986a) appear to have interpreted these data too pessimistically, suggesting that detections close to the vessel are too few because (1) blows are less visible at short distances, (2) whales show vessel avoidance behaviour, or (3) observers did not appreciate the need to record all cues at short distances. Because successive cues are not independent, goodness of fit tests are likely to give spurious significant results. If they are carried out regardless for the hazard-rate model, they are not significant at the 5% level, so Hiby and Ward's conclusion that the data cannot be analysed seems pessimistic. Data sets collected more recently suggest that the method performs adequately.

The fits of the hazard-rate model to the data both with and without left-truncation are shown in Fig. 6.7. In these trials, both blows and sightings of the body of the whale were counted as cues. Hiby and Ward (1986a) estimated the cue rate at 34.98 cues per whale per hour

($\widehat{se} = 4.74$). Supplying this estimate to DISTANCE, together with an estimate of time on effort of 35.8 hours (430 n.m. divided by an average speed of around 12 knots), yields an estimated density of 0.24 whales/n.m.^{2} from untruncated data and 0.26 whales/n.m.^{2} from the truncated data. The goodness of fit statistics are $\chi_6^2 = 11.7$ and $\chi_3^2 = 7.3$ respectively. The p-values for the goodness of fit tests are invalidated by the lack of independence between successive cues from the same animal or animal cluster. Similarly, the analytic estimates of variance are invalid. Without the raw data, it is not possible to apply either the bootstrap or the empirical method to obtain valid variance estimates, because cue counts are not given by cruise leg in Hiby and Ward. In Fig. 6.8, the fits of the Fourier series model to these data, with and without left-truncation, are shown. It yields an estimated density of 0.24 whales/n.m.^{2} without truncation and 0.31 whales/n.m.^{2} with truncation, with respective goodness of fit statistics of $\chi_7^2 = 18.9$ and $\chi_3^2 = 9.1$, indicating a worse fit than the hazard-rate model. Again the p-values corresponding to these statistics are invalid, and we do not present them. The flatter shoulder of the hazard-rate model enables it to fit the counts at short distances more closely. The estimate of density from a line transect survey carried out at the same time as the cue rate trial was 0.37 whales/n.m.^{2}.

6.11 Trapping webs

The estimation of population size (N) from capture data is usually formulated as a capture–recapture problem (e.g. White *et al.* 1982). There, traps are positioned, often at intersections of a rectangular grid, and animals are captured, marked, and released for possible recapture on a subsequent trapping occasion. If the trapping grid is enclosed or the trapped area samples the entire area of interest, then density = number/area can be estimated. However, the usual case is that an area surrounding the trapping grid contains animals that are subject to being captured and thus the effective area being sampled is larger than the area of the grid. One might naïvely estimate density as $\hat{N}/A_g$, where A_g = the area covered by the trapping grid. However, density is then overestimated because grid area is smaller than the area actually sampled by the traps. This problem has been well known for over half a century (Dice 1938). The use of a trapping web (Anderson *et al.* 1983) is an attempt to reformulate the density estimation problem into a distance sampling framework, where density is estimated directly, rather than separately estimating population size and effective area (but see Wilson and Anderson 1985a for an alternative).

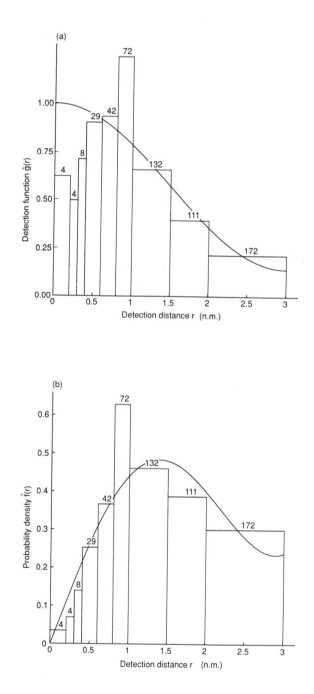

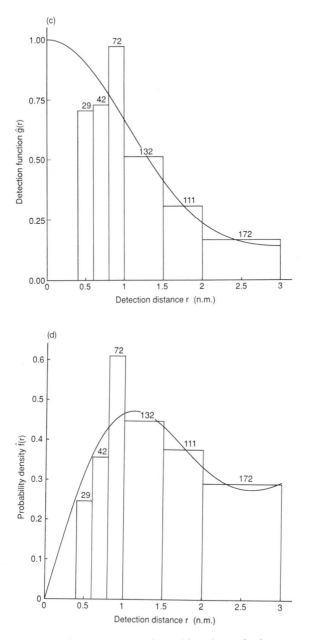

Fig. 6.8. Histograms of the cue count data. Also shown is the one-term Fourier series fit to the data with no left-truncation (a and b) and the two-term fit to the left-truncated data (c and d). The fitted detection functions are shown in (a) and (c) and the corresponding density functions in (b) and (d).

277

Trapping webs are a special case of point transect theory useful in estimating density of animal populations where 'detection' is accomplished by trapping. Animals are trapped in live, snap-trap or pitfall traps. Mist nets or other devices can be employed. Such trap devices are placed in a 'web' such that the density of traps is highest near the centre (Fig. 6.9), thus attempting to assure that $g(0) = 1$.

The web design consists of m lines of equal length, α_T, and each of T traps, radiating from randomly chosen points. A useful rule of thumb is to ensure that $m \times T \geq 200$. The traps are located along each of the m lines, usually (but not necessarily) at some fixed distance interval θ, starting at distance $\alpha_1 = \theta/2$. Points b_i are defined along each line, halfway between traps, for $i = 1, 2, \ldots, T$, with $b_0 = 0$ representing the web centre, and b_T the boundary of the web beyond the last trap. The traps are then at distances $\alpha_i = \theta(i - 0.5)$ for $i = 1, 2, \ldots, T$, and the b_i are at distances $i\theta$ for $i = 0, 1, 2, \ldots, T$.

Thus, traps are placed in rings of increasing radius from the web centre at equal distances along the m lines (Fig. 6.9). All captures in the ith ring of traps are considered to be detections of objects at distance α_i from the centre of the web. The distance data are analysed as grouped data. That is, the total number of captures arising from the ring of traps at distance α_i are treated as grouped data over the interval from distance b_{i-1} to b_i. The total area of the web out to interval i is $c_i = \pi b_i^2$ and the area trapped by the ith ring of traps is then $\Delta_i = c_i - c_{i-1}$. Generally, only first captures (removal data) are recorded and used in the estimation of density. This procedure reduces the impact on estimation of heterogeneity in trap response due to trap-happy or trap-shy animals.

Traps can be placed easily by trained technicians using a stake driven in the ground at the web centre and a rope with knots tied to indicate trap spacing (the α_i). Disturbance of the site should be minimized while traps are being placed in the sampled area. Several trapping webs would be required to sample an area of interest adequately. If only initial captures are of interest, then captured animals can be given a batch mark, to indicate that they have been 'removed', and released back into the population. Sampling is carried out on t occasions (often consecutive days or nights), where typically t is between three and eight.

6.11.1 Assumptions

Analytic theory for the trapping web is an application of point transect sampling theory and the general assumptions apply. The three major assumptions of distance sampling are slightly restated here for the trapping web:

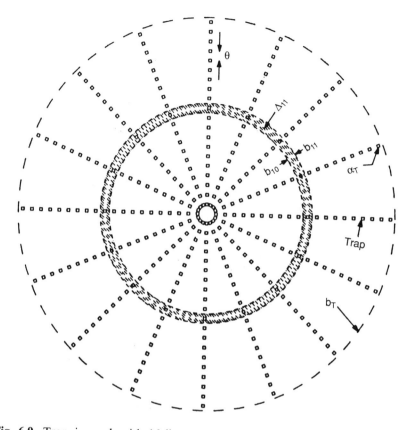

Fig. 6.9. Trapping web with 16 lines ($m = 16$), each of equal length α_T and 20 traps per line ($T = 20$), giving in total 320 traps. The traps are at distances $\alpha_1, \ldots, \alpha_T$ from the centre of the web. The points along each line, halfway between traps, are denoted by b_i, $i = 0, \ldots, T$, where $b_0 = 0$ is the centre and b_T is the boundary of the web, just beyond the last trap. Captures in the eleventh ring of traps are assigned to the shaded ring Δ_{11}, which has area $c_{11} - c_{10}$, where $c_i = \pi \cdot b_i^2$.

1. All animals at the centre of the web are captured at least once during the t occasions. That is, trapping continues until evidence exists that no new animals are being caught near the centre of the web.
2. During the trapping period, animals move over distances that are small relative to the size of each web. Thus, migration through the web is not allowed. Trap spacing is an important consideration and is species-dependent, taking account of the size of home ranges or 'territories'.
3. Distances from the centre of the web to each trap are measured accurately. This assumption is trivial if the trap spacing has been carefully laid out.

Assumption 1 is critical but can be monitored by examining the number of new individuals trapped near the web centre over trapping occasions. Animals near the centre should be captured with probability one. However, if substantial movement occurs over the t occasions (assumption 2), animals that are initially away from the centre of a web may move, eventually to be caught where the trap density is highest. This situation is analogous to point transect sampling where the observation period is long and birds move around the study area. Such movement causes detections near the point to increase and leads to positive bias in the estimator. Bias is worse if animals are attracted to the point or web centre.

6.11.2 Estimation of density

The basic data are the number of first captures in traps in ring j of web i on trapping occasion l, n_{ijl}, where $i = 1, 2, \ldots, k, j = 1, 2, \ldots, T$ and $l = 1, 2, \ldots, t$. Pooling the data over t occasions, the data can be summarized as n_{ij}, where

$$n_{ij} = n_{ij1} + n_{ij2} + \ldots + n_{ijt}$$

Hence n_{ij} is the number of animals trapped in the jth ring of the ith web. Let the total sample size be $n = \sum_i \sum_j n_{ij}$. Then density can be estimated by

$$\hat{D} = \frac{n \cdot \hat{h}(0)}{2\pi k}$$

where the estimate $\hat{h}(0)$ is obtained through standard point transect methods (Chapter 5). The estimator of the sampling variance is

$$\widehat{\text{var}}(\hat{D}) = \hat{D}^2 \cdot [\{\text{cv}(n)\}^2 + \{\text{cv}[\hat{h}(0)]\}^2]$$

If the population is distributed randomly (i.e. Poisson), then Wilson and Anderson recommend $[\text{cv}(n)]^2 = 1/n$. Generally, some degree of spatial aggregation can be expected, and $[\text{cv}(n)]^2 = 2/n$ or $3/n$ might then be more appropriate. If the number of replicate webs, k, is sufficient, it is preferable to estimate the sampling variance of n empirically (Section 3.7.2).

Data analysis is similar to the general theory for point transects, including model selection and inference issues. The challenge with the trapping web is to collect trapping data that mimic the assumptions of point transect sampling and analysis theory. In particular, trap spacing must be related to average home range size or average distance moved and there are presently few guidelines for this decision.

Excessive animal movement near the web centre is problematic. The density of traps near the web centre is high relative to that near the edge of the web. Thus, even if animal movement is random, there is a tendency to trap animals near the web centre, regardless of their original location. If the trap spacing is too small, the problem is made worse and overestimation will likely result. If the animals tend to move in home ranges that are small relative to the size of the web and the trap spacing, then the trapping web may perform well. Alternatively, if animals move somewhat randomly over wide areas in relation to the size of the web and the trap interval chosen, then overestimation may be substantial (see the darkling beetle example, below).

6.11.3 Monte Carlo simulations

Wilson and Anderson (1985b) performed a Monte Carlo study to investigate the robustness of density estimation from trapping web data. Their simulations mimicked small mammal populations whose members were allowed to move in defined home ranges. Home range was simulated from bivariate normal, bivariate uniform and bivariate U–shaped distributions, and from a 'random excursion' model. More details are given by Wilson and Anderson (1985b). A 4 ha area was simulated, 320 traps were positioned in a two-dimensional plane, and animal density was set at two levels, 100/ha and 25/ha. Home range centres were allowed to be spatially random (Poisson), or clumped at three levels of aggregation. Three average probabilities of first capture were simulated at 0.09, 0.16 and 0.24, and these probabilities were allowed to vary by time (trapping occasion), behaviour (trap-shy or trap-happy) and heterogeneity (individual variability); this is model M_{tbh} in Otis et al. (1978: 43). Trapping was simulated for six, eight and ten occasions.

The Monte Carlo results indicated that the combination of a trapping web design and a point transect estimator of density was quite robust. The procedure had typically low bias under a wide variety of realistic situations. Confidence interval coverage was lower than the nominal level, due in part to the use of the Fourier series estimator (Buckland 1987a). The method was recommended in cases where the capture probability was > 0.16 and the number of trapping occasions was at least six. In some extreme situations (e.g. the random excursion model, with low capture probabilities and a clumped spatial distribution), the bias was in the 20–30% range, which might still be substantially less than traditional capture–recapture estimators.

The trapping web does not make any assumptions about geographic closure and is easy to implement in the field. No unique marks or tags are required, and several different types of trap can be used. The results of Wilson and Anderson (1985b) indicated that the trapping web was very promising as an alternative to standard capture–recapture methods. The work of Parmenter *et al.* (1989), summarized in Section 6.11.5, was carried out as a field test of the method where the true density was known.

6.11.4 A simple example

Anderson *et al.* (1983) presented an example of trapping web data from a 4.8 km area south of Los Alamos, New Mexico, where *Peromyscus* spp. were trapped for $t = 4$ nights on a web very similar to that of Fig. 6.9, with trap separation of $\theta = 3$m. The mice were captured in baited live traps and marked using a monel metal tag placed in one ear. Only initial captures were used in the analysis; animals were thus 'removed by marking'. No unmarked mice were caught in the inner area (out to ring 7) on the fourth night and only two new captures were made in this area on the third night. This was taken as evidence that the probability of capture near the web centre was one. A plot of the histogram indicated that mice from beyond the web were being attracted to the baited traps in the web, as the number of captures in rings 19 and 20 (i.e. n_{19} and n_{20}) was somewhat higher than expected. Thus, the distance data were truncated to exclude the two outer rings. This left 76 'detections' in 18 distance groups for analysis; frequencies were 1, 1, 0, 6, 2, 2, 3, 2, 4, 7, 4, 5, 8, 6, 7, 6, 7 and 5, respectively (Anderson *et al.* 1983). Note the lower frequencies in the inner rings, where the area sampled is small relative to that in the outer rings.

Two models were fit to these data: half-normal and hazard-rate, each with cosine adjustment parameters. No adjustment parameters were required and the AIC values were similar for the two models (424.18 and 425.76, respectively). Both models fit the data well as judged by the χ^2 goodness of fit tests ($\chi^2 = 13.2$ with 16 df and 13.6 with 15 df, respectively). Density is estimated at 97.8 mice/ha ($\widehat{se} = 21.3$) under the half-normal model and 86.1 mice/ha ($\widehat{se} = 12.7$) under the hazard-rate model. These estimates compare with 76 animals trapped on the web of area 0.97 ha, suggesting that most animals were caught. Only one web was sampled, hence no inference to a larger area is justified in this simple example.

6.11.5 Darkling beetle surveys

Populations of two species of ground-dwelling darkling beetles were studied in a shrub-steppe ecosystem in southwestern Wyoming to field test the validity of the trapping web on a series of known populations (Parmenter *et al.* 1989). These beetles (10–30 mm body length) attain natural densities so great (> 2000 beetles/ha) that relatively small plots could be surveyed and still have test populations of reasonable size. These beetles are wingless and could be contained by low metal fences. They are easily marked on their elytra with coloured enamel paint, and are relatively long-lived, allowing longer periods of trapping and increased capture success. Pitfall traps were made from small metal cans (80–110 mm), and the web was surrounded by a metal enclosure wall. Traps were placed along 12 lines, each 11 m long, with 1 m trap spacing along the lines. Nine additional traps were placed at the centre of the web, giving 141 in total.

Beetles were captured, marked with enamel and released. These marked beetles constituted the population of known size that was subsequently sampled using the trapping web design. Surveys were done in two different years and different colours were used to denote the year. Several subpopulations, each of known size, were released, allowing analyses to be carried out both separately and in combination, to test the method on a wide range of densities. Additional details were given in Parmenter *et al.* (1989).

Overall the method performed quite well, yielding a correlation coefficient between $\hat{D}$ and D of at least 0.97 for each of four models for $g(r)$. The negative exponential model performed better than the Fourier series, exponential power series and half-normal models. The data exhibited a spike near the web centre, almost certainly caused by considerable movement of beetles and by trap spacing that was too small. All models were fitted after transforming the distance data to areas, a procedure that is no longer recommended. In summary, the results of these field tests were certainly encouraging.

Reanalysis of the darkling beetle data using current theory and program DISTANCE provided a less optimistic impression in that density was substantially overestimated, leading to important insights. The first is that traps were too closely spaced along the lines; trap spacing should have been greater to compensate for the wide area over which beetles of this species move. Second, the beetles had no 'home range' and thus tended to wander widely in relation to the size of the web, which is a function of trap spacing. The trapping web design was envisioned for use with animals that have some form of home range or 'territory'. Third, it is clear that some random movement may result in too many animals being trapped near the web centre. The problem can also arise in bird surveys where some random movement results in the detection of too many birds near the point. This condition leads to a spiked distribution and overestimation of density. Clearly, the additional nine traps placed near the web centre aggravated this problem. Thus, we do not recommend that traps be concentrated at the centre. Research is needed to understand the relationship between spacing and density of traps and the nature of the movement.

If the data are spiked, one might analyse the distance data using some left truncation to eliminate the high numbers trapped near the web centre. Alternatively, one could constrain $\hat{g}(r)$ to be a low order, slowly decreasing function that does not track the spiked nature of the data, but this solution is rather arbitrary and may be ineffective. More experience is needed with sampling from populations of known size to understand better trap spacing and appropriate modelling. Still, this application of point transect sampling has many advantages over capture–recapture. Further studies on populations of known size, using live, snap or pitfall traps or mist nets, could lead to additional insights.

6.12 Migration counts

The main theme of this book is estimation of population abundance by modelling distance data. Counts from migration watch points may be converted into estimates of population size using similar techniques, but by modelling time instead of distance. Typically there will be regular, perhaps daily, counts of numbers of animals passing a watch point. If the animals pass in clusters, then the sampling unit will be the cluster. The basic data are start and end times of watch periods and number of animals or clusters passing during each watch period. Thus the data are in frequency form, being grouped by watch period. There will be gaps between watch periods, corresponding to night or to poor weather. For the basic method, animals are assumed to migrate at the same rate during unwatched periods as during watches. If no migration occurs at night, then time should be defined to end at dusk and start again at dawn. If migration occurs at night, but possibly at a different rate, the rate should be estimated by another method, for example by sonar (active or passive) or radar, or by radio-tagging animals. To model migration time, as distinct from distances in line transect sampling, the following changes are needed to the methodology. First, in line transect sampling the density function is assumed to be symmetric about the line, so only even functions (cosines for the Fourier series model and even powers for polynomial models) are used. For migration counts, odd functions are also needed. Related to this, the key function requires both a location and a scale parameter, whereas only a scale parameter is necessary for line transect sampling, because if sightings to the left of the line are recorded as negative, and distances to the right as positive, the expected distance from the transect is zero under the assumption that the density function is symmetric about the line. Third, allowance must be made for a large number of counts, equal to the number of separate watch periods, whereas in a grouped analysis of line transect data, the number of groups for perpendicular distances seldom exceeds a dozen or so. Finally, having fitted the density to migration times, abundance is estimated by taking the ratio of the area under the entire density to the combined area corresponding to watch periods alone, and multiplying this ratio by the total number of animals counted during watches. Thus, different software is needed to obtain the abundance estimate and its standard error.

We use here as an example the California grey whale census data collected at Monterey, California. The California stock of grey whales migrates from feeding grounds in the Bering and Chukchi Seas to calving areas in Mexican waters every winter, returning north in spring. Aerial and ship surveys confirm that almost the entire population passes close inshore at several points. Counts at coastal migration watch points can therefore be used to estimate population size. Counts at Monterey were annual from 1967–68 through to 1979–80, and further surveys were carried out in 1984–85, 1985–86 and 1987–88. Reilly *et al.* (1980, 1983) gave more information on these surveys, and Buckland and Breiwick (in press) provided abundance estimates corresponding to all surveys. We use analyses of the grey whale count data for 1987–88, extracted from Breiwick *et al.* (unpublished) and Buckland *et al.* (in press), to illustrate analysis of migration count data. In that year, counts were made from two stations (north and south) a few yards apart, to allow estimation of numbers of pods missed during watch periods. The data analysed were numbers of pods passing within each count period, so that the data are grouped, the group endpoints being the start and end of each watch period. Information on duplicate detections was used to reduce the data sets from both stations to a single set of counts of the number of pods detected by at least one station in each watch period. Pods detected travelling north were excluded from the analyses.

The key function selected for fitting the counts was, apart from a scaling factor, the normal density:

$$\alpha(y) = \exp\left\{-\left(\frac{y - \mu}{\sigma}\right)^2\right\}$$

where *y* corresponds to time, measured in days from a predetermined date. Adjustments to the fit of the key were made by adding Hermite polynomial terms sequentially, adjusting the fit first for skewness, then kurtosis, and so on. Four adjustment terms were fitted to the data sets, and likelihood ratios were used to determine which fit was 'best'. If a one-term fit was found to offer no significant improvement over no terms, but a two-term fit gave a significant improvement over no terms at the 5% level, the two-term fit was favoured over both the one-term and the zero-term fits. A three-term (i.e. five-parameter) fit was selected, and this fit is shown in Fig. 6.10. To convert the fitted density to an estimate of population size, it is necessary to evaluate the proportion of the entire untruncated density that corresponds to watch periods. To ensure that the Hermite polynomial fits were sensible in the tails of the migration, zero counts were added for 1 December 1987, before the migration started, and 29 February 1988, after it ended. This had little effect for 1987–88, when counts took place throughout the main migration period (Fig. 6.10), but for some earlier surveys, many pods had passed before the first or after the last count of the season, making the addition of zero counts necessary (Buckland and Breiwick, in press).

In total, and excluding pods travelling north, $n = 3593$ pods were seen from at least one station. The χ^2 goodness of fit statistic corresponding to Fig. 6.10 was $\chi^2_{125} = 334.55$. This value is more indicative of overdispersion of counts than of intrinsic lack of fit of the Hermite polynomial model; in other words, counts in successive watches show greater than

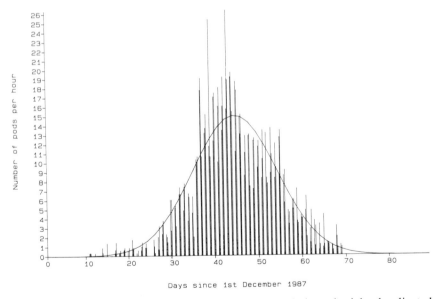

Days since 1st December 1987

Fig. 6.10. Histogram of number of California grey whale pods sighted, adjusted for watch length, by date, 1987–88 survey. Also shown is the Hermite polynomial fit to the histogram.

Poisson variation. The overdispersion was compensated for by multiplying the Poisson variance on the total count by the dispersion parameter, estimated as the χ^2 statistic divided by its degrees of freedom; this multiplicative correction is sometimes termed a variance inflation factor (Cox and Snell 1989). Thus the dispersion parameter estimate is $334.55/125 = 2.676$, giving $\widehat{se}(n) = \sqrt{(3593 \cdot 2.676)} = 98.1$ pods. The fit of the Hermite polynomial model to the counts yields a multiplicative correction for animals passing outside watch periods of $\hat{f_t} = 2.4178$ with standard error 0.0068.

Swartz *et al.* (1987) reported on experiments in which whales were radio-tagged in 1985 and 1986. Of these, 15 were recorded both at night and in daylight. An unpaired t-test on the difference in log day and night speeds revealed no significant difference between Monterey and the Channel Islands ($t_{11} = -1.495$; $p > 0.1$). After pooling the data from both locations, a paired t-test revealed a significant difference in log speeds between day and night ($t_{14} = 2.284$; $p < 0.05$). A back-transformation with bias correction gave a multiplicative correction factor for hours of darkness of 1.100 ($\widehat{se} = 0.045$); thus it is estimated that rate of passage is 10% higher at night, and thus night counts, if they were feasible, would generate counts 10% higher. Counts were carried out for ten hours each day. On average, it is reasonable to add an hour to each end of the day, giving roughly 12 hours of daylight (including twilight) per 24 hours. Thus the multiplicative correction applies approximately to one half of the total number of whales estimated, giving a multiplicative correction factor of $\hat{f_n} = 1.050$ ($\widehat{se} = 0.023$). Swartz (personal communication) notes that the behaviour of the animals off the Channel Islands is very different from when they pass Monterey. If a correction factor is calculated as above from the nine radio-tagged whales off Monterey that were recorded both during the day and at night, we obtain $\hat{f_n} = 1.020$ ($\widehat{se} = 0.023$). Although this does not differ significantly from one, we apply it, so that the variance of the abundance estimate correctly reflects the uncertainty in information on this potentially important parameter.

During the 1987–88 season, counts were carried out independently by observers in identical sheds, 5 m apart. Buckland *et al.* (in press) analysed these double count data using the approach of Huggins (1989, 1991) and Alho (1990), which incorporates covariates to allow for heterogeneity in mark–recapture experiments. We summarize the method here. The procedures for matching detections from the two stations are described by Breiwick *et al.* (unpublished). We assume the matches are made without error.

Assuming that the probability of detection of a pod from one station is independent of whether it is detected from the other, and independent of whether other pods are detected by either station, the full likelihood for all pods passing Monterey during watch periods is

$$L^* = K \cdot \prod_{i=1}^{M} \prod_{j=1}^{2} p_{ij}^{\delta_{ij}} \cdot (1 - p_{ij})^{1 - \delta_{ij}}$$

where M = total number of pods passing during count periods,

p_{ij} = probability that pod i is detected from station j, $i = 1, \ldots, M$, $j = 1, 2$,

$$\delta_{ij} = \begin{cases} 1, & \text{pod } i \text{ is detected from station } j, \\ 0 & \text{otherwise,} \end{cases}$$

and K depends on M, but not on the parameters that define p_{ij}.

Huggins (1989) shows that inference can be based on the conditional likelihood,

$$L = \prod_{i=1}^{n} \prod_{j=1}^{2} \pi_{ij}^{\delta_{ij}} \cdot (1 - \pi_{ij})^{1 - \delta_{ij}}$$

where n = number of pods detected from at least one station,

and $\pi_{ij} = \dfrac{p_{ij}}{p_i}$

with $p_i = 1 - \prod\limits_{j=1}^{2} (1 - p_{ij})$ = probability that pod i is detected from at least one station.

Thus π_{ij} is the probability that pod i is detected from station j given that it is detected from at least one station.

Both Huggins (1989, 1991) and Alho (1990) model the p_{ij} using logistic regression. Algebra yields:

$$\log_e \frac{\pi_{ij}}{1 - \pi_{ij}} = \log_e \frac{p_{ij}}{1 - p_{ij}} - \log_e p_{ij'}, \quad \text{where} \quad j' = 3 - j$$

Hence logistic regression for the p_{ij} can be obtained simply by carrying out logistic regression for the conditional probabilities π_{ij}, and setting an offset variable, equal to $- \log_e \hat{p}_{ij'}$, for each observation. In the first iteration, the offset variable is set to zero (corresponding to normal logistic regression for π_{ij}). Estimates $\hat{p}_{ij}$ are then calculated from the $\hat{\pi}_{ij}$, from which the offset variable is estimated. The model is refitted, and the process is repeated until convergence is achieved. Model fitting was carried out using Genstat (Genstat 5 Committee, 1987).

Potential covariates were date, Beaufort, components of wind direction parallel and perpendicular to the coast, visibility code, distance offshore, pod size and rate of passage (pods per hour); observer, station and watch period were entered as factors. Estimates $\hat{p}_{ij}$ and $\hat{p}_i$ were calculated from the final iteration, from which M was estimated as

$$\hat{M} = \sum_{i=1}^{n} \frac{1}{\hat{p}_i}$$

with

$$\widehat{\text{var}}(\hat{M}) = \sum_{i=1}^{n} \frac{1}{\hat{p}_i^2} \cdot (1 - \hat{p}_i)$$

Thus a correction factor for pods missed by both stations is given by

$$\hat{f}_m = \frac{\hat{M}}{n}$$

with

$$\widehat{\text{se}}(\hat{f}_m | n) = \frac{\sqrt{\widehat{\text{var}}(\hat{M})}}{n}$$

Probability of detection of a pod was adequately modelled as a function of five covariates: pod size; rate of passage; migration date; visibility code; and the component of wind direction parallel to the coast. None of the factors (observer, watch period, station) explained a significant amount of variation. Probability of detection increased with pod size ($p < 0.001$), with rate of passage ($p < 0.001$) and with migration date ($p < 0.05$), and decreased with visibility code ($p < 0.05$). It was also greater when the wind was parallel to the coast from 330° (slightly west of north), and smaller when from 150° (east of south). The correction factor f_m was estimated by $\hat{f}_m = 1.0632$, with standard error 0.00447.

The number of whales passing Monterey is equal to the number of pods multiplied by the average pod size, which was estimated by the average size of pods detected (excluding those moving north). This gave $\bar{s} = 1.959$ ($\widehat{\text{se}} = 0.020$). A correction factor for mean pod size was calculated using data from Reilly et al. (1980), comparing recorded pod sizes with actual pod sizes, determined by observers in an aircraft. For pods of size one, an additive correction of 0.350, with standard error $0.6812/\sqrt{225} = 0.0454$, was used. The correction for pods of size two was

0.178 ($\widehat{se} = 0.9316/\sqrt{101} = 0.0927$), for pods of size three, 0.035 ($\widehat{se} = 1.290/\sqrt{28} = 0.244$), and for pods of size four or greater, the correction was 0.333 ($\widehat{se} = 0.7825/\sqrt{27} = 0.151$). A multiplicative correction factor for mean pod size was then found as:

$$\hat{f_s} = 1 + \frac{0.350n_1 + 0.178n_2 + 0.035n_3 + 0.333n_{4+}}{n \cdot \bar{s}} = 1.131$$

with

$$\widehat{se}(\hat{f_s}|n) \simeq \sqrt{[(0.0454n_1)^2 + (0.0927n_2)^2 + (0.2438n_3)^2 + (0.1506n_{4+})^2}$$
$$+ 0.6812^2 n_1 + 0.9316^2 n_2 + 1.290^2 n_3 + 0.7825^2 n_{4+}]/(n \cdot \bar{s}) = 0.026$$

where n = total number of pods recorded,

n_i = number of pods of size i, $i = 1, 2, 3$,

and n_{4+} = number of pods of size four or more.

The revised abundance estimate was thus found as follows. Counts of numbers of pods by watch period were combined across the two stations, so that each pod detected by at least one station contributed a frequency of one. The Hermite polynomial model was applied to these counts, to obtain a multiplicative correction factor $\hat{f_t}$ to the number of pods detected for whales passing at night or during poor weather. The correction for different rate of passage at night $\hat{f_n}$ was then made. Next, the multiplicative correction $\hat{f_m}$ was applied, to allow for pods passing undetected during watch periods. The estimated number of pods was then multiplied by the mean pod size, and by the correction factor $\hat{f_s}$ for underestimation of pod size, to obtain the estimate of the number

Table 6.7 Estimates of abundance and of intermediate parameters, California grey whales, 1987–88

Parameter	Estimate	Std error	% contribution to $\widehat{var}(\hat{N})$	95% confidence interval
E(Number of pods seen by at least one station) $= E(n)$	3593	98	39	(3406, 3790)
Correction for pods passing outside watch periods, f_t	2.418	0.007	0	(2.405, 2.431)
Correction for night passage rate, f_n	1.020	0.023	27	(0.976, 1.066)
Correction for pods missed during watch periods, f_m	1.063	0.004	1	(1.054, 1.072)
Total number of pods passing Monterey	9419	337		(8781, 10 104)
Mean recorded pod size	1.959	0.020	5	(1.920, 1.999)
Correction for bias in recorded pod size, f_s	1.131	0.026	28	(1.081, 1.183)
Total number of whales passing Monterey	20 869	913		(19 156, 22 736)

Table 6.8 Estimated number of pods, pod size and number of whales by year. (Standard errors in parentheses.) For any given fit, the number of parameters is two greater than the number of terms, corresponding to the two parameters of the normal key

Year	No. of terms	χ^2 [df]	Sample size (pods)	Estimated no. of pods	Estimated average pod size	Relative abundance estimate	Absolute abundance estimate
1967–68	4	83.0 [45]	903	4051 (253)	2.438 (0.063)	9878 (667)	12921 (964)
1968–69	0	70.6 [61]	1079	4321 (134)	2.135 (0.046)	9227 (348)	12070 (594)
1969–70	1	104.5 [67]	1245	4526 (155)	2.128 (0.043)	9630 (383)	12597 (640)
1970–71	2	116.2 [90]	1458	4051 (115)	2.021 (0.033)	8185 (267)	10707 (487)
1971–72	0	71.3 [56]	857	3403 (127)	2.193 (0.048)	7461 (323)	9760 (524)
1972–73	4	91.5 [71]	1539	5279 (152)	2.187 (0.034)	11543 (378)	15099 (688)
1973–74	4	133.7 [66]	1496	5356 (186)	2.098 (0.034)	11235 (431)	14696 (731)
1974–75	0	159.2 [74]	1508	4868 (174)	2.034 (0.035)	9904 (394)	12955 (659)
1975–76	2	101.1 [47]	1187	5354 (218)	2.073 (0.039)	11100 (497)	14520 (796)
1976–77	0	139.7 [87]	1991	5701 (153)	2.052 (0.028)	11700 (353)	15304 (669)
1977–78	0	50.2 [31]	657	7001 (356)	1.843 (0.046)	12904 (731)	16879 (1095)
1978–79	4	152.9 [84]	1730	4970 (159)	2.016 (0.034)	10018 (361)	13104 (629)
1979–80	4	109.3 [55]	1451	6051 (220)	2.068 (0.033)	12510 (498)	16364 (832)
1984–85	3	105.2 [49]	1756	7159 (301)	2.290 (0.038)	16393 (740)	21443 (1182)
1985–86	1	141.4 [104]	1796	6873 (191)	2.237 (0.042)	15376 (515)	20113 (927)
1987–88N	3	205.9 [92]	2426	7756 (221)	2.040 (0.027)	15825 (497)	
1987–88S	3	152.8 [91]	2404	7642 (194)	2.104 (0.029)	16082 (464)	
1987–88 (average)						15954 (481)	20869 (913)

of whales passing Monterey during the 1987–88 migration. Thus the abundance estimate for 1987–88 is given by

$$\hat{N} = n \cdot \hat{f}_t \cdot \hat{f}_n \cdot \hat{f}_m \cdot \bar{s} \cdot \hat{f}_s$$

with

$$\mathrm{cv}(\hat{N}) \simeq \sqrt{\{[\mathrm{cv}(n)]^2 + [\mathrm{cv}(\hat{f_t})]^2 + [\mathrm{cv}(\hat{f_n})]^2 + [\mathrm{cv}(\hat{f_m})]^2 + [\mathrm{cv}(\bar{s})]^2 + [\mathrm{cv}(\hat{f_s})]^2\}}$$

Table 6.7 shows the different components to the estimate $\hat{N}$. Combining them, estimated abundance is 20 869 whales, with $\mathrm{cv}(\hat{N}) = 0.0437$ and approximate 95% confidence interval (19 200, 22 700).

Buckland and Breiwick (in press) scaled their relative abundance estimates for the period 1967–68 to 1987–88 to pass through an absolute abundance estimate for 1987–88. Rescaling them to pass through the revised estimate above yields the estimates of Table 6.8. Figure 6.11 plots the absolute abundance estimates and shows the estimated increase in abundance assuming an exponential model with non-zero asymptote. The estimated mean annual rate of increase is 3.3% per annum ($\widehat{se} = 0.4\%$).

6.13 Point-to-object and nearest neighbour methods

The term 'distance sampling' has been used by botanists in particular to describe methods in which a random point or object is selected, and distances from it to the nearest object(s) are measured. A discussion of these methods was given by Diggle (1983: 42–4). In the simplest case, the distance y to the nearest object is measured; y is a random variable with a pdf, say $f(y)$. However, there is no detection function; the nearest object will be detected with probability one. This is very different from the distance sampling from which this book takes its title, for which there is also a sample of distances y with pdf $f(y)$. The two pdf's can be very similar mathematically, but they are conceptually very different. It is the concept of a detection function that distinguishes the distance sampling of this book. Hence we do not describe point-to-object and nearest neighbour methods in detail here.

For point-to-object and nearest neighbour methods, if the distribution of objects is random, then object density is estimated by

$$\hat{D} = \frac{kn}{\pi \sum_{j=1}^{k} \sum_{i=1}^{n} r_{ij}^2}$$

where k = number of random points or objects,

n = number of point/object-to-object distances measured at each point or object,

r_{ij} = distance of ith nearest object to the jth random point or object, $i = 1, \ldots, n$; $j = 1, \ldots, k$.

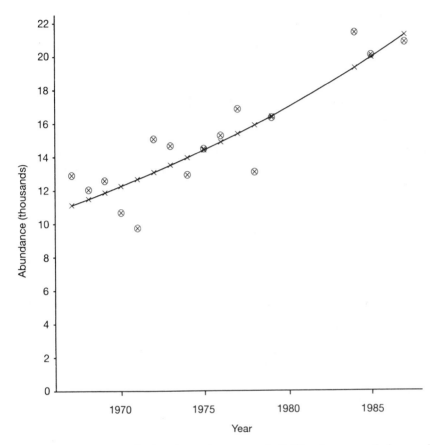

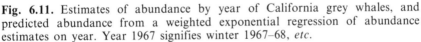

Fig. 6.11. Estimates of abundance by year of California grey whales, and predicted abundance from a weighted exponential regression of abundance estimates on year. Year 1967 signifies winter 1967–68, *etc.*

When the distribution of objects is overdispersed (i.e. aggregated), density is underestimated if distances are measured from a random point, and overestimated if distances are measured from a random object. An average of the two therefore tends to have lower bias than either on its own. Diggle (1983) listed three *ad hoc* estimators of this type.

Some authors have used point-to-object distances only, together with a correction factor for non-Poisson distribution (Batcheler 1975; Cox 1976; Warren and Batcheler 1979), although Byth (1982) showed by simulation that the approach can perform poorly.

Nearest neighbour and point-to-object methods have been used primarily to measure spatial aggregation of objects, and to test the assumption that the spatial distribution is Poisson. Their sensitivity to departures from the Poisson distribution is useful in this context, but renders the methods bias-prone when estimating object density. Except in special cases, such as estimating the density of forest stands (Cox 1976), we do not recommend these methods for density estimation. Their disadvantages are:

1. All objects out to the nth nearest to the selected point or object must be detected. In areas of low density, this may require considerable search effort.
2. It can be time-consuming to identify which are the n nearest objects, and at lower densities it may prove impractical or impossible to determine them.
3. The effective area surveyed cannot be easily predicted in advance, and is highly correlated with object density; a greater area is covered in regions of low object density. By contrast, good design practice in line and point transect surveys ensures that area covered is independent of object density within strata, leading to more robust estimation of average object density.

Point transects and point-to-object methods may both be considered as generalizations of quadrat counts. In both cases, the quadrat may be viewed as circular. For point transects, the area searched, $a = k\pi w^2$, is fixed (and possibly infinite), but the observer is not required to detect all objects in that area. For point-to-object methods, the number of objects n to be detected from each point is fixed, but the radius about each point is variable; all objects within that radius must be detected.

7

Study design and field methods

7.1 Introduction

The analysis methods presented in Chapters 3–5 depend on proper field methods, a valid design, and adequate sample size. This chapter presents broad guidelines for the design of a distance sample survey and outlines appropriate field methods. In general, a statistician or quantitative person experienced in distance methods should be consulted during the initial planning and design of the study. Just as important is the need for a pilot study. Such a preliminary study will provide rough estimates of the encounter rate n/L (line transect sampling) or n/k (point transect sampling), and of variance components from which refined estimates of n and of L or k for the main study are obtained. Additionally, operational considerations can be reviewed and training of participants can occur. **A pilot study is strongly recommended as it can provide insights into how best to meet the important assumptions**.

Careful consideration should be given to the equipment required to allow collection of reliable data. This may include range finders, binoculars with reticles, angle boards or rings, a camera, a compass, and various options for an observation platform, which might vary from none (i.e. one pair of feet) to a sophisticated aircraft or ship, or even a submersible (Fig. 7.1).

The primary purpose of material presented in this chapter is to ensure that the critical assumptions are met. Considerable potential exists for poor field procedures to ruin an otherwise good survey. Survey design should focus on ways to ensure that three key assumptions are true: $g(0) = 1$, no movement in response to the observer prior to detection, and accurate measurements (or accurate allocations to specified distance categories). If the population is clustered, it is important that cluster

Fig. 7.1. Line transect sampling can be carried out from several different types of observation platform. Here, a two-person submersible is being used to survey rockfish off the coast of Alaska. Distances are measured using a small, hand-held sonar gun deployed from inside the submersible.

size be determined accurately. In addition, a minimum sample size (n) in the 60–80 range and $g(y)$ with a broad shoulder are certainly important considerations. Sloppiness in detecting objects near, and measuring their distance from, the line or point has been all too common, as can be seen in Section 8.4. In many line and point transect studies, **the proper design and field protocol have not received the attention deserved**.

Traditional strip transects and circular plots should be considered in early design deliberations. These finite population sampling methods deserve equal consideration with the distance sampling methods. However, if there is any doubt that all objects within the strip or circle are detected, then distances should be taken and analysed (Burnham and Anderson 1984). The tradeoffs of bias and efficiency between strip transects and line transects have been addressed (Burnham *et al.* 1985). Other sampling approaches should also be considered; Seber (1982, 1986) provided a compendium of alternatives and new methods are occasionally developed, such as adaptive sampling (Thompson 1990). A common alternative for animals is capture–recapture sampling, but Shupe *et al.* (1987) found that costs for mark–recapture sampling exceeded those of walking line transects by a factor of three in rangeland studies in Texas. Guthery (1988) presented information on time and cost requirements for line transects of bobwhite quail.

If all other things were equal, one would prefer line transect sampling to point transect sampling. More time is spent sampling in line transect surveys, whereas more time is often spent travelling between and locating sampling points in point transect sampling (Bollinger *et al.* 1988). In addition, it is common to wait several minutes prior to taking data, to allow the animals (usually birds) time to readjust to the disturbance caused by the observer approaching the sample point. Point transect sampling becomes more advantageous if the travel between points can be done by motorized vehicle, or if the points are established along transect lines, with fairly close spacing (i.e. rather than a random distribution of sampling points throughout the study area). If the study area is large, the efficient utilization of effort may be an order of magnitude better for line transect surveys. This principle is reinforced when one considers the fact that it is objects on or near the line or point that are most important in distance sampling. Thus, in distance sampling, the objects seen at considerable distances (i.e. distances y such that $g(y)$ is small, say less than 0.1) from the line or point contain relatively little information about density. In point transect surveys, the count of objects beyond $g(r) = 0.1$ may be relatively large because the area sampled at those distances is so large.

Point transect sampling is advantageous when terrain or other variables make it nearly impossible to traverse a straight line safely while also expending effort to detect and record animals. Multispecies songbird surveys in forest habitats are usually best done using point transect sampling. Point transects may often be more useful in patchy environments, where it may be desirable to estimate density within each habitat type; it is often difficult to allocate line transects to allow efficient and unbiased density estimation by habitat. One could record the length of lines running through each habitat type and obtain estimates of density for each habitat type (Gilbert *et al.* in prep.). However, efficiency may be poor if density is highly variable by habitat type, but length of transect is proportional to the size of habitat area. Additionally, habitat often varies continuously, so that it is more precisely described at a single point than for a line segment. Detection may be enhanced by spending several minutes at each point in a point transect, and this may aid in ensuring that $g(0) = 1$. Remaining at each point for a sufficient length of time is particularly important when cues occur only at discrete

297

times (e.g. bird calls). Some species may move into the sample area if the observer remains at the site too long. Even in line transect sampling the observer may want to stop periodically to search for objects.

7.2 Survey design

Survey design encompasses the placement (allocation) of lines or points across the area to be sampled and across time. The population to be sampled must be clearly defined and its area delimited. A good map or aerial photo of the study area is nearly essential in planning a survey. An adequate survey must always use multiple lines or points (i.e. replication). Consideration must be given to possible gradients in density. If a substantial transition in density is thought or known to exist, it is best to lay the lines parallel to the direction of the gradient (Fig. 1.4). This would also be true if points were to be placed systematically along lines. Alternatively, spatial stratification of the study area might be considered. For example, if two main habitat types occurred in the area of interest, one might want to estimate density in each of the two habitat types. A consideration here is to be sure that adequate sample size is realized in both habitat types. If little is known *a priori*, the strata (i.e. habitat types) should be sampled in proportion to their size. Detection probability often varies with topography, habitat type, and density of objects of interest. Proper design, such as the approaches suggested below, will cope with these realities.

It was often thought that an observer could roam through an area and record only the sighting distances r_i to each object detected. This type of cruising may lead to nearly useless data and unreliable density estimates (Burnham *et al.* 1980; Hayes and Buckland 1983).

7.2.1 Transect layout

Several options exist for the layout of individual lines in a line transect survey or points in a point transect survey. A favoured and practical layout is a systematic design using parallel transects, with a random first start (e.g. Figs 1.4 and 1.6). Then transects extend from boundary to boundary across the study area and are usually of unequal length. Transects are normally placed at a distance great enough apart to avoid an object being detected on two neighbouring transects, although this is not usually critical. Care must be taken such that the transect direction does not parallel some physical or biological feature, giving an unrepresentative sample. For example, if all the lines were on or near fence rows, the sample would be clearly unrepresentative (Guthery 1988).

A common mistake is to have lines follow established roads or corridors. If there is a strong density gradient perpendicular to a linear physical feature, then a design in which lines are parallel to this gradient, and hence perpendicular to the linear feature, should be considered.

A second approach might be to lay out a series of contiguous strips of width $2w$, pick k of these at random, and establish a line or point transect in the centre of each selected strip. Thus, transects would be parallel, but the spacing between transects would be unequal. In some sense, theory would suggest that a valid estimate of the sampling variance could be obtained only with a completely random sample. However, the precision of the systematic sample is often superior to random sampling. There is no compelling reason to use randomly placed lines or points, although a grid of lines or points should be positioned randomly, and oriented either randomly or perpendicular to density contours. Designs that permit overlapping transects should probably be avoided except in specialized cases; this requirement limits the number of possible layouts. Also, designs that require extensive and time-consuming travel between transect lines or points are inefficient.

A third approach is to establish a system of rectangles, whereby the observer travels the perimeter searching for objects along the line or around the points along the line (Fig. 7.2). This allows, for example, an observer on foot to return to a vehicle without losing time walking between transects. This design may be advantageous where a system of roads exists on the study area. The position of the sample rectangles can be selected in several ways (e.g. the southwest corners of the rectangles could be selected at random or they could be placed systematically with a random first start). Many parts of central and western North America have roads on a 1-mile grid, 'section lines', making this design easy to implement in the field.

Transects should not be deliberately placed along roads or trails, as these are very likely to be unrepresentative. Transects following or paralleling ridgetops, hedgerows, powerlines, or stream bottoms are also likely to be unrepresentative of the entire area. We strongly recommend against biasing samples towards such unrepresentative areas. Transects placed subjectively (e.g. 'to avoid dense cover' or 'to be sure the ridge is sampled') are poor practice, and should always be avoided.

The design of point transects would best be done, from a statistical viewpoint, completely randomly (ignoring, for the moment, any need to stratify). This follows from sampling theory whereby the layout of plots (circular or rectangular) should be placed at random. However, in this random design, the amount of time to travel from point to point is likely to be excessive and occasional pairs of points may be quite close

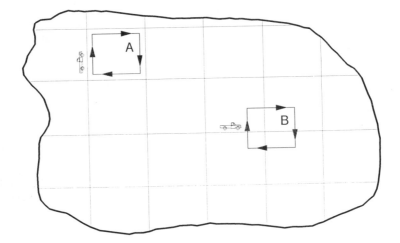

Fig. 7.2. A practical design for line or point transect surveys is to establish a series of rectangles for which the perimeters (or points along the perimeter) are sampled. This design is useful when a network of roads exists on the study area. **A** might be appropriate for surveys where density in undisturbed habitats is of interest, while **B** would be useful in studies of the entire area. Many landscapes have extensive habitat along roads and associated roadsides, fence rows, borrow pits, etc. Perimeter areas to be surveyed can be established at random or systematically with a random first start.

together. This consideration has led ornithologists, in particular, to place a series of points along a transect line. Thus, there might be 20 lines, each having, say, 10 sampling points. These should not be analysed as if they are 200 independent samples; one must be certain that the estimated sampling variance is correctly computed, by taking the transect line of 10 points as the sampling unit. Points could be established at grid intersections of a rectangular grid to achieve a systematic design. Again, the problem here might be the amount of time required to travel from point to point. One might spend 30 minutes walking between successive points and only 5–10 minutes sampling objects at each point.

Detection probability often varies with topography, habitat type, and the density of objects of interest. Proper design, such as the approaches suggested above, will cope with these realities.

If the survey is to be repeated over time to examine time trends in density, then the lines or points should be placed and marked permanently. Sampling of duck nests at the Monte Vista National Wildlife Refuge has been done annually for 27 years using permanent transect markers set up in 1963 (numbered plywood signs atop 2.5 m metal poles). Repeated sampling should be done at time intervals large enough

300

so that the stochastic errors of successive samples are not highly dependent. If an area is to be sampled twice within a short time period, one could consider using a system of transects running north–south on the first occasion and another set of transects running east–west on the second occasion. This scheme, although using overlapping transects, might give improved coverage. However, other schemes might be considered if a strong gradient in density was suspected.

Point transects should also be permanently marked if the survey is to be repeated. One must be cautious that neither the objects of interest nor predators are attracted to the transect markers (e.g. poles and signs would not be appropriate for some studies if raptors used these markers for perching and hunting). A good cover map would aid in establishing sample points and in relocating points in future surveys. In addition, a cover map or false colour infrared image might be useful in defining stratum boundaries.

If there are smooth spatial trends in the large-scale density over an area, then systematically placed lines or points are better than random placement. Ideally, the analysis would fit these trends by some means and derive the variance from the model residuals (Burdick 1979). This topic is addressed in Section 6.3, but is in need of more theoretical development.

No problem arises if a stationary object is detected from two different lines or points. If an animal moves after detection from one line or point to another in a short time period (e.g. the same day), then this may become problematic if it happens frequently and is in response to the presence of the observer. Some sophisticated surveys are designed to obtain double counts of the same object from independent platforms, to allow estimation of $g(0)$ or to correct for the effects of movement (Section 6.4).

7.2.2 Sample size

A basic property of line and point transect sampling theory is that it is the absolute size of the sample that is important when sampling large populations, not the fraction of the population sampled. Thus, if $L = 2400$ m (corresponding to, say, $n = 90$) was sufficient for estimating the density of box turtles on a square kilometre of land, it would also be sufficient for the estimation of density on 25 square kilometres of land (assuming the sampling was done at random with respect to the turtle population). Thus, it would **not** take 25×2400 m of transect to sample the 25 square kilometres area.

The size n of the sample is an important consideration in survey design. If the sample is too small, then little information about density is available and precision is poor. Verner (1985) notes that some surveys have had very small sample sizes ($n \doteq 10$); almost no information about density is contained in so few observations and little can be done regardless of the analysis method used. If the sample is too large, resources might have been used more advantageously elsewhere.

As a practical minimum, n should usually be at least 60–80. Even then, the components of variance associated with both n and $\hat{f}(0)$ (line transects) or $\hat{h}(0)$ (point transects) can be large. If the population is clustered, the sample size (i.e. the number of clusters detected) should be larger to yield similar precision for the abundance estimate of individuals, substantially so if the variance of cluster size is large. If there is a target cv for the density estimate of 25% and $n = 100$ would achieve this for the density of clusters, then a larger n is needed to yield a cv of 25% for the density of individuals. This increase is because variation in cluster size increases the cv of the density estimate for individuals. The variance component associated with cluster size is rarely the largest component.

Sample sizes required are often quite feasible in many survey situations. For example, in aerial surveys of pronghorn (*Antilocapra americana*), it is possible to detect hundreds of clusters in 15–20 hours of survey time. The long-term surveys of duck nesting at the Monte Vista National Wildlife Refuge have detected as few as 41 nests and as many as 248 nests per year over the past 27 years. Effort involved in walking approximately 360 miles per year on the refuge requires about 47 person days per year. Cetacean surveys may need to be large scale to yield adequate sample sizes; in the eastern tropical Pacific, dolphin surveys carried out by the US National Marine Fisheries Service utilize two ships, each housing a cruise leader and two teams of three observers, together with crew members, for 4–5 months annually. Even with this effort, sample sizes are barely sufficient for estimating trends over eight or more years with adequate precision, even for the main stock of interest.

Sample size in point transects can be misleading. One might detect 60 objects from surveying k points and believe this large sample contains a great deal of information about density. However, the area sampled increases with the square of distance, so that many of the observations are actually in the tail of $g(r)$ where detection probability is low. Detections at some distance from the point may be numerous partially because the area sampled is relatively large. Thus, sample size must be somewhat larger for point transect surveys than line transect surveys. As a rough guideline, the sample size for point transects should be approximately 25% larger than that for line transect surveys to attain the same level of precision.

Generally, w should be set large in relation to the expected average distance (either $E(x)$ or $E(r)$). The data can be easily truncated during the analysis, but few (if any) detected objects should be ignored during the actual field survey because they are beyond some preset w, unless distant detections are expensive in terms of resources. For example, dolphin schools may be detected during shipboard surveys at up to 12 km perpendicular distance. These distant sightings add little to estimation and are likely to be truncated before analysis, so that the cost of taking these data is substantial (closing on the school, counting school size, determining species composition) relative to the potential value of the observations. A pilot study would provide a reasonable value for w for planning purposes.

Although we focus discussion here on sample size, the line or point is usually taken as the sampling unit for estimating the variance of encounter rate, and often of other parameter estimates. Thus a sample size of $n = 200$ objects from just one or two lines forces the analyst to make stronger assumptions than a smaller sample from 20 short lines. The strategy of dividing individual lines into segments, and taking these as the sampling units, can lead to considerable underestimation of variance (Section 3.7.4).

(a) *Line transects* The estimation of the line length to be surveyed depends on the precision required from the survey and some knowledge of the encounter rate (n_0/L_0) from a pilot study or from comparable past surveys. Here it is convenient to use the coefficient of variation, $\mathrm{cv}(\hat{D}) = \widehat{\mathrm{se}}(\hat{D})/\hat{D}$, as a measure of precision. One might want to design a survey whereby the estimated density of objects would have a coefficient of variation of 0.10 or 10%; we will denote this target value by $\mathrm{cv}_t(\hat{D})$. Two general approaches to estimating line length are outlined.

First, assume that a small-scale pilot study can be conducted and suppose n_0 objects were detected over the course of a line (or series of lines) of total length L_0. For this example, let $n_0 = 20$ and $L_0 = 5\,\mathrm{km}$. This information allows a rough estimate of the line length and, thus, sample size required to reach the stated level of precision in the estimator of density. The relevant equation is

$$L = \left(\frac{b}{(\mathrm{cv}_t(\hat{D}))^2} \right)\left(\frac{L_0}{n_0} \right) \tag{7.1}$$

where
$$b \doteq \left\{ \frac{\mathrm{var}(n)}{n} + \frac{n \cdot \mathrm{var}\{\hat{f}(0)\}}{\{f(0)\}^2} \right\}$$

While a small pilot survey might be adequate to estimate L_0/n_0 for planning purposes, the estimation of b poses difficulties. However, the value of b appears to be fairly stable and Eberhardt (1978b) provided evidence that b would typically be between 2 and 4. Burnham *et al.* (1980: 36) provided a rationale for values of b in the range 1.5–3. They recommended use of a value of 3 for planning purposes, although 2.5 was tenable. They felt that using a value of 1.5 risks underestimating the necessary line length to achieve the required precision. Another consideration is that b will be larger for surveys where the detection function has a narrow shoulder. Here we use $b = 3$ so that

$$L = \left(\frac{3}{(0.1)^2}\right)\left(\frac{5}{20}\right) = 75.0 \text{ km}$$

Equating the following ratios

$$\left(\frac{L_0}{n_0}\right) = \left(\frac{L}{n}\right)$$

and solving for n gives $n = 300$; the proper interpretation here is that we **estimate** that there will be 300 detections given $L = 75$ km, although the actual sample size will be a random variable. Thus, to achieve a coefficient of variation of 10% one would need to conduct 75 km of transects and expect to detect about 300 objects.

A pilot study to estimate L_0/n_0 can be quite simple. No actual distances are required and n_0 can be as small as, perhaps, 10. Thus, one could traverse randomly placed transects of a predetermined length L_0 and record the number of detections n_0 in estimating (L_0/n_0). The value of w used in the pilot study should be the same as that to be used in the actual survey. Alternatively, the ratio might be taken from the literature or from one's experience with the species of interest. Of course, the results from the first operational survey should always be used to improve the survey design for future surveys.

Second, if the pilot survey is quite extensive, then b can be estimated from the data as $\hat{b} \doteq n_0 \cdot (cv(\hat{D}))^2$ (Burnham *et al.* 1980: 35). From this more intensive pilot survey, the coefficient of variation is computed empirically and denoted as $cv(\hat{D})$. Substituting $\hat{b}$ into Equation 7.1, the line length required to achieve the target precision is given by

$$L = \frac{L_0(cv(\hat{D}))^2}{(cv_t(\hat{D}))^2}$$

For this approach to be reliable, n_0 should be in the 60–80 range; it is perhaps most useful when refining the second year of a study, based on the results from the first survey year.

Many surveys are limited by money or labour restrictions such that the maximum line length is prespecified. Thus, it is advisable to compute the coefficient of variation to assess whether the survey is worth doing. That is, if the $cv(\hat{D})$ is too large, then perhaps the survey will not provide any useful information and, therefore, should not be conducted. The equation to use is

$$cv(\hat{D}) = \left(\frac{b}{L(n_0/L_0)} \right)^{1/2}$$

For the example, if practical limitations allowed only $L = 10$ km,

$$cv(\hat{D}) = \left(\frac{3}{10(20/5)} \right)^{1/2} = 0.274 \text{ or roughly } 27\%$$

The investigator must then decide if this level of precision would adequately meet the survey objectives. If for example $\hat{D} = 100$, then an approximate 95% log-based confidence interval would be [59, 169]. This information might still be useful because the encounter rate is quite high in this example.

If animals occur in clusters, the above calculations apply to precision of the estimated density of clusters. That is, $\hat{D}$ becomes $\hat{D}_s$, the number of animal clusters per unit area. For clustered populations, a pilot survey yields an estimate of the standard deviation of cluster size,

$$\widehat{sd}(s) = \sqrt{\frac{\sum\limits_{i=1}^{n} (s_i - \bar{s})^2}{n-1}}$$

The coefficient of variation of mean cluster size for a survey in which n clusters are detected is then

$$\widehat{se}(\bar{s})/\bar{s} = \widehat{sd}(s)/(\bar{s} \cdot \sqrt{n})$$

For the case of cluster size independent of detection distance, we have

$$\{cv(\hat{D})\}^2 = \{cv(\hat{D}_s)\}^2 + \left[\frac{\widehat{sd}(s)}{\bar{s}} \right]^2 \cdot \frac{1}{n} \tag{7.2}$$

305

Now we substitute $n = L \cdot (n_0/L_0)$ and $\{\mathrm{cv}(\hat{D}_s)\}^2 = \dfrac{b}{L} \cdot \dfrac{L_0}{n_0}$ to get

$$\{\mathrm{cv}(\hat{D})\}^2 = \frac{b}{L} \cdot \frac{L_0}{n_0} + \left[\frac{\widehat{\mathrm{sd}}(s)}{\overline{s}}\right]^2 \cdot \frac{1}{L} \cdot \frac{L_0}{n_0} = \frac{1}{L} \cdot \frac{L_0}{n_0} \cdot \left[b + \left\{\frac{\widehat{\mathrm{sd}}(s)}{\overline{s}}\right\}^2\right]$$

We must select a target precision, say $\mathrm{cv}(\hat{D}) = \mathrm{cv}_t$. Solving for L gives

$$L = \frac{L_0 \, [b + \{\widehat{\mathrm{sd}}(s)/\overline{s}\}^2]}{n_0 \cdot \mathrm{cv}_t^2} \tag{7.3}$$

Suppose that a coefficient of variation of 10% is required, so that $\mathrm{cv}_t = 0.1$. Suppose further that, as above, $n_0 = 20$, $L_0 = 5$ and $b = 3$, and in addition $\widehat{\mathrm{sd}}(s)/\overline{s} = 1$. Then

$$L = \frac{5 \cdot (3+1)}{20 \cdot 0.1^2} = 100 \text{ km}$$

rather than the 75 km calculated earlier for 10% coefficient of variation on $\hat{D}_s$.

Paradoxically, these formulae yield a more precise estimate of population size for a population of (unknown) size $N = 1000$ animals, for which 50 animals are detected in 50 independent detections of single animals, than for a population of 1000 animals, for which 500 animals are detected in 50 animal clusters, averaging 10 animals each. This is partly because finite population sampling theory is not used here. If it was, variance for the latter case would be smaller, as 50% of the population would have been surveyed, compared with just 5% in the first case. A disadvantage of assuming finite population sampling is that it must be assumed that sampling is without replacement, whereas animals may move from one transect leg to another or may be seen from different legs. Use of finite population corrections is described in Section 3.7.5.

In some studies, animals occur in loose agglomerations. In this circumstance, it may be impossible to treat the population as clustered, due to problems associated with defining the position (relative to the centreline) and size of animal clusters. However, if individual animals are treated as the sightings, the usual analytic variance estimates are invalid, as the assumption of independent sightings is seriously violated. Resampling methods such as the bootstrap (Section 3.7.4) allow an analysis based on individual animals together with valid variance estimation.

306

(b) *Point transects* The estimation of sample size and number of points for point transect surveys is similar to that for line transects. The encounter rate can be defined as the expected number of detections per point, estimated in the main survey by n/k. Given a rough estimate n_0/k_0 from a pilot survey and the desired coefficient of variation, the required number of sample points can be estimated as

$$k = \left(\frac{b}{(\mathrm{cv}(\hat{D}))^2} \right) \cdot \left(\frac{k_0}{n_0} \right) \qquad (7.4)$$

As for line transect sampling, b may be approximated by n_0 multiplied by the square of the observed coefficient of variation for $\hat{D}$ from the pilot survey. If the pilot survey is too small to yield a reliable coefficient of variation, a value of 3 for b may again be assumed. If the shoulder of the detection function is very wide, this will tend to be conservative, but if detection falls off rapidly with distance from the point, a larger value for b might be advisable. Some advocates of point transects argue that detection functions for point transect data are inherently wider than for many line transect data sets, because the observer remains at each point for some minutes, ensuring that all birds within a few metres of the observer are recorded, at least for most species. For line transects, the observer seldom remains still for long, so that probability of detection might fall away more rapidly with distance from the line.

Having estimated the required number of points k, the number of objects detected in the main survey should be approximately $k \cdot n_0/k_0$. Suppose a pilot survey of 10 points yields 30 detected objects. Then, if the required coefficient of variation is 10% and b is assumed to be 3, the number of points for the main survey should be $k = (3/0.1^2) \cdot (10/30) = 100$, and roughly 300 objects should be detected.

The above calculations assume that the points are randomly located within the study area, although these procedures are also reasonable if points are regularly spaced on a grid, provided the grid is randomly positioned within the study area. If points are distributed along lines for which separation between neighbouring points on the same line is appreciably smaller than separation between neighbouring lines, precision may prove to be lower than the above equations would suggest, depending on variability in density; if objects are distributed randomly through the study area, precision will be unaffected.

Point transects have seldom been applied to clustered populations, although no problems arise beyond those encountered by line transect sampling. Equation 7.2 still applies, but the expression $n = k \cdot n_0/k_0$ should be substituted, giving

$$\{cv(\hat{D})\}^2 = \{cv(\hat{D}_s)\}^2 + \left[\frac{\widehat{sd}(s)}{\bar{s}}\right]^2 \cdot \frac{1}{k} \cdot \frac{k_0}{n_0}$$

In Equation 7.4, $\{cv(\hat{D})\}^2$ is replaced by $\{cv(\hat{D}_s)\}^2$. Solving for $\{cv(\hat{D}_s)\}^2$ and substituting in the above gives

$$\{cv(\hat{D})\}^2 = \frac{1}{k} \cdot \frac{k_0}{n_0} \cdot \left[b + \left\{\frac{\widehat{sd}(s)}{\bar{s}}\right\}^2\right]$$

Selecting a target precision $cv(\hat{D}) = cv_t$ and solving for k gives

$$k = \frac{k_0\{b + [\widehat{sd}(s)/\bar{s}]^2\}}{n_0 \cdot cv_t^2} \tag{7.5}$$

Continuing the above example, now with clusters replacing individual objects, the number of points to be surveyed is

$$k = \frac{10 \cdot \{3 + [\widehat{sd}(s)/\bar{s}]^2\}}{30 \cdot 0.1^2}$$

If the pilot survey yielded $\widehat{sd}(s)/\bar{s} = 1$ (a plausible value), then

$$k = \frac{10 \cdot (3 + 1)}{30 \cdot 0.1^2} = 133$$

so that roughly 133 points are needed.

7.2.3 Stratification

Sampling effort can be partitioned into several strata in large-scale surveys. This allows separate estimates of density in each stratum (such as different habitat types). Sampling can be partitioned into temporal strata during the day or seasonally. Post-stratification can be used in some cases. For example, the individual lines can be repartitioned by habitat type, based on a large-scale aerial photo on which line locations are drawn accurately. Thus, estimates of density by habitat type can be made. For example, Gilbert *et al.* (in prep.) used a geographic information system (GIS) in this manner for the long-term nesting studies of waterfowl at the Monte Vista National Wildlife Refuge.

For stratified survey designs, the formulae for sample size determination are more complex. The starting point for a given stratum is the formula

$$\text{var}(\hat{D}) = D^2 \left[\{\text{cv}(n)\}^2 + \{\text{cv}(\hat{f}(0))\}^2 \right]$$

Each of the two coefficients of variation is proportional to $1/E(n)$, hence

$$\text{var}(\hat{D}) = D^2 \left[\frac{b_1}{E(n)} + \frac{b_2}{E(n)} \right]$$

where $b_1 = \text{var}(n)/E(n)$ and $b_2 = E(n) \cdot \text{var}\{\hat{f}(0)\}/\{f(0)\}^2$
Now use

$$E(n) = \frac{2LD}{f(0)}$$

to get

$$\text{var}(\hat{D}) = D^2 \left[\frac{(b_1 + b_2) f(0)}{2LD} \right]$$

$$= \left[\frac{D}{L} \right] \left[\frac{(b_1 + b_2) f(0)}{2} \right]$$

If, over the different strata, the detection function is the same, then $f(0)$ and b_2 will be the same over strata. This is often a reasonable assumption. It is plausible that b_1 may be constant over strata; this can be checked by estimating $b_1 = \text{var}(n)/E(n)$ in each stratum. If these conditions hold, then for stratum v,

$$\text{var}(\hat{D}_v) = \left[\frac{D_v}{L_v} \right] K$$

for some K, which can be estimated. To allocate total line length effort, $L = \Sigma L_v$, we want to minimize the sampling variance of the estimated total number of objects in all strata, $\hat{N} = \Sigma A_v \hat{D}_v$, where A_v is the size of area v and summation is over $v = 1, 2, \ldots, V$. If we pretend that each $\hat{D}_v$ is independently derived (it should **not** be under these assumptions), then

$$\mathrm{var}(\hat{N}) = \sum \, [A_v]^2 \, \mathrm{var}(\hat{D}_v)$$

$$= K \sum \, [A_v]^2 \left[\frac{D_v}{L_v} \right] \qquad (7.6)$$

For fixed L, it is easy to minimize Equation 7.6 with respect to the L_v. The answer is expressible as the ratios

$$\frac{L_v}{L} = \frac{A_v \sqrt{D_v}}{\sum A_v \sqrt{D_v}} \qquad (7.7)$$

The total effort L comes from

$$[\mathrm{cv}(\hat{N})]^2 = \left[\frac{K}{L} \right] \left[\frac{\sum A_v \sqrt{D_v}}{\sum A_v \, D_v} \right]^2 \qquad (7.8)$$

Formula 7.7 shows that allocation proportional to $\sqrt{D_v}$ is not unreasonable if stratum sizes are similar. The result in Equation 7.7 is derived under an inconsistency in that given the assumptions made, $f(0)$ should be based on all distance data pooled and the estimators would look like

$$\hat{D}_v = \frac{n_v \cdot \hat{f}(0)}{2 L_v}$$

and

$$\hat{N} = \left[\sum_{v=1}^{V} \frac{A_v \, n_v}{2 L_v} \right] \hat{f}(0) \qquad (7.9)$$

The first order approximate variance of Equation 7.9 is expressible as

$$\mathrm{var}(\hat{N}) = N^2 \left[\frac{f(0)}{2} \right] \left[\frac{b_2}{\sum L_v D_v} + b_1 \sum \frac{(N_v/N)^2}{L_v \, D_v} \right]$$

from which we get an expression for the coefficient of variation of $\hat{N}$ in this case of using pooled distances to get one $\hat{f}(0)$:

$$\mathrm{cv}(\hat{N}) = \left[\frac{b_2 \, f(0)}{2L} \right] \left[\frac{1}{\sum \pi_v \, D_v} + R \sum \frac{p_v^2}{\pi_v \, D_v} \right] \qquad (7.10)$$

310

where

$$R = \frac{b_1}{b_2}$$

$$p_v = \frac{N_v}{N} = \frac{A_v D_v}{\sum A_v D_v}$$

and the relative line lengths by stratum are

$$\pi_v = \frac{L_v}{L}$$

Thus, given L, the allocation problem is to minimize Equation 7.10.

We can use the Lagrange multiplier method to derive the equations to be solved for the optimal $\pi_1, \pi_2, \ldots, \pi_V$. Those equations can be written as

$$\frac{D_j}{\left(\sum \pi_v D_v\right)^2} + R \frac{p_j^2}{\pi_j^2 D_j} = \left[\frac{1}{\sum \pi_v D_v} + R \sum \frac{p_v^2}{\pi_v D_v} \right] \quad j = 1, \ldots, V$$

Fixed point theory can sometimes be used to solve such equations numerically; in this case, it seems to work well. The previous V equations are rewritten below and one must iterate until convergence to compute the π_j. This method is related to the EM algorithm in statistics (Dempster *et al.* 1977; Weir 1990).

$$\pi_j = \sqrt{\frac{R \cdot p_j^2 / D_j}{\left[\frac{1}{\sum \pi_v D_v} + R \sum \frac{p_v^2}{\pi_v D_v} \right] - \frac{D_j}{\left(\sum \pi_v D_v \right)^2}}} \quad j = 1, \ldots, V$$

We programmed these in SAS, explored their behaviour, and concluded that a good approximation to the optimal π_j is to use $\pi_j = p_j$, $j = 1, 2, \ldots, V$. Thus, approximately in this case of pooled distance data,

$$\pi_j = \frac{A_j D_j}{\sum A_v D_v} \tag{7.11}$$

Note the relationship between Equations 7.11 and 7.7. Optimal relative line lengths (i.e. $\pi_1, \ldots, \pi_V$) should fall somewhere between the results

311

of Equations 7.7 and 7.11; the exact values of $\pi_1, \ldots, \pi_V$ are not as critical to the precision of $\hat{D}_1, \ldots, \hat{D}_V$ as the total line length L.

7.2.4 Trapping webs

Trapping webs represent an application of point transect theory (Anderson *et al.* 1983). The method has been evaluated by computer simulation (Wilson and Anderson 1985b) and on known populations of beetles (Parmenter *et al.* 1989) and has performed well. The method was conceived for use in trapping studies of small mammals where the estimation of population density was of interest. The use of distance sampling theory relaxed the assumptions of traditional capture–recapture models (essentially ball and urn models). The method may perform well for populations whose members move relatively little or have somewhat fixed home ranges. Populations of individuals that move randomly over areas large in relation to the trapping web are problematic. In this respect, the positive results found by Parmenter *et al.* (1989) may have been somewhat fortuitous, or at least, require some alternative analysis methods (Section 6.11).

The design of studies using the trapping web approach should be laid out as in Fig. 1.7. As with point transects, some movement of objects through time will result in objects being overrepresented by the traps near the centre of the web, thus leading to overestimation of population density. Placing additional traps near the centre of the web may exacerbate this overestimation, and is not now recommended. Use of at least eight lines is suggested, and 10, 12 or even 16 might be considered. Guidelines for the number of traps are less well defined, although a practical objective is to obtain a sample of trapped animals of at least 60–80, and preferably around $n = 100$. A pilot study using 100–150 traps may often lead to insight on the number required to achieve an adequate sample size. A variety of traps can be used, including snap, live or pitfall traps. Animals, of course, do not have to be marked, unless they are to be returned to the population and thereafter ignored in future samples ('removal by marking'). Simple marking with a felt-tip pen will often suffice.

Trap spacing remains to be studied and we offer only the guideline that traps be spaced along the lines at a distance roughly equal to half the home range diameter of the species being studied. Wilson and Anderson (1985b) suggested 4.5–8 m spacing for mice, voles or kangaroo rats and 8–12 m spacing for larger mammals such as rabbits or ground squirrels. Commonly, captured animals are removed from the population after their initial capture. The field trapping can be done over sequential nights (or days) until it seems clear that no new animals are being caught near the centre. Alternatively, if at the centre of the web most animals that have been marked and released have subsequently been recaptured, then one might conclude that sufficient trapping occasions have been carried out.

Trapping webs can be established using a stake at the web centre and a long rope with knots to denote the trap spacing. Then, the investigator can travel in a circle laying out traps in roughly straight lines radiating from the centre. However, it is not important the traps be on perfectly straight lines. In multiyear surveys, the location of each trap is often marked by a numbered metal stake. We recommend that recaptures of released animals are recorded, and that each trap has a unique number, allowing captures to be assigned to traps. These data allow assumptions to be assessed, and additional analytic methods, such as bootstrap sampling within a web, to be implemented. Further research on the trapping web is needed before more detailed guidelines can be given. DISTANCE can perform analyses on single or multiple trapping web surveys, and provides a useful tool for such research.

7.3 Searching behaviour

Line and point transects are appropriately named because so much that is critical in this class of sampling methods is at or near the line or point. Search behaviour must try to optimize the detection of objects in the vicinity of the line or point, and search effort or efficiency should decrease smoothly with distance. The aims are to ensure that the detection function has a broad shoulder and the probability of detection at the line or point is unity ($g(0) = 1$).

(a) *Line transects* In line transect surveys, the above aims might be enhanced by moving slowly, emphasizing search effort on and near the line, having two or more observers traverse the transects, or using aids to detection such as binoculars. In surveys carried out by foot, the observer is free to use a trained dog, to walk slowly in clumps of heavy cover and faster in low or less suitable cover, or stop frequently to observe. The observer may leave the centreline temporarily, provided he or she records detection distances from the transect line, not from his or her current position. Aerial surveys commonly employ two observers, one covering each side of the aircraft, in addition to the pilot, who might guard the centreline. Shipboard surveys frequently use three or more observers on duty at any one time. In many surveys, it is good practice to look behind occasionally in case an object that was hidden on first approach can be seen.

The survey must be conducted so as to avoid undetected movement away or toward the line or point in response to the observer. To achieve this, most detection distances should exceed the range over which objects might respond to the observer. If a motorized observation platform is used, the range of response might be reduced by using platforms with quiet motors or by travelling faster. In surveys carried out by foot, the observer can ensure more reliable data by moving quietly and unobtrusively. Detection distances can be improved by use of binoculars. If detection cues are continuous, high power binoculars might be used, for example tripod-mounted 25 × binoculars on shipboard surveys of dolphins that typically occur in large schools. If cues are discrete, for example whales surfacing briefly, or songbirds briefly visible amongst foliage, lower magnification is necessary, so that field of view is wider. Indeed, binoculars are often used only to check a possible detection made by the naked eye. In some studies, one observer might scan continuously with binoculars while another searches with the naked eye. Tape cassette players are sometimes used to elicit calls from songbirds, although the observer should avoid attracting birds in towards the transect line. (Note that regular use of tape-cassette players in the territories of some species can cause unacceptable disturbance.)

Certain types of double counting can be problematic. If the objects of interest are immobile objects such as nests, then the fact that a particular nest is detected from two different lines or points is fully allowed under the general theory. Double counting becomes a potential problem only if motile objects are surveyed such that the observer or the observation platform chases animals from one line or point to another or if animals 'roll ahead' of the observer, hence being counted more than once (e.g. 'chain flushes' in surveys of grouse). Movement in response to the observer that leads to double counting should be recognized and avoided in the planning and conduct of a survey.

Although the analysis theory allows the observer to search on only one side of the line (i.e. $\hat{D} = n \cdot \hat{f}(0)/L$), we caution against this practice unless the animal's position relative to the line can be determined reliably and animal movement is relatively minor. If there is a tendency to include animals from the non-surveyed side of the line, then counts near the line will be exaggerated (this is a special type of heaping) and density will be overestimated. Animal movement from one side of the line to the other adds further complications and possible bias in the estimators. No problem would be anticipated if a single observer searches through a side window of an aircraft because there is little chance of including animals from the non-surveyed side of the transect (unless, again, undetected movement is taking place ahead of the observer's view). However, if the aircraft has forward visibility, such as a helicopter, there may be a tendency to include animals on both sides of and very near to the line into the first distance category.

Two alternatives exist for aerial surveys where forward visibility is good but only one observer is available; both involve searching both sides of the line. First, the observer could search a more narrow transect (smaller w) on both sides of the line. This procedure would concentrate most of the searching effort close to the line and this would help ensure that $g(0) = 1$. Second, and perhaps less satisfactory, the width of the transect could be larger on one side of the line than the other side. This would result in an asymmetric detection function and could be more difficult to model. Theory allows asymmetry in $g(x)$, and, if modelling proved too problematic, one could always truncate the distance data and alleviate the problem. In all cases, one should always be cautious to make sure that animals close to the line are not missed. Whenever possible, more than one observer should be used in aerial surveys.

Survey design such as searching only one side of the line illustrates the importance of carefully considering the assumptions of the theory in deciding how best to conduct a survey. Surveying only one side of the line makes the assumptions about movement and measurement error crucial because they will more directly affect the data near the transect centreline. Errors in assigning the detection of an animal to the left or right side of the line are irrelevant if both sides of the line are surveyed, but they are critical if only one side is surveyed. Data near the centreline are most important in obtaining valid estimates of density.

(b) *Point transects* For point transect surveys, the longer the observer remains at each point, the more likely is the probability of detection at the point to be unity, and the broader is the shoulder of the detection function. This advantage is offset by possible movement of objects into the sampled area, which leads to overestimation of density. Optimal time to spend at each point might be assessed from a pilot study. In some cases, it might be useful to observe the point from a short distance and record distances to objects of interest before any disturbance caused by the approach of the observer. Another option is to wait at the point a short period of time before recording, to allow objects to resume normal behaviour. As for line transects, binoculars may be useful for scanning, for checking possible detections, or for identifying the species. Tape cassette recorders may help elicit a response, but as for line transects, great care must be taken not to attract objects towards the observer. After the recording period, the observer may find it necessary to approach an object detected during that period, to identify it.

Detection distances can also be measured out before moving to the next point. If distances are assessed by eye, the task is made easier by use of markers at known distances.

If the radius of each point is fixed at some finite w, one could consider the population 'closed' and use a removal estimator to estimate the population size N (White *et al.* 1982: 101–19). To keep the option of this approach open, the time (measured from the start of the count at the point) at which each object is first detected should be recorded. The count period may then be divided into shorter time intervals, and data for each interval pooled across points. The relevant data would be the number of objects detected in the first time interval, the number of **new** objects detected in the second time interval, and so on. The theory exists, but it has not been used in this type of application. We recommend experimentation with this approach, perhaps with relatively small truncation distances w so that heterogeneity in probability of detection is reduced, as a check on the point transect estimates. Of special interest with such time/distance data is the potential to check that no new detections occur near the point towards the end of the counting period.

(c) *General comments* Ideally, provided $g(0) = 1$, one would like to collect distance data with a very broad shoulder. The choice of an adequate model for $g(y)$ is then relatively unimportant, and D can be estimated with good precision. For many studies, proper conduct of the survey can achieve high detection probabilities out to some distance. Many of the methods employed to ensure $g(0) = 1$ also help to widen the shoulder of the detection function.

Survey data for which the detection function drops off quickly with distance from the line or point, with a narrow shoulder and long tail, are far from ideal (Fig. 2.1). Model selection is far more critical and precision is compromised. Occasionally, little can be done at the design stage to avoid spiked data, but usually, such data indicate poor survey design or conduct (e.g. poor allocation of search effort near the line or point, poor precision in distance or angle estimation, or failure to detect objects prior to responsive movement towards the observer).

In multispecies surveys in diverse or complex habitats, there are likely to be errors in species identification (Bart and Schoultz 1984). As density increases, 'swamping' may occur; accurate data recording might be compromised by the number of sightings, calls, and other cues experienced during a short time interval (Bibby *et al.* 1985). Here, the binomial method of Järvinen and Väisänen (1975; line transects) or Buckland (1987a; point transects), in which distances are assigned to one of just two distance intervals, might be considered, especially if estimates of only **relative** abundance are required.

316

7.4 Measurements

Accurate measurement of distances and angles is quite important. The observer must work carefully and avoid errors in recording or transcribing data. Ancillary data, such as sex, species, and habitat type, are often taken. These data are partitioned by individual line or sample point and recorded. A field form is suggested to structure the recording of data. A field form for recording data is efficient and nearly essential. Figure 7.3 shows two examples; another was presented by Burnham *et al.* (1980: 34). The format for such field forms can usually be improved upon after use during the pilot study. Note-taking on various aspects of the survey should be encouraged and these can be recorded on separate sheets.

Fatigue can compromise accurate data, thus the field effort must consider the time spent surveying each day. Certainly it is unreasonable to believe that an observer can remain at peak searching ability throughout a 7–10-hour day. Fatigue may play a larger role in aerial surveys or foot surveys in difficult terrain. These are important issues and this section provides guidance on data collection.

The careful measurement or estimation of distances near the line or point is critical. **In summary, every possible effort must be made to ensure that accurate measurements are made, prior to any undetected movement, of *all* objects on or near the line or point.** This cannot be overemphasized.

7.4.1 Sighting distance and angle data

For point transects, analyses are based on observer-to-object distances, but for line transects, the widely used methods all require that the shortest distance between a detected object and the line is recorded or estimated. By the time the observer reaches the closest point on the line, the object may not be visible or may have moved in response to the observer's presence. These problems are minor for aircraft surveys in which the speed of the observation platform is sufficient to render movement of the object between detection and the point of closest approach unimportant. For shipboard surveys of marine mammals, sighting distances are frequently several kilometres, and it may take up to half an hour to arrive at the point of closest approach. Further, for many surveys, it is necessary to turn away from the centreline when an animal cluster is detected, both to identify and to count the animals in the cluster. Hence the natural distance to record is the sighting or radial distance r; by recording the sighting angle θ also, the shortest distance between the animal and the line, i.e. the perpendicular distance x, may be calculated as $x = r \cdot \sin(\theta)$ (Fig. 1.5). However, rounding errors in the data cause problems. Angles are seldom recorded to better accuracy

Study area _____ Cloud cover (%) _____ Wind speed _____

Observer name _____ Date _____

Line number _____ Line length (km) _____ Start time _____ End time _____

Sighting number	Perpendicular distance	Covey size	Number of		
			Males	Females	Unknown
1	____	__	__	__	__
2	____	__	__	__	__
3	____	__	__	__	__
⋮					

Sighting number	Covey size	Perpendicular distance interval (m)				
		0 – 50	50 – 100	100 – 150	150 – 250	250 – 400
1	__	__	__	__	__	__
2	__	__	__	__	__	__
3	__	__	__	__	__	__
⋮						

Fig. 7.3. Two examples of a hypothetical recording form for a line transect survey of grouse. The example at the top is for taking ungrouped perpendicular distance data for coveys and the sex of covey mates as ancillary information. The example at the bottom allows for recording of covey sizes and grouped perpendicular distance data. Information on each line, such as its length and the proportion of that length in each habitat type, would be recorded just once on a separate form. Most surveys are somewhat unique, requiring specialized forms for use in the field.

than the nearest 5°, so that an animal recorded to have a sighting distance $r = 8$ km and sighting angle $\theta = 0°$ will have a calculated perpendicular distance of $x = 0$ km, when the true value might be

318

$x = 350$ m or more. Since estimation of abundance depends crucially on the value of the fitted probability density for perpendicular distances evaluated at zero distance, $\hat{f}(0)$ (Burnham and Anderson 1976), the false zeros in the data may adversely affect estimation. The problem is widespread, and more than 10% of distances are commonly recorded as zero, even for land surveys in which distances and angles are apparently measured accurately (e.g. Robinette *et al.* 1974). Possible solutions, roughly in order of effectiveness, are:

1. Record distances and angles more accurately
2. 'Smear' the data (see below)
3. Use models for the detection function that always have a shoulder
4. Group the data before analysis
5. Use radial distance models.

Only the first of these solutions comes under the topic of this chapter, but we cover the others here for completeness, and to emphasize that solution 1, better survey design, is far more effective than the analytic solutions 2–5.

1. Improving accuracy in measuring angles and distances is certainly the most effective solution. It may be achieved by improving technology, for example by using binoculars with reticles (graticules) or range finders, and using angle boards or angle plates on tripods. Most important is that observers must be thoroughly trained, and conscientious in recording data; there is little benefit in using equipment that enables angles to be measured to the nearest degree if observers continue to record to the nearest 5°. Training should include explanation of why accuracy is important, and practice estimates of distances and angles for objects whose exact position is known should be made, under conditions as similar as possible to survey conditions.

2. The concept of 'smearing' the data was introduced by Butterworth (1982b). Although often criticized, for example by Cooke (1985), the technique has become widely used for data from cetacean shipboard surveys. It is an attempt to reduce the effects on the estimates of recording inaccurate locations for detections, through rounding sighting distances and angles to favoured values. When rounding errors occur, the recorded position of an animal may be considered to be at the centre of a sector called the 'smearing sector' (Fig. 7.4); the true position of the animal might be anywhere within the sector. Butterworth and Best (1982) assigned each perpendicular distance a uniform distribution over the interval from the minimum distance between the sector and the centreline to the maximum distance, and selected a distance at random from this distribution to replace the calculated perpendicular distance. Hammond (1984) compared this with assigning a uniform distribution

319

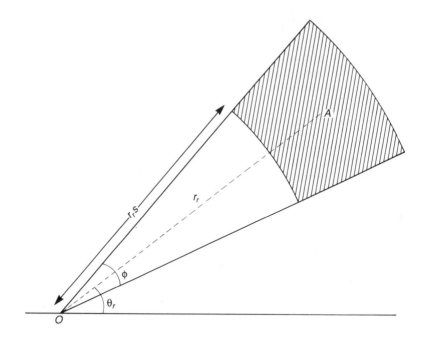

Fig. 7.4. The observer at O records an animal at position A, at radial distance r_r and with sighting angle θ_r. The true position of the animal is considered to be anywhere within the shaded smearing sector. The size of the sector is determined by smearing parameters ϕ and s.

over the sector, selecting a new sighting distance/angle pair at random from the sector and calculating the corresponding perpendicular distance. He also investigated assigning a normal distribution to both the distance and the angle instead of a uniform distribution. He concluded that the degree and method of smearing had relatively little effect on estimation of $f(0)$, but that estimation under either method was improved relative to the case of unsmeared data.

If the data are grouped before analysis, it is unnecessary to sample at random from the assumed distribution within the smearing sector. For example if smearing is uniform over the smearing sector, the sector can be considered to have an area of unity, and the proportion of the sector within each perpendicular distance interval may be calculated. This is carried out for each observation and the resulting proportions are summed within each interval. They can then be rounded to the nearest integer values and line transect models applied in the normal way for grouped data. Alternatively the methods of Section 3.4 (grouped data) follow through when the 'frequencies' are not integer, so that

320

rounding is not required. This approach is described by Buckland and Anganuzzi (1988a).

Values must be assigned to the smearing parameters to control the level of smearing. Butterworth (1982b) incorporated time and vessel speed in his routine, since distance was calculated as speed by time taken to close with a whale or whales. The values for the smearing parameters were selected in a semi-arbitrary manner, by examining the apparent accuracy to which data were recorded. Hammond and Laake (1983) chose the level of smearing in a similar way, although the method of smearing was different; the semicircle ahead of the vessel was divided into smearing sectors so that any point within the semicircle fell in exactly one sector. Objects (in this case, dolphin schools) recorded as being in a given sector were smeared over that sector. Butterworth *et al.* (1984) used data from experiments with a buoy that was fitted with a radar reflector to estimate smearing parameters. None of these offer a routine method for smearing, whereas the angle and distance data contain information on the degree of rounding, suggesting that estimation of the smearing parameters from the data to be smeared should be possible. Buckland and Anganuzzi (1988a) suggested an *ad hoc* method for this. Denote the recorded sighting distance and angle by r_r and θ_r respectively, and the corresponding smearing parameters by s and ϕ (Fig. 7.4), to be estimated. Then the smearing sector is defined between angles $\theta_r - \phi/2$ and $\theta_r + \phi/2$, and between radial distances $r_r \cdot s$ and $r_r \cdot (2 - s)$. Smearing is uniform over the sector, and grouped analysis methods are used, so that Monte Carlo simulation is not required (above). This is the method recommended by Buckland and Anganuzzi, although they also considered two improvements to it. First, rounding error increases with distance from the observer, so that a recorded distance of 1.3 km say is more likely to be rounded down to 1.0 km than 0.7 km is to be rounded up. This may be accounted for by defining the smearing sector between radial distances $r_r \cdot s$ and r_r/s. Second, there are fewer observations at greater perpendicular distances, since the probability of detection falls off. Hence smearing should not be uniform over the smearing sector, but should be weighted by the value of a fitted detection function at each point in the sector. The recommended method therefore has two identifiable sources of bias. One leads to oversmearing, and the other to undersmearing. Buckland and Anganuzzi concluded that the more correct approach did not lead to better performance, apparently because the two sources of bias tend to cancel, and considered that the simpler approach was preferable.

Buckland and Anganuzzi (1988a) estimated the smearing parameters by developing an *ad hoc* measure of the degree of rounding in both the angles and the distances. In common with Butterworth (1982b), they found that errors seemed to be larger in real data than the degree of rounding suggests. They therefore introduced a multiplier to increase the level of smearing and investigated values from 1.0 to 2.5. They noted that undersmearing was potentially more serious than oversmearing, and recommended that the estimated smearing parameters be multiplied by two, which would be correct for example if an angle between 5° and 10° was rounded at random to either endpoint of the interval rather than rounded to the nearest endpoint.

The above methods are all *ad hoc*. Methodological development is needed here to allow the rounding errors to be modelled.

3. If many perpendicular distances are zero, a histogram of perpendicular distances appears spiked; that is the first bar will be appreciably higher than the rest. If, for example, the exponential power series model is fitted to the data, it will fit the spike in the data, leading to a high value for $\hat{f}(0)$ and hence overestimation of abundance. Models for which $g'(0)$ is always zero (i.e. the slope of the detection function at $x = 0$ is zero) are usually less influenced by the erroneous spike, and are therefore more robust. This does not always follow; if distance data fall away very sharply close to zero, then only very slowly at larger distances, the single-parameter negative exponential model is unable to fit the spike, whereas the more flexible two-parameter hazard-rate model can. If the spike is spurious, the negative exponential model can fortuitously provide the more reliable estimation (Buckland 1987b), although its lack of flexibility and implausible shape at small perpendicular distance rule it out as a useful model.

4. If data are grouped such that all perpendicular distances that are likely to be rounded to zero fall in the first interval, the problem of rounding errors should be reduced. This solution is less successful than might be anticipated. First, interval width may be too great, so that the histogram of perpendicular distances appears spiked; in this circumstance, different line transect models can lead to widely differing estimates of object density (Buckland 1985). Second, the accuracy to which sighting angles are recorded often appears to be quite variable. If a detection is made at a large distance, the observer may be more intent on watching the object than recording data; in cetacean surveys the animal may no longer be visible when he/she estimates the angle. Thus for a proportion of sightings, the angle might only be recorded to the nearest 10° or 15°, and 0° is a natural value to round to when there is considerable uncertainty. An attempt to impress upon observers that they should not round angles to 0° in minke whale surveys in the Antarctic led to considerable rounding to a sighting angle of 3° on one vessel!

5. Because rounding errors in the angles are the major cause of heaping at perpendicular distance zero when data are recorded by

sighting angle and distance, it is tempting to use radial distance models to avoid the difficulty. Such models have been developed by Hayne (1949), Eberhardt (1978a), Gates (1969), Overton and Davis (1969), Burnham (1979) and Burnham and Anderson (1976). However, Burnham *et al.* (1980) recommended that radial distance models should not be used, and Hayes and Buckland (1983) gave further reasons to support this recommendation. First, hazard-rate analysis indicates that r and θ are not independently distributed, whereas the models developed by the above authors all assume that they are. Second, hazard-rate analysis also suggests that if detectability is a function of distance r but not of angle θ, then the expected sighting angle could lie anywhere in the interval 32.7° to 45°, whereas available radial distance models imply that it should be one or the other of these extremes, or use an *ad hoc* inter-polation between the extremes. Third, all models utilize the reciprocal of radial distances, which can lead to unstable behaviour of the estimator and large variances if there are a few very small distances. Fourth, despite claims to the contrary, it has not been demonstrated that any existing radial distance models are model robust. A model might be developed from the hazard-rate approach, but it is not clear whether it would be pooling robust, or whether typical data sets would support the number of parameters necessary to model the joint distribution of (r, θ) adequately. We therefore give a strong recommendation to use perpendi-cular distance models rather than any existing radial distance model.

7.4.2 Ungrouped data

The basic data to be recorded and analysed are the n distances. Gener-ally, these are the individual perpendicular distances x_i for line transect surveys or the sighting distances r_i for point transect surveys. Alterna-tively, in line transect surveys, the sighting distances r_i and sighting angles θ_i can be measured, from which $x_i = r_i \cdot \sin(\theta_i)$ (above). These ungrouped data are suitable for analysis, especially if they are accurate. Heaping at zero distance is especially problematic, again illustrating the need to know the exact location of the line or point. Well-marked, straight lines are needed for line transect surveys. Upon detection of an object of interest, the surveyor must be able to determine the exact position of the line or point, so that the proper measurement can be taken and recorded. If sighting angles are being measured, a straight line is needed or the angle will not be well defined.

We recommend the use of a steel tape for measurements for foot surveys of terrestrial populations up to about 30 m. If a stick or lath is used for 'beating', it can be marked off in appropriate measures and used as a measuring stick. If this is to be done, it is wise to use a yard or metre rule. Many surveyors have successfully used a range finder in obtaining estimates of distances out to about 100 m. We discourage the use of visual observation alone in estimating distances and angles. Unless the observers are unusually well trained, such a procedure invites heaping of measurements (at best) or biased estimates of distance with different biases for different observers (at worst). Scott *et al.* (1981) found significant variability in precision among observers and avian species, but no bias in the errors. Often, even simple pacing is superior to ocular estimation.

Observers have a tendency to record objects detected just beyond *w* as within the surveyed area. This might be called 'heaping at *w*' and was noted in the surveys at the Monte Vista National Wildlife Refuge (Chapter 8) where *w* = 8.25 or 12 ft in differing years. In either case, there were more observations in the last distance category than expected for nearly all years.

For shipboard surveys, sighting distances are frequently estimated using reticles or graticules, which are marks on binocular lenses (Fig. 7.5).

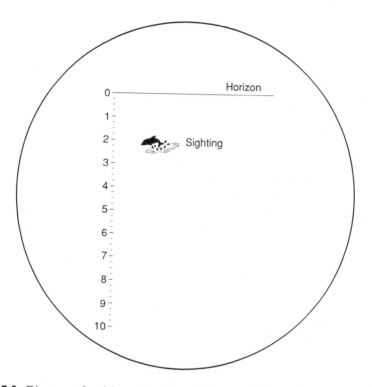

Fig. 7.5. Diagram of reticles used on binoculars on shipboard surveys of marine mammals. Use of these marks allows the computation of sighting distance (text and Fig. 7.6).

The observer records the number of marks down from the horizon to the detected object, and these values are transformed to distances from the observer as follows (Fig. 7.6).

Let R = radius of the Earth $\doteq 6370$ km;

$\quad v$ = vertical height of the binoculars above the sea surface;

$\quad \delta$ = angle of declination between successive divisions on the reticle;

$\quad \phi$ = angle between two radii of the Earth, one passing through the observer and the other passing through any point on the horizon, as seen by the observer

$\quad = \cos^{-1}\{R/(R + v)\}$

Now suppose that the reticle reading is d divisions below the horizon, so that the angle of declination between the horizon and the sighting is $\psi = d \cdot \delta$. Then the sighting distance is approximately

$$r = \frac{v}{\tan(\phi + \psi)}$$

For example, if the observer's eyes are 10 m or 0.01 km above sea level, the angle between successive divisions of the reticle is 0.1°, and the reticle reading is 3.6 divisions below the horizon, then

$$\phi = \cos^{-1}\{6370/(6370 + 0.01)\} = 0.10° \text{ and } \psi = 0.36°$$

so that

$$r = \frac{0.01}{\tan(0.46°)} = 1.25 \text{ km}$$

Note that the horizon is at $h = R \cdot \tan(\phi) = 11.3$ km (Fig. 7.6). These calculations ignore the effects of light refraction, which are generally small for sightings closer than the horizon.

If binoculars are tripod-mounted, sighting angles can be accurately measured from an angle ring on the stem of the tripod, provided observers are properly trained, and the importance of measuring angles accurately is stressed. If binoculars are hand-held, angle boards (Fig. 7.7), perhaps mounted on ship railings, may be found useful; although accuracy is likely to be poor relative to angle rings on tripods, it should still be appreciably greater than for guessed angles. Distance and angle experiments, using buoys at known positions, should be carried out if at all possible, both to estimate bias in measurements and to persuade observers that guesswork can be poor!

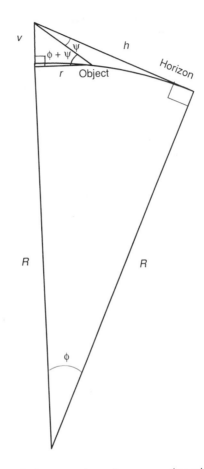

Fig. 7.6. Geometry of the procedure for computing sighting distances from ocular data for shipboard surveys of marine mammals. Reticles provide an estimate of ψ, the angle of declination of the detected object from the horizon, which must be converted into the distance r from the observer to the object.

Ungrouped sightings data are seldom collected for aerial surveys (see below), but if terrain is uneven, so that it is not possible to fly at a fixed altitude, perpendicular distance can be estimated by recording both the angle of declination and the altitude at the time the aircraft passes the detected object.

7.4.3 Grouped data

It is sometimes difficult to measure distances exactly and, therefore, it might be convenient to record data only by distance interval. Then, the

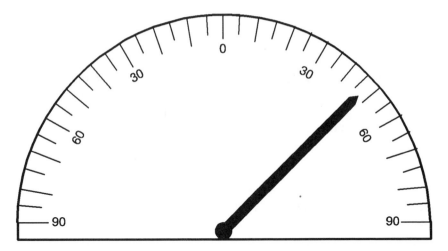

Fig. 7.7. Sighting angles can often be more accurately estimated by the use of an angle board as shown here. Such devices can be hand made and are useful in many applications of distance sampling.

exact distance of an object detected somewhere between, say, 0 and 40 m will not be recorded, but only that it was in the distance interval 0–40. During the course of the survey, a count n_1 will be made of objects in this first distance interval. The survey results will be the frequencies $n_1, n_2, \ldots, n_u$ corresponding to the u distance classes with total sample size $n = \sum n_i$.

In general, let c_i denote the designated distance from the line or point and assume that we have u such distances: $0 = c_0 < c_1 < c_2 \ldots < c_u = w$. In the case of left truncation, $c_0 > 0$. Note, also, that c_u can be finite or infinite. These 'cutpoints' result in u distance intervals and the grouped data are the frequencies n_i of objects detected in the various intervals. Specifically, let n_i = the number of objects in distance interval i corresponding to the interval (c_{i-1}, c_i). If at all possible, there should be at least two intervals in the region of the shoulder. In general, the width of the distance intervals should increase with distance from the line or point, at least at the larger distances. The width of each distance interval might be set so that the n_i would be approximately equal. This rough guideline can be implemented if data from a pilot study are available. Alternatively, if the underlying detection function is assumed to be approximately half-normal, then Table 7.1 indicates a reasonable choice for group interval widths for various u, where Δ must be selected by the biologist. Thus for a line transect survey of terrestrial animals, if $u = 5$ distance intervals were required, and it was thought

327

that roughly 20% of detections would be beyond 500 m, then $\Delta = 100$ m, and the interval cutpoints are 100 m, 200 m, 350 m, 500 m and ∞. The grouped data would be the frequencies $n_1, n_2, \ldots, n_5$. As a guideline, u, the number of distance classes in line transect surveys, should not be less than four and five is much better than four. Too many distance intervals tend to defeat the advantages of such grouping; certainly 7–8 intervals should be sufficient in most cases. Defining too many intervals makes classification of objects into the correct distance interval error-prone and time-consuming. In addition, the use of too many distance intervals distracts attention from the main goal: detections near the line or point.

Table 7.1 Suggested relative interval cutpoints for line and point transects. An appropriate value for Δ must be selected by the user

Number of intervals, u	Suggested relative interval cutpoints for line transects	Suggested relative interval cutpoints for point transects
4	Δ, 2Δ, 4Δ, ∞	2Δ, 3Δ, 4Δ, ∞
5	Δ, 2Δ, 3.5Δ, 5Δ, ∞	2Δ, 3Δ, 4Δ, 5.5Δ, ∞
6	Δ, 2Δ, 3Δ, 5Δ, 7Δ, ∞	2Δ, 3Δ, 4Δ, 5Δ, 6.5Δ, ∞
7	Δ, 2Δ, 3Δ, 4.5Δ, 6Δ, 8Δ, ∞	2Δ, 3Δ, 4Δ, 5Δ, 6Δ, 7.5Δ, ∞
8	Δ, 2Δ, 3Δ, 4Δ, 5.5Δ, 7Δ, 9.5Δ, ∞	2Δ, 3Δ, 4Δ, 5Δ, 6Δ, 7Δ, 8.5Δ, ∞

Collection of grouped data allows a relaxation of the assumption that distances are measured exactly. Instead, the assumption is made only that an object is counted in the correct distance interval. Holt and Powers (1982) reported on an aerial survey of several species of dolphin where counts were made by the following distance intervals: 0.0, 0.05, 0.15, 0.25, ... nautical miles. Terrestrial surveys of jackrabbits might use 0, 50, 100, 175, 250, ∞ m. Note, here the final distance interval is between 250 m and ∞. As few as two distance intervals (i.e. 'binomial' models) are sometimes used in point transect surveys (Buckland 1987a) and in line transect surveys (Järvinen and Väisänen 1975; Beasom *et al.* 1981), although, no goodness of fit test can be made. In general, the use of between five and seven distance intervals will be satisfactory in many line transect surveys.

It is commonly thought that all objects in the first distance interval must be detected (i.e. a census of the first band). This is incorrect; the width of this interval might be 40 m and it is not necessary that all objects be detected in the 0–40 m band. Of course, as the shoulder in the data is broadened, there are significant advantages in estimation. As a guideline, we recommend that the probability of detection should not fall appreciably below unity over at least the first two intervals.

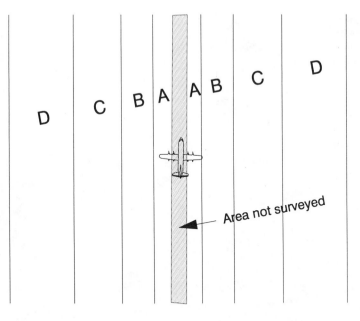

Fig. 7.8. The area below an aircraft can be excluded as shown (shaded area). Here grouped data are recorded in four distance intervals (**A–D**) of increasing widths with distance. Adapted from Johnson and Lindzey (unpublished).

Nearly all aerial surveys collect grouped data. The proper speed and altitude above the ground can be selected after some preliminary survey work. Here it may not be practical to record counts for the first distance interval because visibility below the aircraft is impaired (Fig. 7.8). Ideally the altitude would be high enough so as to leave the objects undisturbed and, thus, avoid movement prior to detection. After this consideration, the aircraft should be flown as low as practical to enhance detection of objects. The altitude should be recorded occasionally during the survey to be sure the pilot is flying at the proper height. The distance intervals may be substantially in error if altitude is not as recorded or if the altitude varies due to terrain. Markers of some type are typically fixed to the aircraft to delineate the distance intervals on the ground for a fixed height above ground (Fig. 7.9). Two sets of markers are required (like the front and rear sight on a rifle); usually markers can be placed on the aircraft windows and wing struts. Observers should be cautioned not to assign objects to distance intervals until they are nearly

Fig. 7.9. Airplane wing struts can be marked to delineate boundaries of the distance intervals on the ground. Other marks on the side window of the airplane are used to assure proper classification of animals to the correct distance interval. Compare with Fig. 7.8. From Johnson and Lindzey (unpublished).

perpendicular to the aircraft. If such assignment is attempted while the object is still far ahead of the aircraft, there is a tendency to assign incorrectly the object to the next largest distance interval (this problem is related to parallax). Occasionally observations are made from only one side of the aircraft, but this is fairly inefficient, often problematic, and should be used only in unusual situations.

Some types of aircraft are far better for biological surveys than others (Fig. 7.10). Ideally, the aircraft should allow good visibility ahead of and directly below the observer. Some helicopters meet these requirements, but are expensive to rent and operate. Airplanes with a high wing and low or concave windows can also make excellent platforms for aerial detection, and craft with clear 'bubbles' at the nose, designed for observation work, are available.

The density of many populations of interest is fairly low, so that recording the counts n_i (and perhaps cluster sizes) can be done by hand

Fig. 7.10. Some aircraft are specifically designed for aerial observation. Note the forward and lateral visibility and the high wing on the aircraft. Helicopters, while more expensive, often provide similar advantages plus the ability to hover or proceed more slowly than a fixed-wing aircraft.

without distracting the observer and, thus, failing to monitor the line. However, it is often best to use a tape recorder, ideally with an automatic time signal, so that the observer can continue searching without distraction. In some cases it might be feasible to use a laptop computer to record data. Some aerial surveys have used the LORAN C navigation system to maintain a course on the centreline, monitor altitude, position and distances (Johnson and Lindzey unpublished). A video camera has been mounted in the aircraft to record the area near the line in pronghorn surveys in Wyoming (F. Lindzey, personal communication). The video can be studied after the flight in an effort to verify that no objects were missed on or near the line. Bergstedt and Anderson (1990) used a video camera mounted on an underwater sled pulled by a research vessel to obtain distance data.

An advantage of collecting grouped data in the field is that exact distances are not required. Instead, one merely classifies an object detected into the proper distance class. Thus, if an object is somewhere near the centre of the distance class, proper classification may be easy. Problems occur only when the object is detected near the boundary between two distance intervals. If this is of concern, one could record the data on two different distance interval sets. Thus, each detection is accurately recorded on one or other of the two sets of intervals. The analysis theory for this situation has been developed but the computer software has not, and we believe that the method may be sensitive to assumptions on how the observer decides to allocate detections to one interval set or the other. A simpler solution is to use a single set of cutpoints, and record which detections are close to a cutpoint. These are then split between the two intervals, so that a frequency of one half is assigned to each (Gates 1979). Of course, a reduction in the number of distance intervals will result in fewer incorrect classifications.

Field studies of measurement error in aerial surveys have been limited. Chafota (1988) placed 59 bales of wood shavings (22.7 kg each) in short grass prairie habitat in northeastern Colorado to mimic pronghorn. A fixed-wing aircraft (Cessna 185) was flown at 145 km/hr at 91.4 m above the ground to investigate detection and measurement errors. Four line transects were flown using existing roads to mark the flight path ($L = 83.8$ km). The centreline of the transect was offset 60 m on both sides of the plane because of the lack of visibility below and near the aircraft (Fig. 7.8). Coloured streamers were attached to the wing struts of the aircraft to help the observer in delineating distance intervals (0–25, 25–50, 50–100 and 100–400 m). No marks were put on the window, thus the observer had only a 'front sight'. Neither the pilot nor the observer had experience in line transect surveys, although both had had experience with aerial strip transect sampling, and neither had knowledge of the number or placement of the bales. The observer did not have training in estimating distances. The performance of the observer on two assessments was reported.

In the first assessment 59 bales were placed in the 0–25 m band to assess the observer's ability to detect objects on or near the centreline (which was offset 60 m). Here the observer detected 58 out of 59 objects in the first band (0–25 m), and the undetected bale was at 22.9 m. However, six of the 58 were recorded as being in the 26–50 band. Worse, two bales were classed in the 50–100 band and an additional two bales were classed in the 100–400 band. Chafota (1988) suggested that possibly the aircraft was flown too low or went off the flight line during part of the survey, thus leading to the large estimation errors.

The second assessment employed 53 bales, including one outside the 400 m distance. The results are shown in Table 7.2. Here, the detection was quite good, as one might expect in the case of relatively large objects placed in short grass prairie habitat. Only one of the 53 bales went undetected (193.9 m). However, the tendency to exaggerate distances is quite clear. Chafota (1988) stressed the need for training in the estimation of distances, an effective pilot study and a carefully designed field protocol. We would concur with these recommendations and add the need for window marking to be used in conjunction with the streamers on the wing struts, an accurate altimeter to maintain the correct altitude, and a navigation system that allows accurate flight lines and positioning (see Johnson and Lindzey unpublished). Chafota (1988) also offered insight into the effects of measurement errors on $\hat{D}$ from the results of Monte Carlo studies.

Table 7.2 Performance of an observer in detecting bales of wood shavings placed at known distances from the centreline in short grass prairie habitat (from Chafota 1988)

Distance interval (m)	Actual frequencies	Observed distance interval (m)				
		0–25	25–50	50–100	100–400	> 400
0–25	21	8	12	1	0	0
25–50	14	1	3	10	0	0
50–100	12	0	1	9	2	0
100–400	5	0	0	1	3	0
> 400	1	0	0	0	0	1
Recorded frequencies:		9	16	21	5	1

It is possible to record sighting distances and sighting angles as grouped. This procedure is not recommended except under unusual circumstances. Transformation of grouped sighting distances and angles into grouped perpendicular distances has several problems and often calls for additional analytic methods to be used prior to the estimation of density. The smearing procedure can be applied to grouped or ungrouped (but heaped) data. It is invariably preferable to collect data that do not require smearing, if at all possible.

7.4.4 Cluster size

Ideally, the size of each cluster observed would be counted accurately, regardless of its distance from the line or point. In practice, one may only be able to estimate the size of the clusters, and such estimates may be biased. Additionally, there may be a tendency to underestimate the size of clusters at the larger distances and small clusters may remain undetected at the larger distances (i.e. size-biased sampling), leading to overestimation of average cluster size if $\bar{s}$ is used. In general, proper estimation of $E(s)$ is possible, but more complicated than use of the simple mean.

Survey design and conduct should attempt to minimize the difficulties in measuring cluster size. More than one observer may aid in getting an accurate count of cluster size. Photography may be useful in some clustered populations, and this has been tried in surveys of dolphin populations. It may be possible to leave the centreline to approach the more difficult clusters, and thereby obtain an accurate count. Sometimes it may be reasonable to obtain estimates of average cluster size from the data in only the first few distance bands for which both size-biased and poor estimation of cluster size are less problematic.

Clusters should be recorded only if the centre of the cluster is within the sampled area (0 to w), but the size of detected clusters should include all individuals within the cluster, even if some of the individuals are beyond w. If the centre of a detected cluster is beyond w, it should not be recorded and no individuals in the cluster should be recorded, even though some individuals might be within the sampled area ($< w$).

Cluster size and the variability among clusters may vary seasonally. For example, Johnson and Lindzey (unpublished) found that pronghorn populations split into small groups of nearly equal size in the spring, whereas much larger and more variable clusters were found during the autumn and winter months. Surveys should be conducted while variability in cluster size is low to avoid a relatively large variance in $\hat{D}$ from the contribution of $\widehat{\mathrm{var}}(\hat{E}(s))$. Small, variable clusters are preferable to large clusters with little variability because the number of detections (i.e. independent observations) will be greater.

7.4.5 Other considerations

In distance sampling it is important to use an objective method in establishing the exact location of the lines or points in the field. Subjective judgement should not play a role here.

If more than one observer is used, the design should allow estimation by individual observer. In line transect surveys, it may be interesting to partition and record the detections and cluster size by whether they are to the left or right of the centreline. Examination of these data may allow a deeper understanding of the detection process that might be useful in the final analysis. For example, in marine surveys, glare may be worse one side of the line than the other, and such data allow the effect of glare to be quantified.

In point transect surveys it might be useful to record a sighting angle θ_i (where $0° \leq \theta_i \leq 360°$) for each detected object. Here, $0°$ would be

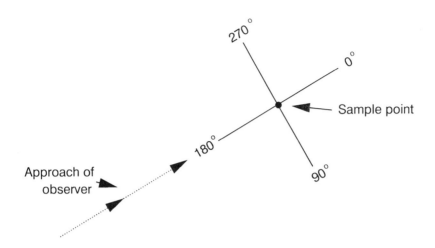

Fig. 7.11. Disturbance by an observer approaching a sample point can often be detected by recording angles θ ($0° \leq θ \leq 360°$) where 0° is directly ahead of the observer's direction of approach. Thus, an angle is recorded for each object detected. These angles might be recorded by group (e.g. 45–135, 136–225, 226–315, and 316–45°).

directly ahead of the direction of approach by the observer (Fig. 7.11). Analysis of such angles could be used to identify a disturbance effect by the observer approaching the sample points. If found to be present, the disturbance effect might be due to animals moving ahead (toward 0°) or merely remaining silent and hidden from the observer.

7.5 Training observers

Large-scale surveys usually employ several technicians to gather the data to be used in estimation of density. This section provides some considerations in preparing technical staff members for their task.

Perhaps the first consideration is to interest the staff in the survey and its objectives and to familiarize them with the study area and its features. Then they must be carefully trained in species identification and become familiar with relevant information about the biology of the species of interest. Particular attention must be given to activity patterns and calls or songs or other cues of the species. Some time in the field with a good biologist is essential. Clear survey instructions must be given and proper data recording forms should be available. Again, a small-scale pilot survey will be highly beneficial. People with prior experience are helpful to a team effort. Discussions held at the end of each day of surveying can be used to answer questions and listen to suggestions. A daily review of histograms of the incoming data will likely reveal possible problems to be corrected (e.g. heaping).

Training of observers is essential if estimates of absolute abundance are required. It is particularly difficult to estimate distances to purely aural cues; Reynolds *et al.* (1980) used an intensive 2-week training period, during which distances to singing or calling birds were first estimated and then checked by pacing them out or by using rangefinders. This is done for different species and for different cues from a single species. The training period should also be used to validate identifications made by each observer. Ramsey and Scott (1981a) recommended that observers' hearing ability be tested, and those with poor hearing be eliminated.

If most objects are located aurally, then the assumption that they are not counted more than once from the same line or point may be problematic. If for example a bird calls or sings at one location, then moves unseen by the observer to another location and again vocalizes, it is likely to be recorded twice. Training of observers, with warnings about more problematic species, can reduce such double counting. In some point transect surveys, points are sufficiently close that a single bird may be recorded from two points. Although this violates the independence assumption, it is of little practical consequence.

In point transect surveys, bias arising from either random or responsive movement and from inadvertent double counting is likely to be less if the time spent at each point is short, but assumptions that the detection function is unity at the point and has a shoulder are then more likely to be violated. Scott and Ramsey (1981) give a useful account of the effect on bias of varying the count period, and in particular warn against longer times, as bird movement can lead to serious overestimation.

Technicians should have instruction and practice in the use of instruments to be used in the survey (e.g. rangefinders, compass, LORAN C, 2-way radios). If distances are to be paced or ocularly estimated, then calibration and checking is recommended.

Basic safety and first aid procedures should be reviewed in planning the logistics of the survey. In particular, aircraft safety is a critical consideration in aerial survey work (e.g. proper safety helmets, fire resistant clothing, fire extinguisher, knowledge of emergency and survival procedures). Radio communication, a good flight plan, and an emergency locator transmitter (ELT) are important for surveys in remote areas or in rugged terrain. Fortunately, many conservation agencies have strict programmes to help ensure aircraft safety for their employees.

336

7.6 Field methods for mobile objects

We listed among the assumptions for line transect sampling that any movement of animals should be slow relative to the observer (Hiby 1986) and independent of the observer. In fact it is possible to relax the requirement that movement is slow. Consider for example seabird surveys. Procedures as laid down by Tasker *et al.* (1984) work well for birds feeding or resting on the sea, or for strip transects for which an instantaneous count of flying birds within a specified area can be made, but traditional line or strip transect estimates can have large upward bias for seabirds in flight. Provided the seabirds do not respond to the observation platform, the following approach allows valid abundance estimation. For birds resting on the sea or feeding in one place, use standard line or strip transect methods; for species that occur in flocks, treat the flock as a detection, and record flock (cluster) size. The position recorded for a flock should be the 'centre of gravity' of the flock, **not** the closest point of the flock to the observer. Record and analyse seabirds in flight separately. Whenever a flying bird (or flock) is detected, wait until it comes abeam of the observation platform, and only then record its position. For line transects, its perpendicular distance is estimated at this point; for strip transects, the bird is recorded only if it is within the strip when it comes abeam. Do not record the bird if it is lost from view before it comes abeam. If it alights on the water, record its position at that time, and record it as resting on the water. Having obtained separate density or abundance estimates for resting/feeding and for flying birds, sum the two estimates. If birds are known to respond to the observation platform, but only when quite close to it, the above procedure may be modified. Determine the smallest distance d beyond which response of flying birds to the platform is likely to be minimal. Instead of waiting for the bird to come abeam, its position is now recorded when its path intersects with a line perpendicular to the transect a distance d ahead of the platform. For this procedure to work, probability of detection at distance d, $g(d)$, should equal, or be close to, one. In this circumstance, flying birds that are first detected after they have crossed the line will have mostly intersected it at relatively large perpendicular distances, and can be ignored.

For point transect surveys, objects that pass straight through the plot, such as birds flying over the plot, should be ignored. Strictly, the count should be instantaneous. If the count is considered to correspond to the start of the count period, objects moving into the plot during the count should be ignored, whereas those that move out of the plot should be recorded at their initial location. If the count is intended to be of all detected objects present at the end of the count period, the converse holds. The first option is the easier to implement in the field. If objects that are moving through the plot are recorded at the location they were detected, density is overestimated (Chapter 5).

7.7 Field methods when detection on the centreline is not certain

A similar strategy to that of Section 7.6 can be adopted for objects that are only visible at well-spaced discrete points in time, so that $g(0) < 1$. Consider for example a species of whale that dives for prolonged periods. Suppose detected whales are only recorded if they are at the surface at the time they come abeam of the observation platform, and their perpendicular distance is estimated at that time. Then a conventional line transect analysis yields an estimate of the density of whales multiplied by the proportion of whales at the surface at any given time. If that proportion can be estimated, then so can population abundance. This strategy is of little use on slow-moving platforms such as ships, since most detected whales will have dived, or moved in response to the vessel, by the time the vessel comes abeam. However, it can be very successful for aerial surveys. Its weakness is that further survey work must be carried out to estimate the proportion of whales at the surface at a given time. This is done by monitoring individual whales over prolonged periods. Possible problems are that it may not be possible to monitor sufficient whales for sufficiently long periods; monitored whales may be affected by the presence of the observer, and may spend an atypical amount of time at the surface; if whales go through periods of short dives followed by longer dives, most of the monitored sequences may be short-dive sequences, since whales are more likely to be lost if they dive for a longer period; whales that habitually spend more time at the surface are more likely to be detected and monitored; it can be difficult to define exactly what is meant by at the surface, especially if monitoring of individual whales is done from a surface vessel, and the line transect surveys from the air.

Methods for estimating $g(0)$ from cetacean shipboard surveys were described in Section 6.4. Discussion of survey design is given there; we do not address the topic here because methodological development is not sufficiently advanced to allow us to make general recommendations with confidence. However, most of the methods described in Section 6.4 rely on detections made from two 'independent observer' platforms. These methods usually require that duplicate detections (whales detected from both platforms) are identified. In this circumstance, general recommendations on field procedures can be made. First, **all** sighting cues

338

should be recorded, for easier assessment of which detections are duplicates. To facilitate this goal further, the exact time of each cue should be noted, preferably using a computerized recording system. Methods are generally sensitive to the judgement of which detections are duplicates, so every attempt should be made to minimize uncertainty, and the uncertainty should be reflected in the estimated variance of $\hat{g}(0)$ (Schweder et al. 1991). Ancillary data such as animal behaviour, cluster size and weather should be recorded for each detection, to allow the analyst to use stratification or covariate modelling to reduce the impact of heterogeneity on $g(0)$ estimation.

7.8 Field comparisons between line transects, point transects and mapping censuses

Several researchers have attempted to evaluate the relative merits of point transect sampling, line transect sampling and mapping censuses through the use of field surveys. We summarize their conclusions here.

7.8.1 Breeding birds in Californian coastal scrub

DeSante (1981) examined densities of eight species of breeding bird in 36 ha of Californian coastal scrub habitat. True densities were established by an intensive programme of colour banding, spot-mapping and nest monitoring. Point transect data were collected by four observers who were ignorant of the true densities. Points were chosen on a grid with roughly 180 m separation between neighbouring points. This gave 13 points, three of which were close to the edge of the study area. Only one-half of those three plots were covered, so that in effect 11.5 points were monitored. The recording time at each point was 8 minutes. Each point was covered four times by each of the four observers. Detection distances were grouped into bands 9.14 m (30 ft) wide out to 182.9 m (600 ft), and into bands twice that width at greater distances. The 'basal radius', within which all birds are assumed to be detected, was estimated as the internal radius of the first band that had a density significantly less than the density of all previous bands. Significance was determined by likelihood ratio testing with a critical value of four (Ramsey and Scott 1979). The density of territorial males was estimated using counts of singing males only, unless twice that number was less than the total count for that species, in which case the density of territorial males was estimated from half the total count. This follows the procedure of Franzreb (1976) and Reynolds et al. (1980). Only experienced observers were used, and they were given four days of intensive training. One day was spent verifying observers' identifications from calls and songs, one day estimating and verifying distances to both visual and aural detections, and two days carrying out simultaneous counts at points. DeSante found that the point transect data yielded underestimates of density, by about 18% when estimates for all eight species are summed. Individual species were underestimated by between under 2% and roughly 70% (Table 7.3; taken from DeSante 1981). Correlation between actual density and estimated density across the eight species was good ($r = 0.982$). Variation in bias between observers was small. The use of the method of Ramsey and Scott (1979) undoubtedly contributed to underestimation of density in DeSante's study; the method assumes that all birds within the basal radius are detected, and the basal radius is estimated here from a small number of points, almost certainly giving rise to estimates of basal radii that are too large. An analysis of the original data by more recent methods might prove worthwhile.

Table 7.3 Actual density and point transect estimates of density of eight bird species in a Californian coastal scrub habitat (from DeSante 1981). The negative errors indicate that the point transect estimates are low, possibly due to poor choice (in the light of recent developments) of point transect detection model

Species	Actual density/36 ha	Point transect estimates		
		Density/36 ha	% error	Basal radius (m)
Scrub jay	3.8	1.1	− 70.0	64.0
Bushtit	2.2	2.1	− 5.0	45.7
Wrentit	36.3	26.9	− 25.9	54.9
Bewick's wren	9.4	8.3	− 11.4	91.4
Rufous-sided towhee	14.0	8.5	− 39.4	91.4
Brown towhee	0.6	0.4	− 36.7	64.0
White-crowned sparrow	32.4	31.8	− 1.9	91.4
Song sparrow	35.5	31.2	− 12.2	64.0
Total	134.2	110.3	− 17.8	

7.8.2 Breeding birds in Sierran subalpine forest

DeSante (1986) carried out a second assessment of the point transect method, in a Sierran subalpine forest habitat. On this occasion a 48 ha study plot was identified in the Inyo National Forest, California. Methods were similar to the above study, with actual densities estimated by intensive spot-mapping and nest monitoring. Twelve points were established with a minimum separation of 200 m, and count time at each point was eight minutes, preceded by one minute to allow bird activity

Table 7.4 Actual density and point transect estimates of density of 11 bird species in a Californian subalpine forest habitat (from DeSante 1986). Densities were estimated from detections of singing males alone, except where indicated by *, for which densities were estimated from counts of all birds. The first row for each species corresponds to late June and the second to mid-July

Species	Actual density/ 48 ha	Point transect estimates		
		Density/ 48 ha	% error	Basal radius (m)
Cassin's finch	26.9	16.5	− 38.6	90
Cassin's finch*	20.6	16.8	− 18.5	60
Dark-eyed junco	23.0	16.5	− 28.2	60
	23.1	10.0	− 56.7	110
Dusky flycatcher*	17.1	28.5	+ 66.8	40
	16.4	19.7	+ 20.3	50
Yellow-rumped warbler	15.3	14.7	− 3.6	100
	15.0	19.4	+ 29.3	80
Mountain chickadee*	12.0	14.9	+ 24.3	40
	11.9	9.7	− 18.3	60
Pine siskin*	12.0	13.6	+ 13.2	50
	13.3	8.9	− 32.7	40
Hermit thrush	5.7	2.8	− 51.6	100
	7.8	12.9	+ 65.3	110
White-crowned sparrow	3.8	1.3	− 67.0	120
	3.0	1.6	− 46.9	100
American robin*	3.4	6.6	+ 95.0	40
	−	−	−	−
Clark's nutcracker*	2.1	4.1	+ 95.9	70
	−	−	−	−
Ruby-crowned kinglet	1.6	1.5	− 3.3	120
	−	−	−	−
Total	122.9	121.1	− 1.5	
	111.1	99.1	− 10.8	

to return to normal after arrival at the point. Counts were carried out on four days in late June and a further four days in the second week of July. Statistical methodology was the same as for the above study. Estimated densities are shown in Table 7.4. Although DeSante gave confidence intervals for point transect density estimates, we do not quote them here, as they were calculated as if 48 points had been counted, when in fact 12 points were each counted four times. His intervals are therefore too narrow; repeat counts on the same point do not provide independent detections, and such data should be entered into DIS-TANCE with sample effort for a point set equal to the number of times the point was covered. The empirical variance option for n combined with the bootstrap option for $\hat{f}(0)$ then gives valid intervals. We further condense DeSante's table to include only common species – those species for which there were more than 25 point transect detections. DeSante concluded that the results were less encouraging than those obtained in his earlier study, and he gave thorough discussion of the possible reasons. These include a higher proportion of birds missed close to the observer, due to the tall canopy, leading to underestimation, and more double-counting of individuals through greater mobility, leading to overestimation. Greater mobility relative to the scrub habitat of his earlier survey occurred because densities were lower and there were more large species, both contributing to larger territories. Further, birds flying over the plot were counted; this is poor practice, leading to overestimation. Birds first detected flying over the point should either be ignored, or counted only if they land within detection distance, and that distance should then be recorded (Section 7.6). The relatively poor performance of the point transect method may be partially attributable to the fact that just 12 points were covered. DeSante considers that an alternative scheme of relocating points each day would not have significantly increased accuracy. Although this may be true for many species, for those species which tend to sing from favoured song posts, four counts from each of 12 points is appreciably less informative than one count from each of 48 points, even when, as here, the study area is too small to accommodate 48 non-overlapping plots.

7.8.3 Bobolink surveys in New York state

Bollinger et al. (1988) compared line and point transect estimates with known densities of bobolinks (*Dolichonyx oryzivorus*) in one 17.6 ha meadow and one 12.6 ha hayfield in New York state. Intensive banding and colour marking established population sizes, and whether each individual was present on a given day. Twelve and ten line transects, respectively, of between 200 and 500 m length were established at the two sites, together with 18 and 14 point transects. One or two transects were covered per day, each transect taking 3–7 minutes. The observer waited four minutes after arriving at the start of a transect to allow birds to return to normal behaviour, and the line(s) was/were covered in both morning and afternoon. A four minute waiting period was also used for point transects, and a four minute counting period was found to be adequate. Two points were covered each morning of the study and two each afternoon. Thus, as for DeSante's studies, adequate sample sizes were obtained by repeatedly sampling the same small number of transects. The Fourier series model was applied to both line and point transect data. For point transects, it was applied to squared detection distances, which can lead to poor performance, as noted earlier.

Bollinger *et al.* found that their point transects took longer to survey than line transects on average, but appreciably less time was spent counting. The number of males counted during point transects was slightly greater on average than during line transects, but substantially fewer females, which are more secretive, were counted. Thus, density estimates were obtained for both males and females from the line transect data, but for males only from the point transect data. The Fourier series was unable to fit adequate non-increasing detection functions to the morning point transect counts, which the authors suggest may be indicative of movement of bobolinks away from the observer. Both methods overestimated male abundance, with the point transect method showing the greater bias (mean relative bias of 140%, compared with 76% for line transects). Bias was found to be lower in general for the afternoon count data. Line transect estimates of female densities were approximately unbiased, although there was a suggestion of underestimation during incubation, countered by overestimation when the young were large nestlings or had fledged. About 25% of bias in male density estimates was attributed to avoidance of field edges by the birds; transects were deliberately positioned so that field edges were not surveyed. Survey design to eliminate or reduce this source of bias is discussed in Section 7.2. Additional bias was considered to be possibly due to 'random' movement of birds, with detection biased towards when the birds were relatively close to the observer, or to attraction to the observer – although this latter explanation is difficult to reconcile with the suggestion that poor fits for point transect data might be due to observer avoidance. It may be that the Fourier series model was inappropriate for the squared distance data, as was found by Buckland (1987a) for the point transect data of Knopf *et al.* (1988), rather than that birds avoided the observer.

7.8.4 Breeding bird surveys in Californian oak–pine woodlands

Verner and Ritter (1985) compared line and point transect counts in Californian oak–pine woodlands. They also considered counts within fixed areas (strip transects and circular plots) and unbounded counts from both lines and points as measures of abundance. They defined four scales for measures of abundance: a 'nominal scale', which requires information only about occurrence; an 'ordinal scale', which requires sufficient information to rank species in order of abundance; a 'ratio scale', which requires relative abundance estimates – bias should be either small or consistent across species and habitats; and an 'absolute scale', which requires unbiased (absolute) estimates of abundance. They assessed the performance of different survey methods in relation to these scales. True bird densities were unknown. Although the area comprised 1875 ha of oak and oak–pine woodlands in the western foothills of the Sierra Nevada, the study plots were just two 19.8 ha plots of comparable relief and canopy cover, one grazed and the other ungrazed. This study, in common with most others of its type, therefore suffers from repeated sampling of the same small area, and hence non-independent detections.

Sampling took place over 8-day periods, with two transects and ten counts covered per day. The transects were 660 m long and were positioned randomly at least 60 m apart each day. The points were located at intervals of 150 m along the transects. The design was randomized and balanced for start time, starting point and count method. All counts were done by a single observer. Four methods of analysis were considered: bounded counts (strip transects of width 60 m and circular plots of radius 60 m); Emlen's (1977) *ad hoc* estimator; Ramsey and Scott's (1978) method; and the exponential polynomial model (Burnham *et al.* 1980). Note that we do not recommend any of these estimators for songbird data. The Fourier series model was found to perform less well than the exponential polynomial model, so results for it were not quoted. Interval estimates were computed for the exponential polynomial model only. Without more rigorous analysis of the data and with no information on true densities, comparisons between the methods of analysis and between line and point transects are severely constrained. However, the authors concluded that line and point transects showed similar efficiency for determining species lists (for point separation of 150 m and 8 min per point); point transects yielded lower counts per unit time, but would be comparable if point separation was 100 m and counting time was 6 min per point; Ramsey and Scott's (1978) method gave widely differing estimates from line transect data relative to those from point transect data; more consistent comparisons between models were obtained from line transects than from point transects; most species showed evidence of movement away from the observer; the exponential polynomial model was thought to be the most promising of the four methods.

7.8.5 *Breeding birds in riparian vegetation along the lower Colorado River*

Anderson and Ohmart (1981) compared line and point transect sampling of bird populations in riparian vegetation along the lower Colorado River. All observers were experienced, and each carried out replicate surveys under both sampling methods in each month from March to

Table 7.5 Density estimates from line and point transect sampling and spot mapping of ten bird species in honey mesquite habitat along the lower Colorado River (from Anderson and Ohmart 1981). Densities are numbers per 40 ha, averaged for March, April and May 1980

Species	Line transect	Point transect, first interval 15 m wide	Point transect, first interval 30 m wide	Territory mapping
Gila woodpecker	2	2	2	2
Ladder-backed woodpecker	4	3	6	8
Ash-throated flycatcher	11	12	10	12
Black-tailed gnatcatcher	14	28	26	24
Verdin	7	8	10	10
Cactus wren	4	4	4	8
Lucy's warbler	37	32	28	41
Northern oriole	12	15	14	13
Crissal thrasher	2	1	2	14
Abert's towhee	21	25	23	21
Total	114	130	125	153

June 1980. The distance walked was identical for each sampling method. The line transect data were analysed using the method of Emlen (1971), and the point transect data using method M1 of Ramsey and Scott (1979). Thus, models that can perform poorly were again used, and this may have compromised some of the authors' conclusions. For example, they sometimes obtained inflated density estimates from the point transect data when detection distances less than 30 m were divided into two or more groups, whereas the method performed well when all observations within 30 m were amalgamated into a single group. A more robust method would be less sensitive to the choice of grouping. Anderson and Ohmart concluded that the point transect surveys took longer to complete when the time spent at each point was 8 minutes, but that times were comparable for recording times of 6 or 7 minutes. More area was covered and more birds detected using the line transect method, because of the dead time between points for the point transect method. The authors tabulated estimated average densities of ten of the more common species, which appear to show relatively little difference between line and point transect estimates, or, for most species, between those estimates and estimates from territory mapping, although overall, line transect estimates were significantly lower than mapping estimates (Table 7.5). Neither method generated average estimates significantly different from the point transect estimates. However, the authors noted that day-to-day variation in point transect estimates was greater than for line transect estimates, and suggested that at least three repeat visits to point transects are necessary, whereas two are sufficient for line transects. They concluded that the line transect method is the more feasible, provided stands of vegetation are large enough to establish transects of 700–800 m in length, and provided that the topography allows ambulation. They indicated that these transects should be adequately cleared and marked. In areas where vegetation occurs in small stands, or where transects cannot be cleared, they suggested that point transects might be preferable.

7.8.6 Bird surveys of Miller Sands Island in the Columbia River, Oregon

Edwards et al. (1981) compared three survey methods, two of which were line and point transect sampling. They described the third as a 'sample plot census', in which an observer records all birds that can be detected within a sample plot. Distances were not recorded, so corrections for undetected birds cannot be made. The method gives estimates of absolute density only if all birds in the sample plot are detected. The study was carried out on Miller Sands Island in the Columbia River, Oregon. Four habitats were surveyed: beach, marsh, upland and tree-shrub. The method of Emlen (1971) was used for the line transect data, and the method of Reynolds et al. (1980) for the point transect data. The authors found that significantly more species were detected using point transects than either line transects or sample plots. However, the truncation point for line transects had been 50 m, and for point transects 150 m, and the sample plots were circles of radius 56.4 m, so the difference is unsurprising. Density estimates were found to be similar for all three methods, although the point transect estimate was significantly higher than the line transect estimate in a handful of cases. The methods were not standardized for observer effort or for time spent in the field, making comparison difficult.

7.8.7 Concluding remarks

More studies, carefully standardized for effort, would be useful. A large study area, too large for all territories to be mapped, is required for a fair comparison of line and point transect sampling and of mapping methods. Within such an area, line and point transect sampling grids could be set up using the recommendations of this chapter, so that both methods require roughly the same time in the field on the part of the observer. In addition, territory mapping should be carried out on a random sample of plots within the area, again so that time in the field

346

is comparable with each of the other methods. The analyses of the line and point transect data should be comparable, for example using the hazard-rate or the Fourier series model in both cases. More than one model should be tried. The precision of each method should then be compared, and an assessment should be made of whether at least one of the methods over- or underestimates relative to the others. If different researchers could agree on a common design, this could be repeated in a variety of habitats, to attempt to establish the conditions and the types of species for which point transect sampling is preferable to line transect sampling or *vice versa*.

The studies described here tend to favour line transects over point transects. This may partly reflect that line transect methodology has had longer to evolve than point transect methodology. It is important to realize that point transect sampling is essentially passive, whereas line transect sampling is active. For birds that are unlikely to be detected unless they are flushed or disturbed, such as many gamebirds or secretive female songbirds, line transect sampling should be preferred. Very mobile species are also likely to be better surveyed by line transects, provided the guidelines for such species given in this chapter are adhered to. For birds that occupy relatively small territories, and which are easily detected at close range, such as male songbirds of many species during the breeding season, point transects may be preferable, especially in patchy or dense habitat. Attempts to estimate abundance of all common species in a community by either method alone are likely to perform poorly for at least some species. If only relative abundance is required, to monitor change in abundance over time, either technique might prove useful. However, bias may differ between species, so great care should be taken if cross-species comparisons are made. Equally, bias may differ between habitats, although well-designed line or point transect studies yield substantially more reliable comparisons across both species and habitats than straight counts of birds without distance data or other corrections for detectability.

Several other authors compare line transect sampling with census mapping. Franzreb (1976, 1981) gives detailed discussion of the merits of each, concluding that census mapping is substantially more labour intensive, but for some species at least, provides better density estimates. Choice of method must take account of the species of interest, whether density estimates for non-breeding birds are required, the habitat of the study area, resources available, and the aims of the study. In the same publication, O'Meara (1981) compares both approaches. His study includes an assessment of the binomial models of Järvinen and Väisänen (1975). These models were found to be more efficient, both in terms of time to record detections into one of just two distance intervals and in terms of variance of the density estimate, than Emlen's (1971, 1977) method, which requires detection distances to be recorded so that they can be assigned to successive bands at increasing distances from the line. Line transect estimates were found to be lower than census mapping estimates, apparently due to imperfect detection of birds at or near the line (Emlen 1971; Järvinen and Väisänen 1975), but estimates could be obtained for twice as many species from the line transect data. Redmond *et al.* (1981) also compared census mapping with the line transect methods of Emlen (1971) and Järvinen and Väisänen (1975), for assessing densities of long-billed curlews (*Numenius americanus*). They also found that the method of Järvinen and Väisänen was easier to apply than that of Emlen, because it requires just two distance intervals, and was far more efficient than census mapping in terms of resources in the case of territorial male curlews. Female curlews were not reliably surveyed using line transects, nor were males during brood rearing.

Several field evaluations have been made of distance sampling theory in which a population of known size or density is sampled and estimates of density made (Laake 1978; Parmenter *et al.* 1989; White *et al.* 1989; Bergstedt and Anderson 1990; Otto and Pollock 1990). Strictly speaking, these are not evaluations of the distance sampling methods, but rather an assessment of the degree to which the critical assumptions have been met under certain field conditions. We encourage more studies of this type as such results often provide insights into various issues. **We strongly recommend that the person performing the data analysis should not know the value of the parameter being estimated.**

7.9 Summary

Line and point transect sampling are well named because it is the area near the line or point that is critical in nearly all respects. In many ways the statistical theory and computational software are now more adequately developed than the practical field sampling methods. **The proper design and field protocol have not received the attention deserved prior to data collection.**

Having determined that line or point transect sampling is an appropriate method for a study, the planning of the sampling programme must focus on the three major assumptions and attempt to ensure their validity: (1) $g(0) = 1$, (2) no undetected movement, and (3) accurate measurements or counts (e.g. no heaping, especially at zero distance). Furthermore, if the population exists in clusters, then accurate counts of cluster size must be taken. Sample size, as a rough guideline, should be at least 60–80; formulae for determining appropriate sample size are

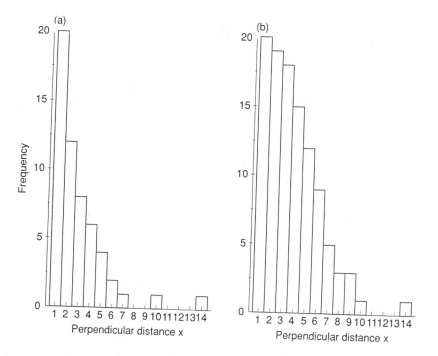

Fig. 7.12. (a) These line transect data are spiked, difficult to model, subject to imprecision in estimating density, and usually the result of poor survey design or conduct. Proper design and field procedures should result in data more nearly as depicted in (b). These data exhibit a shoulder and can be analysed effectively if no undetected movement occurred and distances were measured accurately. Some truncation prior to analysis is suggested in both cases.

given in Section 7.2. The distance data should be taken such that the detection function $g(y)$ has a shoulder (Fig. 7.12). Transect width w should be large enough so relatively few detections are left unrecorded; plan on data truncation as part of the analysis strategy.

A pilot study is highly recommended. A preliminary survey provides the opportunity to assess a large number of important issues, assumptions, practicalities, and logistical problems. Failure to complete a pilot programme often means wasted resources at a later stage of the survey. Consultation with a biometrician familiar with sampling biological populations is strongly advised.

8

Illustrative examples

8.1 Introduction

Several analyses of real data are presented here to illustrate line and point transect analysis, and use of program DISTANCE. Units of measurement of the original study are adhered to, to avoid quoting cutpoints as 'every 0.305 m' instead of 'every foot'. Five of the examples are line transect surveys and three are point transect studies. The first two examples are case studies in which the true density is known. The first of these is a set of line transects to estimate the number of bricks that had been placed on the bed of a region of Lake Huron, as part of a programme to assess the viability of using an underwater video system to monitor numbers of dead lake trout. This is followed by a reanalysis of the wooden stake surveys carried out as part of a student project in sagebrush–grass near Logan, Utah (Laake 1978). The third example is a comprehensive analysis of the data from annual surveys of duck nest density that have been carried out for 27 years at the Monte Vista National Wildlife Refuge in Colorado. An analysis of fin whale data from ship sightings surveys in the North Atlantic illustrates use of stratification, and the final line transect example is one for which sightings data on dolphins are collected by observers placed on board tuna fishing vessels. Thus, there is no control over the cruise track, and methods must be used that reduce the effects of the potentially strong biases in the data.

The first point transect example is an illustration of model selection and stratification on house wren data collected during surveys of riparian vegetation in South Platte River bottomland in Colorado. The second example is an analysis of data on the six species most commonly recorded in songbird surveys of the Arapaho National Wildlife Refuge, also in Colorado. In the final example, an approach for assessing effects of habitat on density using point transect sampling is described, and an analysis of binomial point transect data is carried out in which bird density and detectability in restocked forest plantations in Wales are related to habitat succession.

8.2 Lake Huron brick data

Bergstedt and Anderson (1990) presented the results of 13 independent, line transect surveys of known populations of bricks placed approximately at random on the floor of portions of Lake Huron. The bricks were to simulate dead lake trout, and the purpose of the survey was to explore the feasibility of large-scale surveys of dead lake trout on the lake bottom. The distance data were taken remotely via a video camera mounted on an underwater sled and pulled along the lake bottom by a surface vessel. The data were taken in five distance categories (0, 0.333, 0.667, 1, 1.333, 1.667 m), and these cutpoints were marked on a cathode ray tube on board the vessel. The relationship between the cutpoints on the screen and the perpendicular distance on the lake bottom was defined by calibration studies before the first survey. The true density in all surveys was 89.8 bricks/ha, and this allows an evaluation of the utility of line transect sampling theory. Full details of this survey are given by Bergstedt and Anderson (1990). Our purpose here is to reanalyse the data from the 13 surveys using the methods presented in earlier chapters.

Table 8.1 Summary of density estimates and coefficients of variation for the Lake Huron brick data under five models of the detection function (data from Bergstedt and Anderson 1990). True density = 89.8/ha. The final means are weighted by line length

Survey	HN + cos		Uni + cos		Uni + poly		HN + Herm		Haz + poly	
1	138.5	19.5	144.4	15.7	137.1	19.2	126.8	16.6	104.0	16.2
2	96.4	18.7	92.6	32.5	100.0	15.6	93.1	27.1	88.2	15.6
3	88.8	20.4	91.4	19.2	86.2	17.0	88.8	20.4	70.0	17.7
4	84.0	19.7	85.5	15.9	83.2	19.6	77.1	16.0	63.0	16.2
5	78.3	22.0	81.6	20.0	72.9	18.4	78.3	22.0	61.3	18.9
6	84.5	33.8	93.6	14.6	90.4	17.7	84.4	27.5	75.9	16.4
7	80.7	38.2	76.9	26.3	92.4	19.3	85.4	16.6	71.1	16.0
8	122.4	19.1	107.0	15.7	104.1	19.2	122.4	19.1	86.2	16.7
9	96.2	19.5	99.6	15.7	95.7	19.2	96.2	19.5	73.6	16.0
10	72.8	20.2	70.2	26.7	67.7	16.9	67.6	30.0	61.5	16.0
11	89.4	20.9	93.4	19.2	86.1	17.3	89.4	20.9	70.6	18.0
12	92.7	18.7	83.9	15.3	81.6	18.0	81.9	29.2	76.5	18.0
13	56.8	20.7	58.2	20.0	55.1	17.8	56.8	20.7	55.5	27.2
Wt. Ave.	88.7	6.2	88.3	6.0	86.5	6.0	86.2	6.8	72.4	4.7

Table 8.1 provides a summary of the estimates of density for each of the 13 surveys under each of five models: half-normal + cosine, uniform + cosine, uniform + simple polynomial, half-normal + Hermite polynomial, and the hazard-rate model + simple polynomial. Information related to model selection for the pooled data is presented in Table 8.2. AIC selects the hazard-rate model with no adjustment terms, and this model fits the grouped distance data well ($p = 0.70$). The remaining four models (half-normal + cosine, uniform + cosine, uniform + polynomial and half-normal + Hermite polynomial) have similar AIC values, but fit the data less well ($0.09 \leq p \leq 0.29$).

Table 8.2 Model selection statistics for the pooled Lake Huron brick data

Key function	Adjustment	Total parameters	AIC	p-value[*]
Half-normal	Cosine	4	1671.2	0.09
Uniform	Cosine	3	1670.3	0.29
Uniform	Polynomial	3	1670.3	0.29
Half-normal	Hermite	3	1672.7	0.06
Hazard-rate	Polynomial	2	1667.9	0.70

[*] Goodness of fit test.

Surprisingly, the hazard-rate model provides the poorest estimates of mean density based on the weighted mean of the estimates (Table 8.1). The hazard-rate model fitted a very flat shoulder to the pooled data and produced estimates that were low in 12 of the 13 surveys. Performance of the other four estimators is quite good (all were slightly low due, in part, to the results of survey 13) (Table 8.1). Estimated confidence intervals for mean density covered the true value for all but the hazard-rate model.

The weighted mean of the estimates was very close to the true parameter, except for the hazard-rate model (88.7, 88.3, 86.5, 86.2 and 72.4, respectively). Any of the first four models performs well, especially when one considers that the average sample size per survey was only 45. The hazard-rate key + polynomial adjustment does more poorly in this example, and is the only model whose 95% confidence interval fails to cover the true density. Troublesome is the fact that the hazard-rate model was selected as the best of the five models by AIC in nine of the 13 data sets (surveys).

The reader is encouraged to compare these results with the original paper by Bergstedt and Anderson (1990), which includes discussion of various points relating to possible measurement errors, missing bricks in the first distance category, and potential problems with survey 13.

Bergstedt and Anderson (1990) note that some bricks were missed near the centreline, and that the cutpoints drawn on the cathode ray tube, although accurate in the initial calibration, were perhaps compromised by the uneven lake bottom.

8.3 Wooden stake data

Laake (1978) set out 150 unpainted wooden stakes ($2.5 \times 5 \times 46$ cm) within a rectangular area of sagebrush–grass near Logan, Utah, in the spring of 1978 to examine the performance of the line transect method.

Table 8.3 Summary of stake data taken in 1978 in a sagebrush–grass field near Logan, Utah (Laake 1978). Density in each of the 11 surveys was 37.5 stakes/ha. Cosine adjustments were added as required in modelling $f(x)$. For each survey, $L = 1000$ m and $w = 20$ m

Survey no.	Key function	Sample size	Density estimate	cv(%)	Log-based 95% confidence interval	
1	Half-normal	72	37.11	19.3	(25.51,	53.99)
	Uniform		30.00	14.5	(22.63,	39.77)
2	Half-normal	48	35.18	19.9	(23.90,	51.78)
	Uniform		36.01	20.1	(24.36,	53.23)
3	Half-normal	74	28.76	15.8	(21.16,	39.10)
	Uniform		29.26	14.8	(21.92,	39.07)
4	Half-normal	59	38.31	19.1	(26.42,	55.52)
	Uniform		33.30	17.5	(23.68,	46.81)
5	Half-normal	59	34.41	19.9	(23.37,	50.66)
	Uniform		29.58	19.0	(20.47,	42.76)
6	Half-normal	72	26.38	16.2	(19.24,	36.17)
	Uniform		27.08	15.6	(19.98,	36.69)
7	Half-normal	55	34.48	19.9	(23.44,	50.72)
	Uniform		34.69	21.1	(23.06,	52.18)
8	Half-normal	61	33.31	20.2	(22.51,	49.30)
	Uniform		34.48	21.3	(22.82,	52.09)
9	Half-normal	46	28.32	21.9	(18.51,	43.31)
	Uniform		23.52	21.2	(15.60,	35.48)
10	Half-normal	43	34.16	20.1	(23.15,	50.42)
	Uniform		32.69	21.1	(21.71,	49.22)
11	Half-normal	53	29.80	17.4	(21.25,	41.80)
	Uniform		31.45	17.8	(22.23,	44.50)
Mean	Half-normal	642	32.75	3.6	(30.54,	35.11)
	Uniform		31.10	3.6	(28.99,	33.36)
Pooled	Half-normal	642	34.37	7.2	(29.86,	39.55)
	Uniform		34.60	7.5	(29.86,	40.08)

The stakes were placed in a restricted random spatial pattern such that the number of stakes was distributed uniformly as a function of distance from the line. In fact, each 2 m distance category had 15 stakes present. Stakes were driven in the ground until about 37 cm remained above ground. One stake was placed about every 7 m of transect and alternated between left and right sides of the line. Exact placement was generated randomly within the 7 m section. True density was 37.5 stakes per hectare.

A single, well-marked line ($L = 1000$ m, $w = 20$ m) was traversed by 11 different, independent observers. The observers were carefully instructed and supervised and fatigue was probably a minor factor as each survey could be completed by the observer in approximately 2 h. Observers traversed the line at different times, thus the data for each of the 11 surveys are independent. The number of stakes detected (n) varied from 43 to 74, corresponding to $\hat{P}_a$ ranging from 0.29 to 0.49 (Table 8.3). Histograms and estimated detection functions ($\hat{g}(x)$) differed greatly among observers. In field studies, the detection function would also be affected by habitat type and species being surveyed. These factors affect n and make it, alone, unreliable as an index of density.

Two strategies were used for the analysis of these data. First, the data were pooled over the 11 surveys ($n = 642$) and AIC was computed for five models,

Model	AIC
Half-normal + cosine	2412.9
Uniform + cosine	2415.2
Uniform + polynomial	2417.4
Half-normal + Hermite	2450.8
Hazard-rate + cosine	2416.4

Thus, the half-normal + cosine is selected as the best model. In fact, all models seem fairly satisfactory for data analysis except the half-normal + Hermite polynomial model where the LR test indicated that the first adjustment term (for kurtosis) was not required ($p = 0.617$). There are options in DISTANCE, such as SELECT = **forward** (which yields AIC = 2415.1 for this model) or LOOKAHEAD = 2 (also giving AIC = 2415.1), that allow the user to avoid such poor model fits (Section 8.7). Estimates of density are shown in Table 8.3 under the two best models, as suggested by the AIC criterion. The estimates of density are quite similar between these two models.

While n varied widely, estimates of density varied from only 26.38 to 38.31, using two models for the detection function (half-normal and uniform key functions with cosine adjustment terms). Confidence interval coverage cannot be accurately judged from only 11 replicates, but examination of the intervals in Table 8.3 shows no particular indication of poor coverage. Estimates of density are low in all cases, except for survey 4 for the half-normal estimator. Averaging the density estimates over the 11 surveys indicates a negative bias of approximately 13–17%. Pooling the data over the 11 surveys provides an approximate estimate of bias of about − 7 to − 8%. The main reason for the negative bias seemed to be some lack of model fit near the line. Examination of the histograms and estimated detection functions seems to indicate that models commonly fit poorly near zero distance. Some of the negative bias is due to measurement error for stakes near the line. The exact location of each stake was known, and errors in measurement could be assessed. For example, the information for three stakes (i.e. stake numbers 45, 93 and 103) is shown:

Stake no.	103	45	93
True distance (m)	0.92	5.03	14.96
Ave. distance ($\bar{x}$)	0.73	4.77	14.63
$\widehat{sd}(x)$	0.139	0.151	0.277
$100\,(\widehat{sd}(x)/\bar{x})$	18.9	3.2	1.9

This suggests the measurement error is largely due to improper determination of the centreline. Finally, the negative bias is partially the result of observers missing about 4% of the stakes in the first metre and 13% of the stakes in the first two metres. However, this is offset by the tendency to underestimate distances near the line. One observer was seen actually tripping over a stake on the centreline, but still the stake was not detected. Stakes do not move and do not respond to the observer; for field surveys of animals, the relative importance of the different assumptions may be very different. If this survey was to be repeated, we would enlarge the study area and lengthen the line such that $E(n) \doteq 80$. Also, observers would be shown the evidence that stakes near the centreline were occasionally missed, and that measurements were often in error, in the hope that these problems could be lessened.

A second strategy will be illustrated that is less mechanical, requires a deeper level of understanding of the theory, and is somewhat more subjective. The pooled data are displayed as a histogram using 1 m cutpoints (Fig. 8.1). This information suggests several aspects that should influence the analysis. First, the distance data have a long tail, a small mode around 15 m, and considerable heaping at 10 m at the expense of 9 and 11 m. Some heaping at 0, 5 and 15 m is also seen. A histogram based on 0.5 m cutpoints indicated that detection declined rapidly near the line. Thus, the data have some of the characteristics

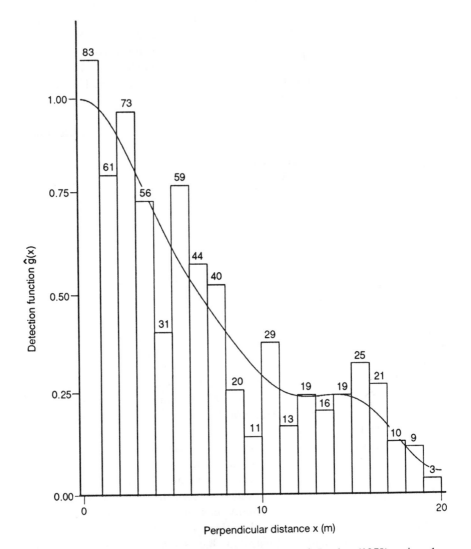

Fig. 8.1. Histogram of the wooden stake data of Laake (1978) using 1 m cutpoints. Also shown is a 3-term Fourier series fitted to these grouped data. Note the heaping at 0, 5 and 10 m, the relatively long tail in the distribution, and the additional mode near 15–16 m.

illustrated in Fig. 2.1. The modelling of $g(x)$ will require additional terms to model the extra mode and long tail (Fig. 8.1 shows a model with three cosine terms, and this still seems to underestimate the data near the line). Thus, a reasonable approach might be to truncate the data at

356

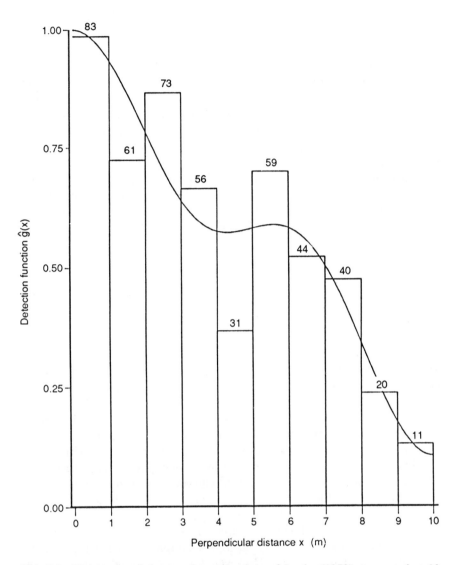

Fig. 8.2. Histogram of the wooden stake data of Laake (1978), truncated at 10 m. The detection function is modelled using a uniform key function and three cosine adjustment terms.

$w = 10$ m in the hope of obtaining a better fit near the line and alleviating problems in the tail of the distribution.

At this point, one could choose a robust model for $g(x)$ and proceed; here we will first use the uniform key and cosine adjustment function (Fourier series). Fitting these truncated data using a uniform key

357

function and a cosine adjustment function still required three terms, but provided a good fit near the line (Fig. 8.2). Some heaping at 5 m now shows more clearly. This model of the pooled data provided $\hat{D} = 37.71$ ($\widehat{se}(\hat{D}) + 2.82$) with a log-based confidence interval of (32.58, 43.64). These calculations assume var(n) = 0, as the stakes were placed uniformly. In reality, the value of var(n) in this study would be small but non-zero; zero was used only for illustration. Now, the data from each separate survey (i.e. individual person) can be modelled using a 3-term cosine series. The average estimate of density from the 11 surveys was 36.99 ($\widehat{sd}(\hat{D}) = 5.04$), again close to the true density of 37.5.

Second, having selected a truncation point ($w = 10$ m), one could select a model using AIC, based either on the individual data sets or on the pooled data sets. Here, we will examine only the pooled data. The results are summarized below.

Model	AIC
Half-normal + cosine	2118.7
Uniform + cosine	2112.4
Uniform + polynomial	2114.7
Half-normal + Hermite	2118.7
Hazard-rate + cosine	2126.8

The AIC selects the uniform + cosine model (the Fourier series) as the best model and, as shown above, provides an excellent estimate of density. The other models provided estimates that were inferior (half-normal + cosine = 33.60, uniform + polynomial = 31.47, half-normal + Hermite = 33.59, and hazard-rate + cosine = 31.90), and this could have been judged from the poor model fits near the line (Fig. 8.3). Likelihood ratio tests chose models with one parameter, except the hazard-rate model (two parameters) and the uniform + cosine (three parameters). Only the uniform + cosine model had confidence intervals that covered the true parameter.

Because the authors knew the value of D during these analyses, we cannot allege total objectivity in this example, which happened to provide an excellent estimate. However, the point is that careful review of the distance data can suggest anomalies (e.g. spiked data, long tails, heaping, a second mode), and these can suggest analysis approaches that should be considered equally with the more mechanical approach of using AIC or only likelihood ratio tests. We advocate some truncation, especially in cases where there are clear outliers in the distance data. Poor fit of the model to the data near the line should always be of concern. Generally the guidelines outlined in Section 2.5 will serve the analyst well in planning the analysis of distance data.

8.4 Studies of nest density

Studies of duck nest density have been conducted annually since 1964 at the Monte Vista National Wildlife Refuge in Colorado. During the 27 years of these studies, 10 041.6 miles of transect were walked and 4156 duck nests were found. Here we will examine the data for individual species and years to illustrate various points, approaches and difficulties. No attempt is made here to provide a final, comprehensive analysis of these data. Further details are found in Gilbert *et al.* (in prep.).

Strip transects were originally established on the refuge in a systematic design, running north–south, with 100 yds between transects and $w = 8.25$ ft, giving a 5.5% sample of the entire 18.4 square mile refuge. Each transect centreline was marked by a series of numbered plywood signs attached to a 2.5 m pole (Burnham *et al.* 1980: 32). Only one-half of the original transects were surveyed during the 1969–90 period, except in 1971 when only one-quarter of the original transects were run and in 1977 when no survey was conducted. In 1969–79 w was increased to 12 ft, but this was changed back to 8.25 ft during the 1980–85 period. Strip width was increased again to 12 ft in 1986–87 and finally changed back to 8.25 ft during 1988–90. Distances to detected nests were not recorded during 1964–66 and 1975–79. These erratic changes were often due to personnel or budget limitations. Transects were searched twice each year to monitor nest density of both early and later nesting species. A third search was made in a few years, but few nests were found and these data are not included in any of the examples given here. The mallard (*Anas platyrhynchos*) was the most common nesting duck, but substantial numbers of northern pintail (*Anas acuta*), gadwall (*Anas strepera*), northern shoveler (*Anas clypeata*), and teal (*Anas cyanoptera, A. discors*, and *A. carolinensis*) nests were also found. Species identity could not be determined for many nests; these were classed as 'unknown'.

The refuge, at an elevation of 7500 ft, is characterized by level terrain and high desert vegetation in relatively simple communities. The drier, alkaline sites contain greasewood (*Sarcobatus vermiculatus*) and rabbit-brush (*Chrysothamnus* spp.) on the higher sites, while saltgrass (*Distichlis stricta*) dominates the lower sites. The wetter sites are dominated by baltic rush (*Juncus balticus*), but other species include cattail (*Typha latifolia*), spikerush (*Eleocharis macrosachya*), bullrush (*Scirpus validus*), and sedges (*Carex* spp.). Water is managed from pumped and artesian wells and irrigation sources and a system of dikes and borrow pits allow open water to be interspersed with vegetation cover to create good waterfowl nesting habitat. This area has one of the highest duck nest densities of any in North America.

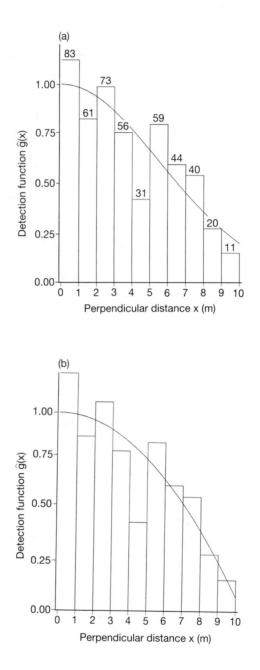

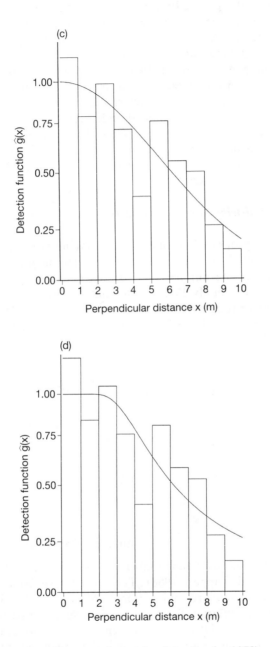

Fig. 8.3. Histograms of the wooden stake data (Laake 1978) and model fits for (a) the half-normal + cosine, (b) uniform + polynomial, (c) half-normal + Hermite, and (d) hazard-rate + cosine model.

The original design used strip transects, and it was assumed that all duck nests within the strip were detected. Perpendicular distance data were collected in 1967 and 1968, and it was clear by 1970 that some nests near w remained undetected, even with the narrow width of 8.25 ft (Anderson and Pospahala 1970). Perpendicular distances were recorded in 1969–74 and 1986–90, primarily as a means to relocate nests so that the nest fate could be determined. In years when $w = 12$ ft, it was likely that more nests remained undetected. Thus, distance sampling theory is appropriate to obtain estimates of density, account for different sampling intensities, resolve differences in transect width w, and provide a basis for correction for undetected nests in years when no distance measurements were taken.

8.4.1 Spatial distribution of duck nests

On biological grounds it seemed likely that nests were somewhat randomly distributed on the refuge. Thus, one might expect that the variation among the number of nests (n_i) detected by transect line (l_i) would be approximately Poisson, so that the variance in total sample size (n) might be roughly Poisson (i.e. $\widehat{\text{var}}(n) \doteq n$). The variance of n was computed empirically for mallard and non-mallard nests for each of 26 years using

$$\widehat{\text{var}}(n) = L \sum_{i=1}^{k} l_i \left(\frac{n_i}{l_i} - \frac{n}{L} \right)^2 / (k - 1)$$

where

$$L = \sum_{i=1}^{k} l_i$$

and

$$k = \text{number of replicate lines}$$

Modelling of var(n) as a function of n is common in statistical application (Carroll and Ruppert 1988). The ratio $\hat{b} = \widehat{\text{var}}(n)/n$ was computed for each year for mallard and non-mallard nests (Table 8.4); b is often called a variance inflation factor. A random spatial distribution of nests yields $\hat{b} \doteq 1.0$. The relationship between n and $\widehat{\text{var}}(n)$ can be estimated by a weighted linear regression through the origin, where the weight is the sample size. The point estimate under this approach is equivalent to a ratio estimator. The estimated value of b for mallards was 1.630

362

Table 8.4 Sample size (n), estimates of the empirical variance of n ($\widehat{var}(n)$), and their ratio ($\hat{b}$) for mallard and non-mallard duck nests

Year	Mallards			Non-mallards		
	n	$\widehat{var}(n)$	$\hat{b}$	n	$\widehat{var}(n)$	$\hat{b}$
1964	142	210.20	1.48	119	201.27	1.69
1965	140	188.79	1.34	72	94.41	1.31
1966	151	238.01	1.58	109	146.69	1.35
1967	150	255.58	1.70	94	156.29	1.66
1968	176	281.84	1.60	139	240.67	1.73
1969	112	168.79	1.51	118	166.61	1.41
1970	103	166.44	1.62	126	225.92	1.79
1971	66	179.61	2.72	48	139.98	2.92
1972	63	99.10	1.57	64	102.08	1.60
1973	102	160.18	1.57	137	450.11	3.28
1974	69	157.18	2.28	40	54.43	1.36
1975	37	45.49	1.23	28	57.86	2.07
1976	49	61.89	1.26	46	72.48	1.58
1978	20	27.61	1.38	35	49.33	1.41
1979	13	9.73	0.74	40	96.06	2.40
1980	34	23.46	0.69	55	82.93	1.51
1981	50	105.63	2.11	54	79.27	1.47
1982	57	125.91	2.21	53	88.53	1.67
1983	79	123.10	1.56	35	48.69	1.39
1984	112	180.20	1.61	38	50.16	1.32
1985	70	112.88	1.61	82	105.49	1.29
1986	114	205.24	1.80	83	105.90	1.27
1987	130	164.00	1.26	132	195.82	1.48
1988	93	182.94	1.97	57	65.66	1.15
1989	74	103.67	1.40	56	58.63	1.05
1990	56	117.97	2.11	32	37.63	1.18
Wt. Ave.			1.630			1.677

($\widehat{se} = 0.068$) and 1.677 ($\widehat{se} = 0.111$) for non-mallards. These estimates are not significantly different ($z = 0.36$) and, thus, a pooled estimate of b for all species was computed as $\hat{b} = 1.651$ ($\widehat{se} = 0.063$). This weighted regression has an adjusted correlation of $r^2 = 0.93$ (Fig. 8.4), and provides $\widehat{var}(n) \doteq 1.7n$ which will be used in the remaining material for this example. This reflects some contagion in the distribution of duck nests, related, no doubt, to the variable distribution and quality of nesting habitat on the refuge. Minor species such as the redhead (*Aythya americana*), which nest in specialized habitat types, probably had a very non-random spatial distribution of nests (i.e. $b \gg 1.0$). Note that var(n) and b can be estimated for all years, even those where perpendicular distances were not recorded.

8.4.2 Estimation of density

This material focuses primarily on the grouped distance data collected during 1969–74 and 1986–87, years when $w = 12$ ft. Histograms of the distance data are shown in Fig. 8.5 for the mallard, pintail, gadwall, teal, shoveler and unknown species. The first interval includes nests from 0 to 11 inches, the second interval includes nests detected from 12 to 23 inches, etc. Relatively few pintail nests were recorded within 2–3 ft of the centreline, whereas for other species, there appears to have been preferential heaping of nests close to the transect centreline. In each case, except the pintail, many nests found in the second interval (12–23 inches) were probably heaped into the first interval and perhaps into the third interval. Heaping at zero distance is common, and has been reported several times in distance sampling literature (e.g. Robinette *et al.* 1974), especially for data taken as sighting distance (r_i) and angle (θ_i) and then the perpendicular distances (x_i) computed as $x_i = r_i \cdot \sin(\theta_i)$. Also, there was a strong tendency to record nests somewhat outside the transect boundary (w) as exactly 12 ft from the centreline. This latter form of heaping was right-truncated in all the analyses of the Monte Vista data in this chapter by setting w to 11.9 ft, thus excluding nests beyond 12 ft.

Estimated detection functions $\hat{g}(x)$ are also plotted in Fig 8.5, assuming the half-normal key function with Hermite polynomial adjustments, if required. The half-normal key function seems quite reasonable for modelling these data. While there was substantial variation and some obvious heaping in the counts (n_i), the fit appeared fairly good, with the clear exception of that for the pintail (discussed further in Section 8.4.3). Nests of the mallard seem to be most easily detected, as shown by the estimated unconditional probability of detection in the surveyed strip of area $a = 2wL$ $(\hat{P}_a = 0.80, \widehat{se} = 0.03$; Table 8.5). Gadwall nests were also easy to detect $(\hat{P}_a = 0.76, \widehat{se} = 0.09)$, whereas teal, shoveler, and unknown species nests were less detectable $(\hat{P}_a = 0.64, \widehat{se} = 0.04$; $\hat{P}_a = 0.60, \widehat{se} = 0.07$; and $\hat{P}_a = 0.63, \widehat{se} = 0.04$, respectively). Pooling all nests for all species results in an unconditional probability of detecting a nest within the transect of 0.78, $\widehat{se} = 0.02$. Mallards (and pintails) nest early in the season when most vegetation has little new growth and detection might be easier. Other species tend to nest later and may experience more concealment in the vegetation.

Estimates of P_a for nests of all species combined were higher in years when $w = 8.25$ ft $(\hat{P}_a = 0.84, \widehat{se} = 0.07)$ compared to years when $w = 12$ ft $(\hat{P}_a = 0.78, \widehat{se} = 0.02)$. In general, P_a is a function of w in that $P_a = 1/\{w \cdot f(0)\}$; under large sample approximations, $cv(\hat{P}_a) = cv\{\hat{f}(0)\}$. However, fewer nests were found using the narrower transect. The

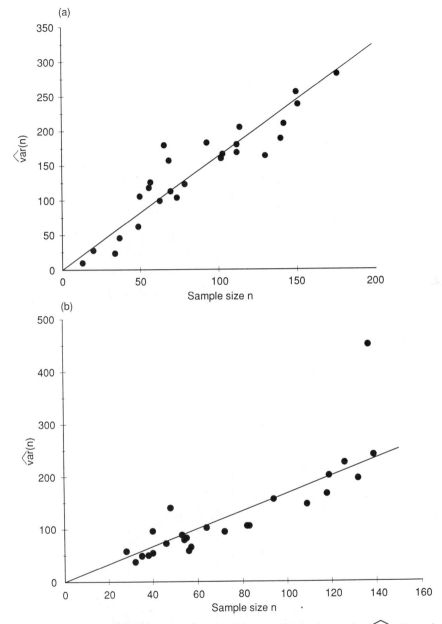

Fig. 8.4. Relationship between the empirical variance in *n* (i.e. $\widehat{\mathrm{var}}(n)$) and sample size (*n*) for (a) mallard nests and (b) non-mallard nests.

narrow transects are inefficient as shown by the mean encounter rate (e.g. 0.867 and 1.137, respectively, for total nests). The values of *w*

365

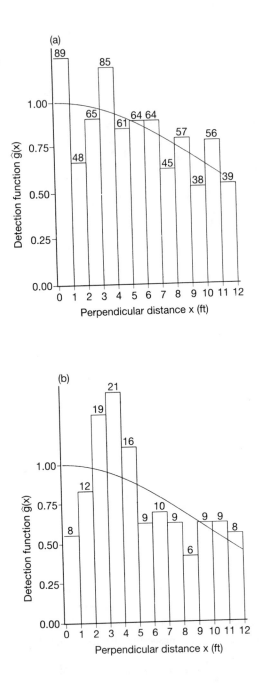

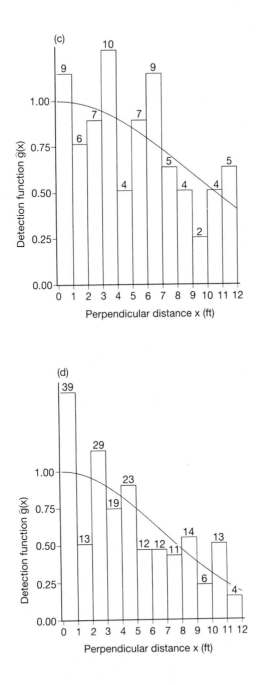

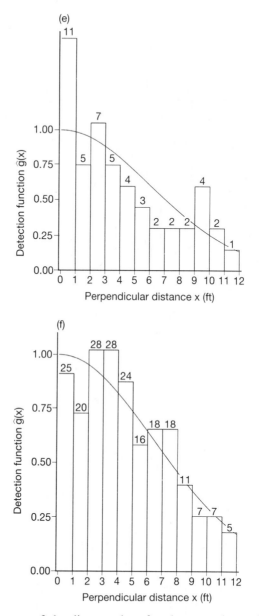

Fig. 8.5. Histograms of the distance data for the nests detected at the Monte Vista National Wildlife Refuge in Colorado, USA, during 1969–74 and 1986–87. In addition, estimated detection functions using the half-normal key function and Hermite polynomial adjustments are shown for nests of (a) mallard ($n = 711$), (b) pintail ($n = 136$), (c) gadwall ($n = 72$), (d) teal ($n = 195$), (e) shoveler ($n = 48$), and (f) unknown species ($n = 207$).

(either 8.25 or 12 ft) were certainly too large to meet the assumptions of strip transect sampling and too small for good efficiency in line transect sampling. Using the data for mallard nests, pooled for the 1969–74 and 1986–87 period, $\hat{g}(8.25) \doteq 0.8$, thus the observer undoubtedly detected many nests beyond w and could not record them. Detection near w is still good even for the wider transects [$\hat{g}(12) \doteq 0.6$], thus many nests were still readily detectable beyond the transect boundary. As only about 0.58 mallard nests were found per transect mile (Table 8.5), it is clear that increasing w would allow sample size to increase with little additional survey effort. Finding a nest is a relatively rare event and if more nests could be found, little additional time would be required to take and record the relevant measurements. It would be interesting to increase w to perhaps 15, 18, or even 20 ft in future surveys to improve efficiency for both the observers and the estimators. An additional adjustment term might be required in modelling $g(x)$ but an overall gain in the estimation process is likely. Observers would have to be cautioned to emphasize search on and near the centreline and not divert too much attention near w. Of course, it would be advantageous if heaping could be lessened and more accurate measurements were taken. Perpendicular distances might be remeasured as a test of accuracy when the fate of the nests is checked at a later time.

Table 8.5 Summary of statistics and estimates for survey of duck nests at the Monte Vista National Wildlife Refuge in Colorado during 1969–74 and 1986–87. See Fig. 8.5 for histograms of these data and estimated detection functions using the half-normal model and Hermite polynomial adjustment terms. Density is in nests/mile^2

Species	n	$\hat{P}_a$	n/L	$\hat{f}(0)$	cv$\{\hat{f}(0)\}$ (%)	$\hat{D}$	cv$(\hat{D})$ (%)[*]
All duck	1415	0.77	1.137	0.1079	2.6	323.8	3.7
Mallard	711	0.85	0.580	0.0901	3.8	149.6	5.4
Teal	195	0.64	0.157	0.1305	6.4	54.0	9.6
Pintail	136	0.78	0.109	0.1063	8.5	30.7	12.1
Gadwall	72	0.76	0.058	0.1094	11.5	16.7	16.5
Shoveler	48	0.60	0.039	0.1388	12.4	14.1	19.0
Unknown	207	0.63	0.166	0.1325	6.1	58.2	9.3

[*] Assuming $\widehat{\text{var}}(n) = 1.7\, n$.

Mallard nests were found in adequate numbers to allow annual estimates of nest density to be made. Histograms of the distance data are shown in Fig. 8.6, with estimated detection functions $\hat{g}(x)$. Although the annual sample sizes were generally fairly adequate ($n_i > 60$ in every

ILLUSTRATIVE EXAMPLES

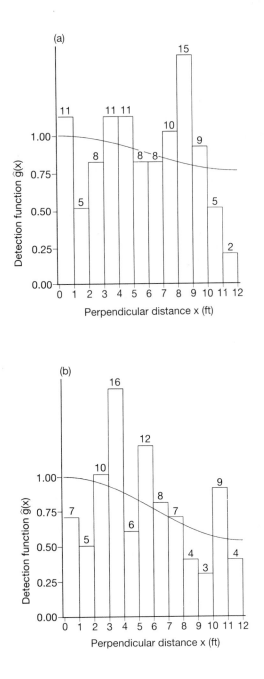

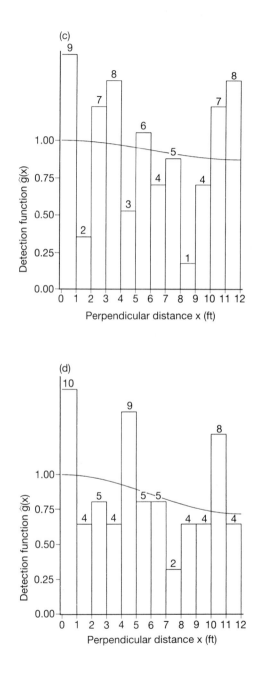

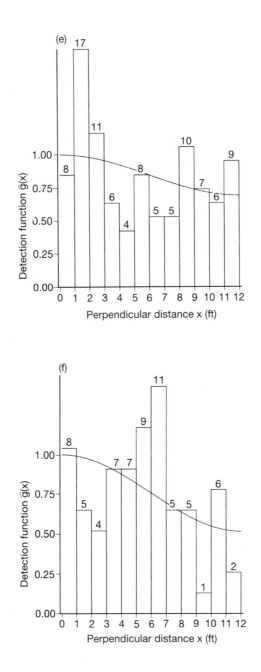

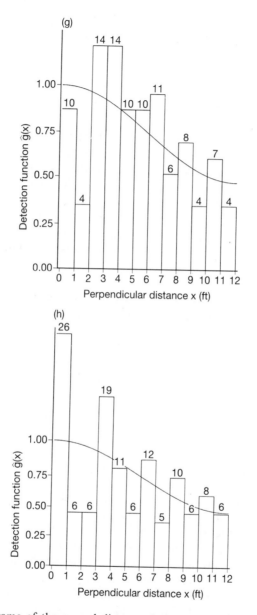

Fig. 8.6. Histograms of the annual distance data for mallard nests detected at the Monte Vista National Wildlife Refuge in Colorado, USA, during 1969–74 and 1986–87. The estimated detection function is also shown. The respective sample sizes (n) are (a) 103, (b) 91, (c) 64, (d) 64, (e) 96, (f) 70, (g) 102 and (h) 121.

year), the histograms look 'rough' and certainly exhibit heaping and perhaps some careless measurements. Except in the early years of the survey, observers generally had little training and were possibly not that motivated.

Estimation of annual nest density from data such as shown in Fig. 8.6 can be achieved using

$$\hat{D}_i = n_i \cdot \hat{f}_i(0)/2L_i, \quad \text{where } i = \text{year}$$

An attractive alternative exists if one is willing to make the assumption that the form of $g_i(x)$ or $f_i(x)$ is the same, or nearly the same, each year. This assumption seems biologically reasonable as the vegetation is low and somewhat sparse, nests of the various species do not move in response to the observer, the transects are rather narrow, and nests appear somewhat alike (but may vary somewhat depending on the stage of incubation). These reasons seem fairly compelling, but it is advisable to test the null hypothesis that the grouped distance data arose from a common detection function. Here a reasonable strategy of analysis is to fit the distance data, pooled over the eight years, to several good candidate models and select the model with the smallest AIC. These values for the mallard data under five models were:

Model	AIC
Half-normal + cosine	3513.8
Uniform + cosine	3513.9
Uniform + polynomial	3514.3
Half-normal + Hermite	3513.8
Hazard-rate + cosine	3515.4

Any of the five models could be used in this case, with little difference among the models. AIC selects the single-parameter half-normal key function with no adjustment terms, although the uniform key function plus a cosine term (i.e. Fourier series) is second best by a trivial margin. The uniform + polynomial model might also be a satisfactory model, followed by the 2-term hazard-rate model. For this example, the uniform + cosine (Fourier series) model will be used. Then, the eight individual distance data sets are analysed using the same model (i.e. the number of adjustment terms fixed) (Fig. 8.6), to give the log-likelihood value for the model based on the pooled data ($\log_e(\mathcal{L}_p)$) and the eight log-likelihood values for the individual data sets ($\log_e(\mathcal{L}_i)$). A likelihood ratio test may now be used:

$$\chi^2 = 2[\sum_{i=1}^{8} \log_e(\mathcal{L}_i) - \log_e(\mathcal{L}_p)]$$

where the test statistic is asymptotically distributed as χ^2 with $n \cdot (r - 1)$ degrees of freedom (n = number of model parameters and r = number of data sets). With the mallard data,

$$\chi^2 = 2[-1753.64 + 1755.93]$$
$$= 4.58$$

with $1(8 - 1) = 7$ df, giving a p-value of 0.711. Thus, use of a common $\hat{f}(x)$ for all years seems appropriate. One can pool the distance data over the eight years to obtain $\hat{f}(0)$ and its standard error. Then, yearly estimates of density can be computed using the yearly sample size (n_i) and

$$\hat{D}_i = n_i \cdot \hat{f}(0)/2L_i, \quad \text{where } i = \text{year}$$

The estimate of $f(0)$ is specific to a given transect width, i.e. the value of $\hat{f}(0)$ for data from a transect with $w = 12$ ft cannot be used for years when the transect width was 8.25 ft. This approach seems appropriate for the mallard data, and it is nearly essential for species such as teal where yearly sample size would not support a reliable estimate of the year-specific detection function $g_i(x)$ (or, equivalently, $f_i(x)$). Data for the shoveler ($n = 48$) and gadwall ($n = 72$) support only an estimate of an average $f(0)$ over the eight-year period, but this analysis approach allows annual estimates of density by using the year-specific sample sizes n_i. This general approach can be used for the analysis of other years of data where different transect widths were used.

Estimates of density under the five models were quite similar, ranging from 149.3 to 157.0 mallard nests per square mile. This might have been expected because the AIC values were all of similar magnitude. Models for $g(x)$ contained only one parameter (either a key or an adjustment parameter), except the hazard-rate model, with two parameters. Thus, two of the five models used only the half-normal key function.

8.4.3 Nest detection in differing habitat types

Despite large sample sizes, well distributed in time and space, the ability to examine differences in detectability by vegetation type was limited by the fact that baltic rush and greasewood made up approximately 68% and 15% of the vegetation on the refuge, respectively. Initially it was hypothesized that nest detectability would decline more rapidly with distance from the centreline in the tall, but often sparse, stands of greasewood when compared to the lower, more dense areas of rush. Instead, it became clear that the histogram of grouped distance data for nests found in greasewood indicated a mode well away from the transect centreline. It was hypothesized that observers would avoid the thorny

375

Fig. 8.7. Stand of greasewood with extensive areas of bare ground typify many upland sites on the Monte Vista National Wildlife Refuge. Observers may tend to avoid walking through these thorny shrubs, thus biasing the data.

greasewood (see Fig. 8.7) by walking off line and around these shrubs. Thus, nests at the base of these shrubs tended to go undetected near the transect centreline. Nests detected at the edge of greasewood clumps would be detected with near certainty while the observer was temporarily off the centreline (and thus avoiding the greasewood). Once such a nest was found its distance to the centreline was measured and recorded. Such temporary departures from the transect centreline could explain the odd distance data for the pintail nests (Fig. 8.5). Perhaps pintail were common nesters in greasewood types and, thus, many were missed near the centreline. Indeed, 24.2% of the pintail nests were found in greasewood; surely this percentage would be still higher if nests near the centreline in greasewood were all detected. Other species nested in greasewood types less frequently: mallard 15.6%, gadwall 19.5%, teal 6.9%, and shoveler 2.6%. We tentatively conclude that observers were reluctant to enter the thorny greasewood type, and this resulted in nests being missed near the centreline.

An alternative explanation is that the observer measured the distance from his or her position to the nest and that pintail tended to nest at least two feet into the greasewood type. Then Fig. 8.5b would arise without missing any nests near the centreline; instead, the data would arise because the observer's path would go through habitat with a low pintail nest density. In any event, the presence of obstacles such as greasewood on the line must be dealt with effectively in the field survey or the analysis of the data can be problematic. We do not always advocate that the observer plunge through such cover types; instead, extra care in searching must be taken when an easier path is temporarily followed. For example, the observer could go around clumps of such vegetation both to the left and then to the right, searching the centreline more carefully. In any event, the measurements must be taken from the transect centreline, not to the observer who may be away from the centreline.

A definitive analysis of data such as those for the pintail nests is not possible. Approximate analyses that might be useful could be considered. First, one could fit a monotonically constrained function for $g(x)$ as is shown in Fig. 8.5b for the half-normal key function with Hermite polynomial adjustments. This is likely to result in an underestimate of density if a substantial number of nests near the centreline was undetected. However, in this particular case, one knows from several other, similar species in this survey that the shape of $g(x)$ has a broad shoulder, so that the procedure might be acceptable.

Second, one could use some arbitrary left-truncation and then estimate $f(0)$ and D using, for example, the uniform + cosine or half-normal + Hermite model. First, one could decide on a truncation point; 3 ft might be reasonable for the pintail nest data. Here the grouped distance data less than 3 ft could be discarded, the remaining data rescaled as if the third interval was actually the first interval, and proceed to estimate density in the usual way (Fig. 8.8a). This is likely to be similar to the first procedure because we have reason to suspect that the detection function for pintail nests is fairly flat. Still, in this case, some underestimation

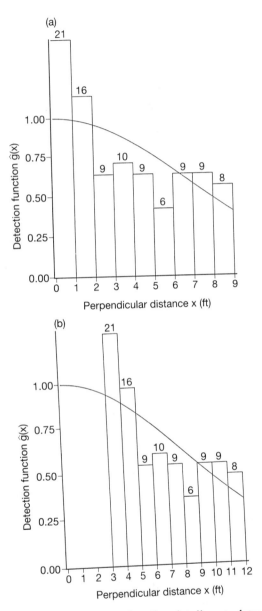

Fig. 8.8. Histograms of the distance data for pintail nests detected at the Monte Vista National Wildlife Refuge in Colorado, USA, during 1969–74 and 1986–87. Two estimates of $g(x)$ are shown using alternative ways to left-truncate the data and minimize the problems observed in the first 3–4 distance categories. Left-truncation and rescaling are shown in (a), and the Alldredge and Gates (1985) approach is shown in (b)

might be expected (unless $g(0) \doteq 1.0$ but nests close to zero tended to be recorded at around 3 ft; then overestimation might result).

Third, the left truncation procedure of Alldredge and Gates (1985) could be employed, using the same truncation point. The result of this procedure is very dependent upon the model chosen and is often imprecise (Fig. 8.8b). In this example, where something is known about the distribution of distances of nests of other species of ducks, it seems likely that density of pintail nests is overestimated using this approach. Of course, any left truncation decreases sample size. The results of using the three approaches for the pintail nest data are summarized below for the half-normal key function and Hermite polynomial adjustments:

Method	n	$\hat{D}$	cv(%)
Full data	136	30.7	12.1
Left-truncate and rescale	97	29.8	14.2
Full left-truncation	97	35.0	17.9

The three estimates seem fairly reasonable for the pintail nest data, although one might prefer a density estimate near 30–32, rather than 35, unless the observer's path around greasewood types tended to sample areas of low pintail nest density. Considerable precision is lost in efforts to alleviate this problem; this is to be expected given the uncertainty introduced.

8.4.4 Models for the detection function g(x)

Various combinations of the key and adjustment functions provide flexibility in modelling the detection function $g(x)$. For data sets exhibiting a reasonable shoulder and meeting the other assumptions of

Table 8.6 Summary of density estimates (above) and coefficients of variation (below) for five models of $g(x)$ and four duck species, 1969–74 and 1986–87

Key function	Adjustment function	Gadwall ($n = 72$)	Teal ($n = 195$)	Shoveler ($n = 48$)	Mallard ($n = 711$)
Half-normal	Cosine	16.7	63.1	19.0	149.6
		16.5	12.4	22.4	5.4
Half-normal	Hermite	16.7	54.0	14.1	149.6
		16.5	9.6	19.0	5.4
Uniform	Cosine	17.4	54.8	14.0	155.2
		16.5	9.0	17.7	5.7
Uniform	Polynomial	15.9	61.5	12.2	147.3
		15.0	10.9	16.2	5.0
Hazard-rate	Cosine	16.2	67.0	17.9	147.6
		19.4	18.0	23.1	6.6

distance sampling, the choice of model, among those recommended here, is relatively unimportant. Estimates of density and estimated standard errors are summarized in Table 8.6 for several reasonable models for nest data on mallard, gadwall, teal and shoveler for the Monte Vista data. The differences in estimated density are small relative to the estimated standard errors. The standard errors given relate to the estimates made from data pooled over eight years.

If the distance data are distributed in a more spiked form, the choice of model is more difficult and the estimate of density more tenuous. The models recommended here are likely to perform reasonably well, except in pathological cases. A model with an appreciably smaller χ^2 goodness of fit value, if constrained to be non-increasing, will tend to be better than other models with the same number of estimated parameters. However, in general, goodness of fit tests are of relatively little help in model selection. In particular, some lack of fit near w is of little consequence in comparing model fit among several models.

8.5 Fin whale abundance in the North Atlantic

Large-scale line transect surveys of the North Atlantic to assess whale abundance were carried out in 1987 and 1989 (North Atlantic Sightings Surveys, NASS-87 and NASS-89). We analyse here the fin whale (*Balaenoptera physalus*) data from the 1989 survey collected by Icelandic vessels to illustrate the use of stratification. The analyses are extracted from Buckland *et al.* (1992b).

In 1989 four Icelandic vessels surveyed Icelandic and adjacent waters during July and August. The area covered was mostly within the East Greenland/Iceland stock boundaries for fin whales, and we consider here abundance estimation for that stock alone.

Sighting distances and angles were smeared and assigned to perpendicular distance intervals, using smearing method 2 of Buckland and Anganuzzi (1988a), and the hazard-rate model was fitted to the group frequencies. Detections were often of more than one animal, so an analysis of clusters was carried out; average cluster (school) size was roughly 1.5 whales. Several potential stratification factors were identified: geographic block, Beaufort (a scale for wind speed, generally determined from sea state), cloud cover, vessel and school size. Ideally stratification should be by all of these factors, but sample size considerations preclude this. Variables Beaufort, cloud cover and school size could be entered as covariates to avoid sample size difficulties, although it is then necessary to define a linear or generalized linear model between these effects and say effective strip width or encounter rate, and confounding between say Beaufort and geographic location, and hence between Beaufort and whale density, is inevitable. For analysing minke whale data, Gunnlaugsson and Sigurjónsson (1990) used generalized linear modelling to estimate sighting efficiency in different Beaufort states during NASS-87. This approach also has shortcomings when Beaufort varies strongly with geographic location, if whale density also varies geographically. For example, encounter rate may be lower in high Beaufort simply because a disproportionate amount of rough weather encountered by survey vessels was in an area with low animal density. Geographic stratification reduces but does not eliminate this effect. The problem of estimating fin whale abundance is easier than that of estimating North Atlantic minke whale abundance since cues are more visible. We adopt a simpler approach here to determine stratification factors.

To assess say the effect of Beaufort, average school size, encounter rate and effective strip width were estimated for each Beaufort (0–6) in turn, pooling across all other possible stratification factors. Standard errors were calculated for each estimate, and z-tests carried out to assess whether there are significant differences in estimates at different Beauforts. Standard error for school size was calculated as sample standard deviation divided by square root of sample size; for encounter rate, the rate per day was calculated, and the sample variance of these rates, weighted by daily effort, used as described for the empirical method of Section 3.7.2; and the standard error for effective strip width was obtained from likelihood methods, *via* the information matrix. The stratification factors are confounded with each other, and the above approach ignores interactions between them; analyses are supplemented here by knowledge of likely effects of the different factors on the three components of estimation to determine an appropriate analysis. Thus results from z-tests are not used blindly; if a pairwise test indicates that effective strip width is wider at Beaufort 4 than Beaufort 1, it would be considered spurious, because it is counter to the knowledge that detection is easier in low Beaufort, whereas if there was a trend towards narrower effective strip widths as Beaufort increases, stratification would be deemed necessary.

Suppose mean size of schools detected during Beaufort 0 is $\bar{s}_0$, and during Beaufort 1, $\bar{s}_1$. Denote their standard errors by $\widehat{se}(\bar{s}_0)$ and $\widehat{se}(\bar{s}_1)$ respectively. Then a z-test is carried out by calculating

$$z = \frac{\bar{s}_0 - \bar{s}_1}{\sqrt{[\{\widehat{se}(\bar{s}_0)\}^2 + \{\widehat{se}(\bar{s}_1)\}^2]}}$$

The distribution of z is approximately normal. Thus if $z > 1.96$ or $z < -1.96$, the mean school sizes differ significantly at the 5% level

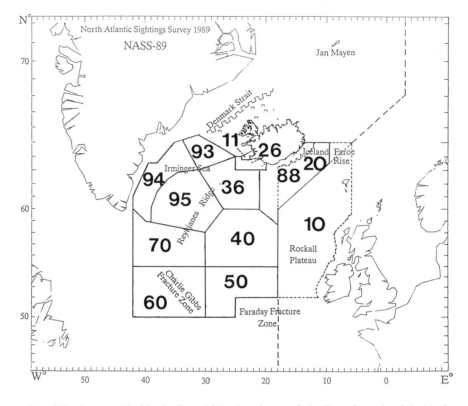

Fig. 8.9. Geographic blocks for which abundance of the East Greenland/Iceland stock of fin whales is estimated from Icelandic 1989 data.

($p < 0.05$). Evidence against the null hypothesis that the mean school size is the same in both sea states is strong if p is small, whereas if p is large, the data are consistent with the null hypothesis.

No significant differences in encounter rates by sea state (Beaufort 0–6) were found (Table 8.7). Mean school size did not differ significantly for Beauforts 0–3 or for Beauforts 4–6, but there was strong evidence that the mean of recorded school sizes in Beaufort $\geqslant 4$ is smaller than for Beaufort 2 or 3. The effective strip width was significantly smaller for Beaufort 0 than for all other Beauforts except 5, and significantly larger for Beaufort 1 than for Beauforts 3, 4, 5 or 6 ($p < 0.05$). No other differences were significant at the 5% level, although the effective strip width was significantly smaller at Beaufort 5 than at Beauforts 2, 3 and 6 at the 10% level. The unexpected result for Beaufort 0 corresponds to a very small sample size (13); otherwise, there is an indication that effective strip width decreases with Beaufort, which is what we would expect. We estimate densities separately by low Beaufort (0–3) and high Beaufort (4–6), and average resulting estimates across Beaufort categories, weighting by effort. This analysis is valid provided the probability of detection on the centreline, $g(0)$, is unity for both Beaufort categories; the effective strip width need not be the same for both categories.

A similar analysis of cloud cover produced no significant differences, except that the encounter rate at cloud cover 3 was significantly higher than at cloud cover 2 ($p \doteq 0.01$), probably because relatively more cloud cover 3 occurred in areas of high fin whale density. If cloud cover 3 did increase detectability, effective strip width might be expected to increase, yet no pairwise comparisons provided any evidence of this ($p > 0.2$ for all six pairwise tests).

Table 8.7 Number of sightings (after truncation but before smearing), effective strip width, encounter rate and mean school size by sea state, Icelandic fin whale data, NASS-89. Standard errors in parentheses. Values in the same column with different superscript letters differ significantly ($p < 0.05$)

Beaufort	Number of sightings n	Effective strip width (n.m.)	Encounter rate (schools/100 n.m.)	Mean school size $\bar{s}$
0	13	0.55 $(0.05)^a$	3.08 $(1.90)^a$	1.69 $(0.24)^{ab}$
1	42	2.37 $(0.22)^b$	3.47 $(0.54)^a$	1.48 $(0.15)^{ab}$
2	83	2.00 $(0.23)^{bc}$	4.19 $(1.12)^a$	1.54 $(0.08)^a$
3	78	1.60 $(0.20)^c$	4.13 $(0.47)^a$	1.63 $(0.11)^a$
4	44	1.17 $(0.29)^c$	2.55 $(0.38)^a$	1.25 $(0.10)^b$
5	33	0.49 $(0.19)^{ac}$	2.42 $(0.92)^a$	1.21 $(0.10)^b$
6	18	1.61 $(0.19)^c$	2.51 $(1.73)^a$	1.22 $(0.14)^b$

Table 8.8 Number of sightings (after truncation but before smearing), effective strip width, encounter rate and mean school size by area, Icelandic fin whale data, NASS-89. Standard errors in parentheses. Blocks 11 (no sightings) and 26 (one sighting) are ignored. Values in the same column with different superscript letters differ significantly ($p < 0.05$)

Block	Number of sightings n	Effective strip width (n.m.)	Encounter rate (schools/100 n.m.)	Mean school size $\bar{s}$
36	54	0.94 $(0.51)^{abc}$	4.86 $(1.31)^{ab}$	1.35 $(0.10)^{ab}$
40	15	1.88 $(0.14)^a$	1.30 $(0.75)^{cd}$	1.13 $(0.14)^a$
50	23	2.07 $(0.35)^a$	2.03 $(1.06)^{bcd}$	1.35 $(0.13)^{ab}$
60	36	1.31 $(0.32)^{abc}$	3.26 $(1.40)^{abc}$	1.36 $(0.10)^{ab}$
70	9	0.68 $(0.18)^c$	1.18 $(0.51)^{cd}$	1.11 $(0.12)^a$
88	32	1.68 $(0.44)^{ab}$	2.57 $(0.45)^{bc}$	1.56 $(0.13)^{bc}$
93	70	1.87 $(0.18)^a$	16.39 $(1.96)^e$	1.69 $(0.12)^c$
94	66	1.14 $(0.21)^{bc}$	7.84 $(2.28)^{ae}$	1.53 $(0.12)^{bc}$
95	5	0.75 $(0.31)^{abc}$	0.24 $(0.20)^d$	1.20 $(0.22)^{abc}$

The geographic blocks defined for Icelandic surveys in 1989 are shown in Fig. 8.9. Highly significant differences between some blocks in encounter rate and mean school size are unsurprising, and we stratify by block for each of these components of estimation. There are also several pairwise comparisons between blocks that indicate significant differences between effective strip widths. Blocks 40, 50 and 93 yield wide estimated effective strip widths, whereas the estimates for blocks 70 and 95 are small (Table 8.8). Given adequate sample size, stratification could be by block, as for encounter rate and mean school size. However, effective strip width estimation is unreliable for small samples. There were only nine sightings in block 70 and five in block 95, rendering comparisons between them and other blocks of little value. Thus just the differences between block 94 and blocks 40 ($p \doteq 0.04$), 50 ($p \doteq 0.03$) and 93 ($p \doteq 0.05$) are genuine cause for concern. For estimating effective strip width, we choose here to stratify the area into two parts: south (blocks 40, 50, 60 and 70) and north, since this also effectively stratifies by vessel type (below).

Table 8.9 Number of sightings (after truncation but before smearing), effective strip width, encounter rate and mean school size by vessel, Icelandic fin whale data, NASS-89. Standard errors in parentheses. Values in the same column with different superscript letters differ significantly ($p < 0.05$)

Vessel	Number of sightings n	Effective strip width (n.m.)	Encounter rate (schools/100 n.m.)	Mean school size $\bar{s}$
Sk	43	1.09 $(0.43)^a$	1.79 $(0.47)^a$	1.26 $(0.08)^a$
AF	49	1.15 $(0.26)^a$	1.45 $(0.21)^a$	1.31 $(0.08)^a$
$Hv8$	83	1.43 $(0.33)^a$	4.58 $(0.47)^b$	1.43 $(0.08)^{ab}$
$Hv9$	136	1.37 $(0.17)^a$	8.02 $(2.46)^b$	1.61 $(0.08)^b$

The three components of estimation were also considered by vessel (Table 8.9). Most pairwise comparisons between vessels for encounter rate were significant, as were many of those for mean school size. These differences arise largely because vessels operated in different blocks; there is strong confounding between vessel differences and block differences. If vessel differences in encounter rate in particular occurred because different vessels have different searching efficiencies, significant differences in effective strip width between vessels might be anticipated, yet none were close to significance ($p > 0.2$ in all pairwise tests). Given the similarity in effective strip widths across vessels, we pool distance data across vessels prior to analysis. Effective strip widths for the two research vessels (1.09 n.m. and 1.15 n.m.) were slightly smaller than for the whaling vessels (1.43 n.m. and 1.37 n.m.). Although these differences are not significant, the impact on the analyses of estimating the effective strip width for research vessels separately from that for whaling vessels was assessed, and found to be slight. Because all effort in southern blocks was carried out by research vessels and most effort in the northern blocks was by whaling vessels, the decision to estimate the effective strip width separately for the northern and southern blocks, and to estimate encounter rate and mean school size by individual block, in effect gives stratification by vessel type.

Effective strip width did not show significant differences by size of school at the 5% level (Table 8.10), although there was a weak indication that the effective width was greater for schools of four or more animals than for single animals ($p \doteq 0.1$). Since 68% of sightings were of single animals, and a further 22% were of pairs, the effect of variation in detectability due to school size on abundance estimates will be slight. However, stratification by school size is likely to be more valid and was adopted. Estimated effective strip width is almost identical for single animals and for pairs, and very few schools of more than three animals were detected, so two strata were defined: small schools (one or two animals) and large schools (three or more). Small sample sizes forced one modification to the preferred method of analysis: the number of large schools was too small to allow estimation of effective strip width separately for high and low Beaufort, so that for large schools only, a pooled estimate of effective strip width across Beaufort categories was calculated.

Table 8.10 Number of sightings (after truncation but before smearing), effective strip width and encounter rate by school size, Icelandic fin whale data, NASS-89. Standard errors in parentheses. Values in the same column with different superscript letters differ significantly ($p < 0.05$)

School size	Number of sightings n	Effective strip width (n.m.)	Encounter rate (schools/100 n.m.)
1	211	1.27 $(0.16)^a$	2.27 $(0.16)^a$
2	68	1.25 $(0.28)^a$	0.71 $(0.13)^b$
3	22	1.50 $(0.25)^a$	0.23 $(0.05)^c$
4	8	1.71 $(0.23)^a$	0.09 $(0.03)^d$
> 4	2	–	–

To assess the impact of the decision to stratify by school size on estimates, two further analyses were carried out. The first of these was exactly as above, except data were not stratified by school size. In the second, the data were reanalysed with individual animals as the sighting unit. Thus a school of size three between 0.75 and 1.0 n.m. perpendicular distance contributes a frequency count of three to that distance interval (before smearing). This method of analysis is used in the southern hemisphere minke whale subcommittee of the International Whaling Commission, so that school size can be estimated as the ratio of animal density, from this analysis, to school density, estimated conventionally. We do not recommend this approach in general, although it can be effective, if variances are estimated by robust methods. When estimates were summed across geographic blocks prior to combining across Beaufort categories, the preferred method of analysis (stratifying by school size) gave a total estimate of 11 054 whales ($\widehat{se} = 1670$). Without stratification by school size, the estimate was 11 702 whales ($\widehat{se} = 1896$). When individual whales were taken as the sampling unit, an estimate of 11 758 whales ($\widehat{se} = 1736$) was obtained. Note that this latter strategy gave a very similar standard error to the other methods, even though sightings of individual whales were not independent events, and sample size is thus artificially increased. This occurs because of the robust method of estimating the variance in encounter rate, found by calculating the sample variance of the rate per day, weighted by daily effort, used as described for the empirical method of Section 3.7.2. Nevertheless, this approach underestimates the variance in effective strip width, unless it is obtained by resampling methods.

Abundance estimates for the East Greenland/Iceland stock of fin whales are given by block in Table 8.11. The sum of these estimates does not equal the corresponding estimate of 11 054 whales given above, due to the effects of calculating a weighted average of high and low Beaufort estimates within each block instead of first combining across blocks. The two estimates would be equal if the proportion of effort at low Beaufort was the same in every block. Suppose that 50% of effort occurred at low Beaufort overall, but in a given block, just 5% of effort occurred at low Beaufort. The method of summing estimates across blocks before averaging across Beaufort categories would give equal weight to the low and high Beaufort estimates in this block, whereas the method of Table 8.11 would give the high Beaufort estimate 19 times the weight of the low Beaufort estimate. The latter method is more appropriate, so the final abundance estimate of 10 378 whales ($\widehat{se} = 1655$) is obtained by weighting the low and high Beaufort estimates by respective effort in individual blocks.

This example shows that reliable abundance estimates may be obtained by geographic block even when sample size within a block is very small.

Table 8.11 Abundance estimates by block, East Greenland/Iceland fin whale stock, 1989

Block	Number of sightings n	$\hat{D}$ (whales/ 10 000 n.m.^{2})	$\widehat{se}(\hat{D})$	Size of block (n.m.^{2})	Abundance estimate $\hat{N}$	$\widehat{se}(\hat{N})$
36	54	270	72	44 172	1195	316
40	15	68	39	107 842	735	421
50	23	87	57	99 750	865	569
60	36	158	67	131 458	2071	879
70	9	74	58	88 571	658	517
88	32	129	38	59 848	770	230
93	70	873	220	21 761	1900	480
94	66	450	101	46 092	2073	467
95	5	16	14	69 396	111	95
All	323	155	25	668 891	10 378	1655

In two of the blocks of Table 8.11, sample size was under 10, yet analysis was possible stratifying not only by block but also by school size and Beaufort category. Of the three components of estimation, only effective strip width (or equivalently, $f(0)$) cannot be reliably estimated when samples are small. If this parameter can be assumed to be constant across at least some of the stratification categories, small sample size problems are avoided. The method is far superior to prorating a total estimate, obtained by pooling data across blocks, between blocks according to their respective areas, which requires that density of animals is uniform across the entire surveyed area. Variance estimation requires some care, since the individual block estimates are not independent. Provided the common component ($\hat{f}(0)$) of the respective estimates is removed when calculating the variance of their sum, then incorporated in the variance estimate using the delta method for approximating the variance of a product, as described in Section 3.7.1, valid variance estimates can be obtained quite simply. For the relatively complex fin whale analyses, variances are found as follows.

Within a stratum, abundance N is estimated by

$$\hat{N} = \frac{n \cdot \hat{f}(0) \cdot \bar{s} \cdot A}{2L}$$

with

$$\widehat{var}(\hat{N}) = \hat{N}^2 \cdot \left[\frac{\widehat{var}(n)}{n^2} + \frac{\widehat{var}\{\hat{f}(0)\}}{\{\hat{f}(0)\}^2} + \frac{\widehat{var}(\bar{s})}{\bar{s}^2} \right]$$

where n = number of sightings within 3 n.m. of the centreline in the stratum,

$\hat{f}(0)$ = estimated probability density of perpendicular distances, evaluated at zero,

$\bar{s}$ = mean school size,

L = distance covered while on effort,

A = size of the area containing the population of N animals.

For a given block, the above yields independent estimates of $f(0)$, and hence of animal abundance corresponding to small schools in low Beaufort $\hat{N}_{sm,\,lo}$, small schools in high Beaufort $\hat{N}_{sm,\,hi}$, and large schools $\hat{N}_{la}$ (unstratified by Beaufort). Then an estimate of abundance for animals in small schools is obtained by taking an average, weighted by effort carried out at low Beaufort (L_{lo}) and high (L_{hi}):

$$\hat{N}_{sm} = \frac{L_{lo} \cdot \hat{N}_{sm,\,lo} + L_{hi} \cdot \hat{N}_{sm,\,hi}}{L_{lo} + L_{hi}}$$

and

$$\widehat{\mathrm{var}}(\hat{N}_{sm}) = \frac{L_{lo}^2 \cdot \widehat{\mathrm{var}}(\hat{N}_{sm,\,lo}) + L_{hi}^2 \cdot \widehat{\mathrm{var}}(\hat{N}_{sm,\,hi})}{(L_{lo} + L_{hi})^2}$$

An abundance estimate for the block is then

$$\hat{N}_{bl} = \hat{N}_{sm} + \hat{N}_{la}$$

with

$$\widehat{\mathrm{var}}(\hat{N}_{bl}) = \widehat{\mathrm{var}}(\hat{N}_{sm}) + \widehat{\mathrm{var}}(\hat{N}_{la})$$

Within say the northern blocks, for which $\hat{f}(0)$ estimates are in common, total abundance is estimated by

$$\hat{N}_N = \sum \hat{N}_{bl}$$

where summation is over all northern blocks. To estimate the variance of this estimate, note that $\hat{N}_{bl}$ may be expressed as

$$\hat{N}_{bl} = \frac{L_{lo} \cdot \hat{N}_{sm,\,lo} + L_{hi} \cdot \hat{N}_{sm,\,hi}}{L_{lo} + L_{hi}} + \hat{N}_{la}$$

$$= \frac{L_{lo} \cdot \hat{f}_{sm,\,lo}(0) \cdot \hat{M}_{sm,\,lo} + L_{hi} \cdot \hat{f}_{sm,\,hi}(0) \cdot \hat{M}_{sm,\,hi}}{L_{lo} + L_{hi}} + \hat{f}_{la}(0) \cdot \hat{M}_{la}$$

$$= l_{lo} \cdot \hat{f}_{sm,\,lo}(0) \cdot \hat{M}_{sm,\,lo} + l_{hi} \cdot \hat{f}_{sm,\,hi}(0) \cdot \hat{M}_{sm,\,hi} + \hat{f}_{la}(0) \cdot \hat{M}_{la}$$

where

$$l_{lo} = \frac{L_{lo}}{L_{lo} + L_{hi}}$$

$$l_{hi} = \frac{L_{hi}}{L_{lo} + L_{hi}}$$

$$\hat{M}_{sm,\, lo} = \frac{n_{sm,\, lo} \cdot \bar{s}_{sm,\, lo} \cdot A}{2 L_{lo}}$$

evaluated for that block, and similarly for $\hat{M}_{sm,\, hi}$ and $\hat{M}_{la}$.
The component $\hat{f}(0)$ is common across blocks, whereas the other components of the abundance estimate are not. Thus

$$\hat{N}_N = \hat{f}_{sm,\, lo}(0) \cdot \sum [l_{lo} \cdot \hat{M}_{sm,\, lo}] + \hat{f}_{sm,\, hi}(0) \cdot \sum [l_{hi} \cdot \hat{M}_{sm,\, hi}] + \hat{f}_{la}(0) \cdot \sum \hat{M}_{la}$$

where summation is over blocks. Denote the three terms in this expression by T_i, $i = 1, 2, 3$; these three terms are independent. Consider the final term,

$$T_3 = \hat{f}_{la}(0) \cdot \sum \hat{M}_{la}$$

This has variance

$$\widehat{\text{var}}(T_3) = T_3^2 \cdot \left[\frac{\widehat{\text{var}}\{\hat{f}_{la}(0)\}}{\{\hat{f}_{la}(0)\}^2} + \frac{\sum \widehat{\text{var}}(\hat{M}_{la})}{\{\sum \hat{M}_{la}\}^2} \right]$$

where

$$\widehat{\text{var}}(\hat{M}_{la}) = \hat{M}_{la}^2 \cdot \left[\frac{\widehat{\text{var}}(n_{la})}{n_{la}^2} + \frac{\widehat{\text{var}}(\bar{s}_{la})}{\bar{s}_{la}^2} \right] \text{ evaluated in each block}$$

Similarly,

$$\widehat{\text{var}}(T_1) = T_1^2 \cdot \left[\frac{\widehat{\text{var}}\{\hat{f}_{sm,\, lo}(0)\}}{\{\hat{f}_{sm,\, lo}(0)\}^2} + \frac{\sum l_{lo}^2 \cdot \widehat{\text{var}}(\hat{M}_{sm,\, lo})}{\{\sum l_{lo} \cdot \hat{M}_{sm,\, lo}\}^2} \right]$$

and likewise for $\widehat{\text{var}}(T_2)$

389

Finally, $\widehat{\text{var}}(\hat{N}_N) = \sum_{i=1}^{3} \widehat{\text{var}}(T_i)$

If total abundance in the southern blocks is estimated by $\hat{N}_S$, it and its variance are estimated in the same way as for $\hat{N}_N$. Total abundance over the whole area is then the sum of these estimates, with variance equal to the sum of the respective variances, since $f(0)$ is estimated independently in the two areas. Applying the above methods, we obtain an abundance estimate of 10 378 fin whales, with standard error 1655. Assuming $\hat{N}$ is lognormal, the estimated 95% confidence interval for N is (7607, 14 158) animals (Section 3.7.1).

8.6 Use of tuna vessel observer data to assess trends in abundance of dolphins

In some circumstances, the scientist has little control over the design of line transect surveys, so that robust analysis techniques must be used to cope with potentially biased data. An example is the extensive database gathered by observers placed on board tuna vessels in the eastern tropical Pacific. Large tuna, and in particular the yellowfin tuna (*Thunnus albacares*), associate with dolphin schools in this region, and fishermen use speed boats to herd dolphins into large purse-seine nets to catch the tuna under them. Although most dolphins are released again, high mortality, largely through entanglement, can occur. To estimate this mortality, and to assess its impact on dolphin stock size, the observers record data on many variables, including sighting angles and distances to detected schools. We use these data for one of the affected stocks, the northern offshore stock of spotted dolphin (*Stenella attenuata*), to illustrate the techniques of Buckland and Anganuzzi (1988b) and Anganuzzi and Buckland (1989), which attempt to estimate trends in abundance that are robust to the many biases inherent in sightings data from fishing vessels. This had been attempted earlier, notably by Hammond and Laake (1983).

The database comprises hundreds of cruises and thousands of detections of dolphin schools, which have accumulated annually since 1975. Thus, the analyst can concentrate on reducing the effects of biases in the data, to yield smoother trends in relative abundance estimates, at the expense of precision.

Equation 3.4 gave the general formula for estimating animal density from line transect data. For tuna vessel dolphin sightings data, the constant $c = 1$, and g_0 may be assumed to be unity, since school sizes are large and the sighting cue is continuous. Thus density may be expressed as

$$D = \frac{E(n) \cdot f(0) \cdot E(s)}{2L}$$

If the stock area is of size A, the stock size may be expressed as

$$N = \frac{E(n)}{L} \cdot f(0) \cdot E(s) \cdot \frac{A}{2} \tag{8.1}$$

The first term in this expression is the encounter rate, the second term is the density function evaluated at zero, which is the reciprocal of the effective strip width, the third term is the true mean school size for the stock, and the remainder is a known constant. Thus there are three components to estimate. Because tuna vessel search effort is concentrated where captains expect to catch tuna, correction is required to adjust for search effort. Since the methods described here are to monitor changes in abundance, absolute abundance estimates are not required, but we must assume that bias in the relative abundance estimates is consistent between years, i.e. percentage relative bias should be constant. However, in a year of abundance of yellowfin tuna, effort is concentrated on dolphin schools, whereas in a year of scarcity, it becomes more economic to catch tuna too small to associate with dolphins. Thus if search effort is not corrected for, we cannot expect bias to be consistent. We correct for search effort in the following way.

Suppose for a random point in the ocean we can estimate the expectation of each of the three components, encounter rate, $f(0)$ and school size. Then their product, multiplied by the stock area divided by two, provides an estimate of abundance. The estimates need not be independent, provided variance estimation allows for correlation. We can therefore reduce the estimation problem to three simpler problems, each comprising a single parameter. Note that it is the expectation of each component for a random point in the ocean that is required. Thus an estimate of the true mean school size of the stock is inappropriate, since in areas of high school density, average school size may also be high. Such estimation would not be problematic if each component showed very little variation through the stock area. In practice variation is large, but if strata are defined such that the variation in a component within each stratum is small relative to variation between strata, the component can be estimated within each stratum, and averaged across strata, weighting by the area of each stratum. A method of stratifying the area is therefore required. The stratification can be different for each component. (Another option is to stratify by search effort, and estimate all three components from this single stratification. However, the low effort stratum proves to be very heterogeneous, and typically contains a high proportion of the stock, and the method lacks robustness.)

The stock area is post-stratified using the tuna vessel data. For each component, a variable that correlates well with that component must be identified. We use crude encounter rate, average school size and average sighting distance, calculated by degree square, as variables that should correlate well with encounter rate, school size and $f(0)$ (or effective strip width, $1/f(0)$). If these values are used unsmoothed, post-stratification leads to substantial bias, and no estimates are available for degree squares in which there was no effort. It is therefore necessary to use a smoothing routine on each of these variables. This implicitly assumes that each component of Equation 8.1 varies smoothly throughout the stock area. For a given component, the smoothed values for each degree square are ordered from smallest to largest, the number of strata is determined as a function of sample size, and the ordered list divided so that the sample size in each stratum is as equal as possible. The relevant component is then estimated within each stratum, and an average of the stratum estimates, weighted for stratum areas, is calculated. Strata are then determined for the next component, and the process is repeated. Once all component estimates are calculated, they are combined using Equation 8.1. To estimate $f(0)$ within each stratum, the hazard-rate model was assumed, and data were truncated at five nautical miles.

Criteria were set up and data failing to meet them were discarded. Many complete cruises and substantial parts of many more were deleted in this way, to reduce sensitivity of the methods to bias. Full details are given by Buckland and Anganuzzi (1988) and by Anganuzzi and Buckland (1989). The bootstrap was used to generate robust precision estimates. The bootstrapping unit was taken to be a cruise, to take account of variation between different observers, crews and vessels, while maintaining a fair degree of independence between units. Estimates for 1975–89 for the northern offshore stock of spotted dolphins, taken from Anganuzzi and Buckland (1989) and Anganuzzi et al. (1991), are given in Table 8.12.

Buckland et al. (1992a) estimated the underlying trends in dolphin abundance by smoothing the estimates of Table 8.12. They considered various smoothing methods such as moving averages, running medians and polynomial regression. Their chosen method was a compound running median known as 4253H, twice (Velleman and Hoaglin 1981).

Suppose that $\{X(t)\}$, $t = 1, \ldots, N$, is a time series of length N, and let $\{S_i(t)\}$ be a smoothed version of it, found by calculating an i-period running median. We can construct compound smoothing methods such as $\{S_{ij}(t)\}$, which is simply $\{S_j(S_i(t))\}$. Thus, a 4253 running median method smooths a time series using a 4-period running median, which

Table 8.12 Estimates of relative abundance and related parameters for the northern offshore stock of spotted dolphin. [ ] indicate that the 1977 estimate was used. Bootstrap standard errors (estimated from $B = 79$ replicates) are given in parentheses. $n =$ number of sightings after deletion and truncation; $\hat{\mu} = 1/\hat{f}(0)$ (nautical miles); $n/L =$ estimated encounter rate (schools/1000 n.m.); $\hat{E}(s) =$ estimated average school size for a random point in the stock area; $\hat{N}_s =$ estimated number of schools in stock (thousands); $\hat{N}_d =$ estimated number of dolphins in stock (millions)

Year	n	$\hat{\mu}$	n/L	$\hat{E}(s)$	$\hat{N}_s$	$\hat{N}_d$
1975	761	[2.74]	9.00	634	6.23	3.95
		(0.30)	(1.28)	(71)	(1.14)	(1.00)
1976	876	[2.74]	7.40	830	5.13	4.25
		(0.30)	(0.92)	(92)	(0.88)	(0.91)
1977	1700	2.74	6.46	855	4.48	3.83
		(0.30)	(0.52)	(73)	(0.69)	(0.75)
1978	720	2.65	6.58	680	4.72	3.21
		(0.23)	(0.53)	(69)	(0.55)	(0.54)
1979	516	2.36	7.03	521	5.66	2.95
		(0.32)	(0.67)	(54)	(0.92)	(0.56)
1980	1460	2.37	6.36	654	5.10	3.34
		(0.28)	(0.39)	(80)	(0.71)	(0.58)
1981	1593	2.62	6.39	547	4.63	2.54
		(0.24)	(0.37)	(44)	(0.56)	(0.44)
1982	1383	2.37	6.32	503	5.07	2.55
		(0.57)	(0.34)	(54)	(1.34)	(0.56)
1983	731	2.92	5.94	316	3.86	1.22
		(0.37)	(0.64)	(44)	(0.53)	(0.25)
1984	636	3.19	8.81	411	5.25	2.16
		(0.34)	(0.78)	(51)	(0.74)	(0.36)
1985	1976	2.78	8.96	471	6.12	2.88
		(0.18)	(0.63)	(41)	(0.56)	(0.35)
1986	2197	2.76	9.11	504	6.28	3.16
		(0.17)	(0.41)	(30)	(0.50)	(0.30)
1987	3529	2.85	8.82	502	5.89	2.95
		(0.13)	(0.41)	(36)	(0.42)	(0.29)
1988	2259	3.15	8.10	550	4.89	2.69
		(0.16)	(0.62)	(35)	(0.45)	(0.33)
1989	3569	3.42	8.39	624	4.66	2.91
		(0.18)	(0.30)	(40)	(0.32)	(0.28)

is in turn smoothed by a 2-period running median, smoothed again by a 5-period running median, and then by a 3-period running median (i.e. $\{S_{4253}(t)\} = \{S_3(S_5(S_2(S_4(t))))\}$). Near the endpoints, where there are not enough values surrounding a point to be smoothed using the specified running median, a shorter period running median may be used. The endpoints of the resultant time series are calculated by estimating $X(0)$ and $X(N + 1)$, the 'observed' values at $t = 0$ and $t = N + 1$, and then calculating

$$S_{4253}(1) = \text{median } \{\hat{X}(0), X(1), S_{4253}(2)\}$$

and $\qquad S_{4253}(N) = \text{median } \{S_{4253}(N - 1), X(N), \hat{X}(N + 1)\}$

$\hat{X}(0)$ is found by extrapolating from the straight line which passes through the smoothed values at $t = 2$ and $t = 3$, i.e. $\hat{X}(0) = 3 \cdot S_{4253}(2) - 2 \cdot S_{4253}(3)$; similarly,

$$\hat{X}(N + 1) = 3 \cdot S_{4253}(N - 1) - 2 \cdot S_{4253}(N - 2)$$

The H in 4253H, twice denotes a linear smoothing method commonly used with running medians, which is known as Hanning. It is a 3-period weighted moving average for $t = 2, \ldots, N - 1$, with weights $\{0.25, 0.5, 0.25\}$. The endpoints remain unchanged.

The pattern of the time series may be recovered be calculating the residuals of the series (i.e. the differences between the smoothed and unsmoothed estimates), smoothing the residual series by the same method as the time series was smoothed by, and then adding the smoothed values of the residuals to the smoothed values of the series. This is known as smoothing 'twice'. For example, if we define the residuals of the time series smoothed by 4253H to be $\{E(t)\} = \{X(t) - S_{4253}(t)\}$, then the values of the times series smoothed by 4253H, twice can be defined by

$$\{S_{4253H, \text{ twice}}(t)\} = \{S_{4253H}(t) + S_{4253H}(E(t))\}$$

Thus, the 4253H, twice running median method uses a 4253 running median to smooth the time series, estimates the endpoints of the smoothed series and then smooths the resultant series by Hanning. The residuals of the series are calculated and are also smoothed, using the same method as above. The smoothed values of the residuals are then added to the smoothed values of the time series to produce a time series smoothed by 4253H, twice. The advantage of using running medians is that the magnitude of an extreme estimate does not affect the resultant smoothed time series.

To assess changes in abundance over time, the bootstrap was again used, and 85% confidence intervals for relative abundance in each year

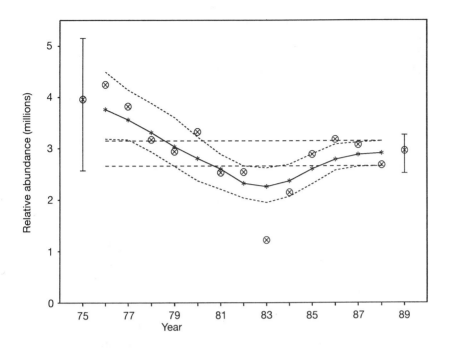

Fig. 8.10. Smoothed trends in abundance of the northern offshore stock of spotted dolphin. The broken lines indicate approximate 85% confidence limits. The horizontal lines correspond to 85% confidence limits for the 1988 estimate. If these limits both lie above the upper limit for an earlier year, abundance has increased significantly between that year and 1988 ($p < 0.05$); if the limits both lie below the lower limit for an earlier year, abundance has decreased significantly.

were estimated using the percentile method. The rationale for the choice of confidence level is that if two 85% confidence intervals do not overlap, the difference between the corresponding relative abundance estimates is significant at roughly the 5% level ($p \leq 0.05$), whereas if they do, the difference is not significant ($p > 0.05$). One bootstrap replication was carried out for each year, and the bootstrap estimates were smoothed using the running median routine. This process was repeated 79 times, and for each year, the sixth smallest and sixth largest smoothed estimates were taken as approximate 85% confidence limits. The median of the smoothed bootstrap estimates (i.e. the 40th estimate of each ordered set of 79) was used as the 'best' estimate of trend. Figure 8.10 shows the estimates of underlying trend for the northern offshore stock of spotted dolphins. The broken horizontal lines correspond to the upper and lower 85% confidence limits for the 1988 relative abundance estimate. Years for which the entire confidence interval lies outside the region between the broken horizontal lines show a significantly different relative abundance from that for 1988. The estimated trend is downwards until around 1983. Estimated abundance in 1976 was significantly higher than in 1988 ($p < 0.05$), but there is some evidence of a recovery between 1983 and 1988 ($p \doteq 0.05$). Thus northern offshore spotted dolphins appeared to decrease through the 1970s and early 1980s, with numbers remaining stable or increasing since.

8.7 House wren densities in South Platte River bottomland

We use data on house wrens (*Troglodytes aedon*) to illustrate model selection. The data were collected from 155 points, with between 14 and 16 points in each of ten 16 ha study blocks. The blocks were established in riparian vegetation along 30 km of South Platte River bottomland near Crook, Colorado. The study was described by Knopf (1986) and Sedgwick and Knopf (1987). The house wren was the most frequently recorded bird, and sample sizes were sufficient to allow estimation by block as well as across blocks. Thus, the option to stratify can also be examined.

The following models were tried: Fourier series (uniform key and up to four cosine adjustments); Hermite polynomial (half-normal key and up to four Hermite polynomial adjustments); half-normal and up to four cosine adjustments; and hazard-rate with at most two simple polynomial adjustments. Terms were tested for inclusion using the likelihood ratio test with a p-value of 0.05 and DISTANCE option LOOKAHEAD set to two. Intervals for goodness of fit tests were set at 0.0, 7.5, 12.5, 17.5, 22.5, 27.5, 32.5, 42.5, 62.5 and 92.5 m. The largest detection distances were at 90 m. To assess the impact of truncation, the last two intervals were discarded. Thus, the truncation point was 42.5 m, corresponding to 10% truncation of observations. The intervals were chosen to avoid possible favoured rounding distances, such as 10 m or 25 m. We recommend that goodness of fit is not used for model selection, but if it is, we recommend strongly that intervals are set using the option GOF within the ESTIMATE procedure of DISTANCE. The default intervals used by DISTANCE do not take account of rounding to favoured values, and may frequently give spurious significant test statistics.

A summary of results is given in Table 8.13. Note that the log-likelihood and the Akaike Information Criterion (AIC) are of no use for determining whether data should be truncated; values for different models are only comparable if the same truncation point is selected. Density estimates from untruncated data in Table 8.13 are mostly smaller and more precise than those from truncated data. They are probably also more biased given that fits to the untruncated data are less good. The exception is the Hermite polynomial model, which provides the best of the four fits to the untruncated data, and the worst fit to the truncated data. The Fourier series and hazard-rate models perform particularly poorly on the untruncated data. The Fourier series model is not robust to poor choice of truncation point for both line and point transects, whereas the hazard-rate model appears to be robust when data are untruncated for line transects but not for point transects (Buckland 1985, 1987a). The fit of all but the Hermite polynomial model is improved by truncation, and density estimates are more similar under different models for truncated data. We therefore select the model that gives the largest log-likelihood and the smallest AIC value when applied to truncated data for further analyses. This model is the hazard-rate with simple polynomial adjustments.

Table 8.13 Summary of results from fitting different models to house wren data. FS = Fourier series model (uniform key and cosine adjustments); HP = Hermite polynomial model (half-normal key and Hermite polynomial adjustments); HC = half normal key and cosine adjustments; Hz = hazard-rate key and simple polynomial adjustments. The truncation distance of $w = 92.5$ m is larger than the largest recorded distance, so no data are truncated, and the value $w = 42.5$ m corresponds to truncation of 10% of detection distances

Model	Number of adjustments	Log-likelihood	χ^2 statistic (df)		p-value	AIC	$\hat{D}$	Log-based 95% confidence interval
Data untruncated ($w = 92.5$ m)								
FS	4	-3312.7	18.8	(3)	< 0.001	6633.4	6.72	(5.95, 7.58)
HP	3	-3308.4	10.8	(4)	0.03	6624.8	8.28	(6.98, 9.82)
HC	3	-3308.4	10.7	(4)	0.03	6629.9	8.47	(7.24, 9.91)
Hz	1	-3329.7	39.3	(4)	< 0.001	6665.5	6.05	(5.28, 6.93)
Data truncated at $w = 42.5$ m								
FS	3	-2760.0	7.0	(3)	0.07	5526.0	9.05	(7.48, 10.95)
HP	1	-2762.0	12.1	(4)	0.02	5528.0	7.84	(6.77, 9.07)
HC	1	-2760.4	7.6	(4)	0.11	5524.8	9.01	(7.43, 10.92)
Hz	1	-2758.9	7.1	(3)	0.07	5523.8	8.14	(6.44, 10.30)

Estimates stratified by block and by observer are shown in Table 8.14. Goodness of fit tests indicate that fits to eight of the ten blocks are very good, although the data from blocks 0 and 6 are less well modelled. The effective detection radius is high for blocks 0 and 5, but is similar for all other blocks, at around 20 m. Densities vary appreciably between blocks. The final abundance estimate from the analysis stratified by block is very similar to that obtained from an unstratified analysis (Table 8.13, last row). The confidence interval is rather wider, reflecting

Table 8.14 Analyses of house wren data using the hazard-rate model with truncation at 42.5 m. Standard errors are given in parentheses; confidence intervals were calculated assuming a log-normal distribution for $\hat{D}$, and the Satterthwaite correction was not applied. *Estimated density found as the average of the block estimates; the corresponding standard error is found as (square root of the sum of squared standard errors for each block) divided by the number of blocks

Observer	Block	Effective detection radius $\hat{\rho}$	Encounter rate n/k	Estimated density	Log-based 95% confidence interval	Goodness of fit test p-value
All	0	30.4	0.64	2.21	(1.41, 3.47)	0.01
		(1.3)	(0.14)	(0.52)		
All	1	21.0	1.88	13.60	(7.28, 25.41)	0.42
		(3.2)	(0.24)	(4.45)		
All	2	22.7	2.17	13.35	(8.81, 20.24)	0.52
		(2.1)	(0.24)	(2.87)		
All	3	19.1	1.05	9.10	(1.52, 54.46)	0.46
		(10.8)	(0.14)	(10.40)		
All	4	20.4	1.71	13.05	(7.89, 21.61)	0.35
		(2.2)	(0.25)	(3.41)		
All	5	31.1	1.20	3.96	(2.64, 5.94)	0.39
		(1.9)	(0.20)	(0.83)		
All	6	23.5	1.22	7.00	(4.66, 10.51)	0.01
		(2.0)	(0.15)	(1.47)		
All	7	23.4	1.15	6.70	(2.86, 15.72)	0.35
		(5.1)	(0.16)	(3.06)		
All	8	23.5	1.16	6.69	(3.98, 11.25)	0.97
		(2.7)	(0.17)	(1.80)		
All	9	19.3	1.38	11.76	(4.44, 31.15)	0.31
		(4.9)	(0.22)	(6.22)		
All	All			8.38*	(5.94, 11.83)	
				(1.48)		
1	All	21.8	1.26	8.44	(5.28, 13.49)	< 0.001
		(2.6)	(0.08)	(2.05)		
2	All	18.8	1.06	9.56	(6.42, 14.24)	0.84
		(1.8)	(0.08)	(1.96)		
3	All	19.7	1.12	9.16	(6.43, 13.07)	0.15
		(1.6)	(0.09)	(1.67)		
4	All	33.0	1.40	4.11	(3.19, 5.30)	0.27
		(1.8)	(0.10)	(0.53)		
2 and 3	All	19.3	1.09	9.30	(7.10, 12.18)	0.31
		(1.2)	(0.07)	(1.28)		
1–3	All	20.1	1.15	9.06	(6.96, 11.79)	< 0.001
		(1.3)	(0.06)	(1.22)		

the larger number of parameters that have been estimated. There seems little advantage here to stratification, unless estimates are required by block; this is likely to be true generally when effort per unit area is the same in all strata.

Of more interest is the stratification by observer. Data from observers 1, 2 and 3 yield remarkably similar estimates. However, the first observer's data are modelled poorly. Inspection of the data and output from DISTANCE shows that observer 1 preferentially rounded distances around, or rather over, 10 m to exactly 10 m, and distances between 25 m and 40 m were predominantly recorded as, or close to, 30 m. Such rounding generally generates little bias, but intervals for goodness of fit testing need to be widened and reduced in number to obtain a reliable test when it is present. More serious is the apparent bias in the data of observer 4. The number of detections per point is rather greater than for the other observers, which is consistent with the higher effective detection radius, yet density is estimated to be well under half of that estimated from the data of each of the other observers. It is possible that observer 4 concentrated on detecting birds at greater distances, at the expense of missing many birds close to the point. More likely perhaps is that the effective detection radius was similar to that for the other observers, but that distances were overestimated by observer 4 by roughly 50%. This would be sufficient to explain the large difference in density estimates between observer 4 and the others. Whatever the cause, it seems clear that the data for observer 4 should be viewed with suspicion, whereas those for observers 2 and 3 appear to be most reliable. Our preferred analyses for these data use the hazard-rate model with up to two simple polynomial adjustments and truncation at 42.5 m, are unstratified by block, and discard data from observer 4. If there is concern about the poor fit of this model, the data of observer 1 should also be deleted. The resulting estimates with and without the data of observer 1 are shown in Table 8.14.

We have shown that poor model fits can be improved by truncating data. We now use the untruncated data to illustrate other strategies for improving the fit of a model. Other than truncation, the user of DISTANCE has several options. First, the analyst has control over how many adjustment terms are tested before DISTANCE concludes that no significant improvement in the fit has been obtained. It is not uncommon that a single adjustment term does not improve the fit of a model significantly, whereas the combined effect of it and a further term does yield a significant improvement. If LOOKAHEAD is set equal to one, the better model will not be found, whereas LOOKAHEAD = 2 will allow DISTANCE to select it, at the expense of slower run times. Second, the user can change the method by which DISTANCE selects models using SELECT. For SELECT = **sequential** (the default), it fits the lowest order term first, then adds successively higher order terms sequentially. If SELECT = **forward** is specified, DISTANCE will test for inclusion of each term not yet in the model. If the term that gives the largest increase in the value of the likelihood yields a significant improvement in the fit, it is included, and DISTANCE tests for inclusion of another term. This is analogous to a forward stepwise procedure in multiple regression. SELECT = **all** allows all possible combinations of adjustment terms to be tested, and that giving the minimum Akaike Information Criterion value is selected. Third, the key function (half-normal, hazard-rate, uniform or negative exponential) may be changed, and fourth, a different type of adjustment term (simple or Hermite polynomial, or cosine) can be selected. The combinations of these options that were applied to the house wren data are listed in Table 8.15. The

Table 8.15 Models and model options used for fitting house wren data. Distance data were pooled across observers and blocks, and were untruncated ($w = 92.5$ m)

Model	LOOK-AHEAD	Selection method	Key function	Adjustment terms
1	1	Sequential	Hazard-rate	Cosine
2	2	Sequential	Hazard-rate	Cosine
3	1	Forward	Hazard-rate	Cosine
4	–	All	Hazard-rate	Cosine
5	1	Sequential	Hazard-rate	Simple polynomial
6	2	Sequential	Hazard-rate	Simple polynomial
7	1	Forward	Hazard-rate	Simple polynomial
8	–	All	Hazard-rate	Simple polynomial
9	1	Sequential	Uniform	Cosine
10	2	Sequential	Uniform	Cosine
11	1	Forward	Uniform	Cosine
12	–	All	Uniform	Cosine
13	1	Sequential	Uniform	Simple polynomial
14	2	Sequential	Uniform	Simple polynomial
15	1	Forward	Uniform	Simple polynomial
16	–	All	Uniform	Simple polynomial
17	1	Sequential	Half-normal	Cosine
18	2	Sequential	Half-normal	Cosine
19	1	Forward	Half-normal	Cosine
20	–	All	Half-normal	Cosine
21	1	Sequential	Half-normal	Hermite polynomial
22	2	Sequential	Half-normal	Hermite polynomial
23	1	Forward	Half-normal	Hermite polynomial
24	–	All	Half-normal	Hermite polynomial

user may also specify the p-value for selecting between fits in likelihood ratio tests. For the runs considered in this example, only PVALUE = **0.05** was used, but if this tended to fit too few terms, a larger value (e.g. 0.15) might be preferred.

Table 8.16 Summary of results from fitting different models to house wren data. The models are defined in Table 8.15

Model	Number of adjustments	Log-likelihood	χ^2	(df)	p-value	AIC	$\hat{D}$	Log-based 95% confidence interval	
1	0	-3337.1	61.2	(6)	< 0.001	6678.3	5.56	(4.89,	6.32)
2	3	-3305.9	8.2	(3)	0.04	6621.9	8.58	(7.33,	10.03)
3 and 4	1	-3306.7	9.4	(5)	0.10	6619.3	8.74	(7.48,	10.22)
5 and 6	2	-3321.4	30.1	(4)	< 0.001	6650.8	6.72	(5.79,	7.79)
7 and 8	3	-3309.8	15.7	(3)	0.001	6629.9	7.51	(6.31,	8.93)
9	3	-3332.0	59.9	(5)	< 0.001	6670.0	5.28	(4.75,	•5.88)
10–12	5	-3308.1	13.2	(3)	0.004	6626.3	7.46	(6.52,	8.54)
13	3	-3529.3	501.1	(5)	< 0.001	7064.6	2.65	(2.34,	3.00)
14	5	-3510.3	306.2	(3)	< 0.001	7030.7	3.24	(2.93,	3.58)
15	3	-3582.8	515.9	(5)	< 0.001	7171.5	2.56	(2.32,	2.82)
16	4	-3456.8	271.5	(4)	< 0.001	6921.7	3.39	(3.00,	3.82)
17 and 18	1	-3313.0	13.8	(6)	0.03	6630.1	7.33	(6.48,	8.30)
19	2	-3310.4	11.2	(5)	0.05	6626.7	8.38	(7.13,	9.84)
20	3	-3308.6	10.8	(4)	0.03	6625.2	8.47	(7.24,	9.92)
21 and 23	0	-3327.3	39.6	(7)	< 0.001	6658.5	6.16	(5.51,	6.89)
22 and 24	2	-3312.8	13.9	(5)	0.02	6631.7	7.33	(6.39,	8.40)

The results of Table 8.16 indicate a clear 'winner' among the models. The hazard-rate model with cosine adjustments and using selection methods forward and all both lead to a hazard-rate model with a single cosine adjustment of order 4. Only this model yields a goodness of fit statistic that is not significant at the 5% level, and its Akaike Information Criterion value is 2.6 lower than the next best model. The density estimate is rather higher than that obtained from the favoured model on truncated data from above. Figure 8.11 shows that the fitted detection function has only a very narrow shoulder. For these data, use of cosine adjustment terms leads generally to a narrow shoulder, whereas polynomial adjustments to the hazard-rate model tend to preserve a wider shoulder. In Fig. 8.12, the fitted detection function obtained by making simple polynomial adjustments to the hazard-rate key, together with selection option forward or all, is shown. Although this model fits the data less well, its wider shoulder may be a better reflection of reality. It yields a density estimate rather lower than that from the favoured method for truncated data from above.

8.8 Songbird surveys in Arapaho National Wildlife Refuge

For this example, we use data supplied by F.L. Knopf from extensive songbird surveys of parts of the Arapaho National Wildlife Refuge, Colorado (Knopf *et al.* 1988). We consider counts carried out in June of 1980 and 1981, and analyse the six most numerous species, namely the yellow warbler (*Dendroica petechia*), brown-headed cowbird (*Molothrus ater*), savannah sparrow (*Passerculus sandwichensis*), song sparrow (*Melospiza melodia*), red-winged blackbird (*Agelaius phoeniceus*) and American robin (*Turdus migratorius*). In 1980, three pastures, labelled Pastures 1, 2 and 3, were surveyed by one visit to each of 124, 126 and 123 points respectively. In 1981, four pastures, 0, 1, 2 and 3, were surveyed during one visit to each of 100 points per pasture. All birds detected within 100 m of the point were noted and their locations were flagged, so that their distances could be measured to the nearest 10 cm. Although pastures varied in size, for the purposes of illustration, we assume that each was the same size.

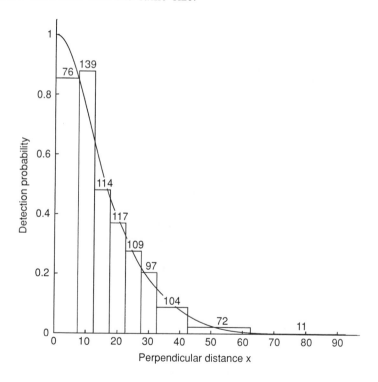

Fig. 8.11. The detection function obtained by fitting the hazard-rate key with cosine adjustments to untruncated house wren data.

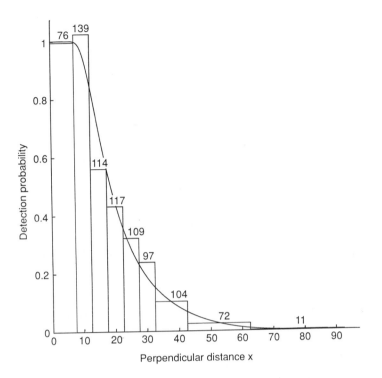

Fig. 8.12. The detection function obtained by fitting the hazard-rate key with simple polynomial adjustments to untruncated house wren data.

Analyses were carried out adopting the half-normal key with cosine adjustments. This model combines the key of the Hermite polynomial model with the adjustments of the Fourier series model. It is computationally more efficient than the former model, and uses a more plausible key function than the latter. For yellow warbler, savannah sparrow and song sparrow, some fits were found to be poor, so the detection distances were truncated at 52.5 m. Other analyses are untruncated. The variance of the number of detections was found using the empirical option within DISTANCE.

Yellow warbler analyses are summarized in Table 8.17. Separate estimates were obtained by stratum (pasture). In all, 205 detections were made in 1980 and 342 in 1981. Analyses of the brown-headed cowbird counts (Table 8.18) are less straightforward. First, count frequencies by distance from the point were highly variable in some pastures. Examination of the data showed that this was caused by detections of groups of birds. If more than one bird was recorded at exactly the same distance from the same point, we assume here that the birds comprised a single flock (cluster). In the analyses of Table 8.18, the option OBJECT = cluster was selected. For the first analyses, detection distance data were pooled across pastures, and a single set of estimates per year determined. The second set of analyses are by pasture. The estimates of $h(0)$ (or equivalently, of the effective radius of detection ρ) are imprecise, because the number of detections (as distinct from individual birds) per pasture was low, ranging from 21 (Pasture 1, 1980) to 50 (Pasture 3, 1980). Potentially more serious, bias may be high, as there is little information from which to select an appropriate model when sample size is small. Indeed, the effective radii for 1981 (Table 8.18) indicate appreciably more variability between pastures than can be explained by the standard errors. Either detectability of brown-headed cowbirds varied substantially between pastures or sample size was inadequate at least in some pastures for estimating the effective detection radius with low bias. If the latter explanation is more likely, then pasture estimates with higher precision and lower bias may be obtained by estimating the

Table 8.17 Analyses of yellow warbler point transect data, Arapaho National Wildlife Refuge. Standard errors are given in parentheses. Estimated density for category 'all' is found as the average of the pasture estimates; the corresponding standard error is found as (square root of the sum of squared standard errors for each pasture) divided by the number of pastures

Year	Pasture	Effective detection radius $\hat{\rho}$	Encounter rate n/k	Estimated density	Log-based 95% confidence interval		Goodness of fit test p-value
1980	1	26.2	0.73	3.40	(1.97,	5.87)	0.67
		(3.5)	(0.06)	(0.97)			
	2	27.1	0.45	1.96	(1.38,	2.79)	0.42
		(1.7)	(0.06)	(0.36)			
	3	17.0	0.37	4.02	(2.42,	6.67)	0.80
		(1.6)	(0.07)	(1.06)			
	All			3.12	(2.30,	4.23)	
				(0.49)			
1981	0	17.7	0.71	7.18	(4.87,	10.58)	0.60
		(1.4)	(0.09)	(1.44)			
	1	20.8	0.98	7.22	(4.52,	11.54)	0.07
		(2.4)	(0.08)	(1.75)			
	2	26.0	0.69	3.25	(2.36,	4.47)	0.13
		(1.5)	(0.08)	(0.53)			
	3	16.4	0.62	7.34	(5.03,	10.71)	0.19
		(1.2)	(0.08)	(1.43)			
	All			6.25	(5.05,	7.74)	
				(0.68)			

effective radius from data pooled across pastures (first section of Table 8.18), and estimating other parameters individually by pasture (second section of Table 8.18). This assumes that detectability does not vary with pasture, and utilizes the fact that average cluster size and expected

Table 8.18 Analyses of brown-headed cowbird point transect data, Arapaho National Wildlife Refuge. The first set of results was obtained by carrying out an unstratified analysis, the second set by stratifying by pasture, the third set by estimating $h(0)$ from unstratified distance data and other components separately by pasture, the fourth set by stratifying by cluster size, and the fifth set by correcting for size bias in mean cluster size. Standard errors are given in parentheses

Year	Pasture	Effective detection radius $\hat{\rho}$	Encounter rate n/k	Mean cluster size $\bar{s}$	Estimated density	Log-based 95% confidence interval	Goodness of fit test p-value
Unstratified							
1980	All	38.2	0.26	1.72	0.99	(0.73, 1.35)	0.44
		(2.0)	(0.03)	(0.11)	(0.16)		
1981	All	34.8	0.35	1.73	1.59	(1.18, 2.13)	0.44
		(1.9)	(0.03)	(0.11)	(0.24)		
Stratified by pasture							
1980	1	36.6	0.17	1.71	0.69	(0.34, 1.41)	0.46
		(5.0)	(0.04)	(0.21)	(0.26)		
	2	38.7	0.21	2.04	0.93	(0.50, 1.70)	0.60
		(4.2)	(0.04)	(0.28)	(0.29)		
	3	38.5	0.41	1.56	1.36	(0.92, 2.02)	0.55
		(2.5)	(0.06)	(0.12)	(0.28)		
	All				0.99	(0.72, 1.36)	
					(0.16)		
1981	0	25.5	0.28	1.39	1.91	(1.03, 3.54)	0.23
		(3.2)	(0.05)	(0.14)	(0.62)		
	1	56.9	0.33	2.03	0.66	(0.38, 1.14)	0.59
		(5.1)	(0.06)	(0.26)	(0.19)		
	2	42.1	0.42	1.76	1.33	(0.86, 2.06)	0.44
		(2.8)	(0.06)	(0.18)	(0.30)		
	3	24.9	0.36	1.69	3.12	(1.64, 5.94)	0.66
		(3.4)	(0.05)	(0.25)	(1.05)		
	All				1.75	(1.23, 2.49)	
					(0.32)		
Stratified by pasture, except for $h(0)$							
1980	1				0.63	(0.39, 1.13)	
					(0.18)		
	2				0.95	(0.61, 1.59)	
					(0.24)		
	3				1.39	(0.97, 2.00)	
					(0.26)		
	All				0.99	(0.73, 1.35)	
					(0.16)		

Table 8.18 *(Contd.)*

Year	Pasture	Effective detection radius $\hat{\rho}$	Encounter rate n/k	Mean cluster size $\bar{s}$	Estimated density	Log-based 95% confidence interval	Goodness of fit test p-value
1981	0				1.03 (0.25)	(0.67, 1.68)	
	1				1.76 (0.44)	(1.11, 2.90)	
	2				1.95 (0.43)	(1.28, 2.98)	
	3				1.61 (0.38)	(1.07, 2.63)	
	All				1.59 (0.24)	(1.18, 2.13)	
Stratified by cluster size							
Single birds							
1980	All	36.5 (2.9)	0.16 (0.02)	1.00	0.37 (0.08)	(0.25, 0.56)	0.07
1981	All	32.5 (2.1)	0.22 (0.02)	1.00	0.65 (0.11)	(0.47, 0.90)	0.19
Clusters ($\geq$ two birds)							
1980	All	40.5 (3.0)	0.11 (0.02)	2.78 (0.15)	0.58 (0.13)	(0.37, 0.89)	0.49
1981	All	39.4 (4.3)	0.13 (0.02)	2.92 (0.20)	0.80 (0.21)	(0.48, 1.32)	0.30
All birds							
1980	All				0.95 (0.15)	(0.70, 1.29)	
1981	All				1.44 (0.23)	(1.05, 1.97)	
Mean cluster size corrected for size bias							
1980	All	38.2 (2.0)	0.26 (0.03)	1.68 (0.10)	0.97 (0.16)	(0.71, 1.33)	0.44
1981	All	34.8 (1.9)	0.35 (0.03)	1.49 (0.07)	1.36 (0.20)	(1.02, 1.81)	0.44

number of detections can be estimated with low bias from small samples, whereas the effective radius often cannot. The approach is described in Section 3.8. Note that care must be taken when estimating variances.

The third section of Table 8.18 shows the estimates obtained from the above approach, assuming the pastures were equal in area. Note how much some of the pasture estimates differ from those found by estimating the effective radius of detection within each pasture. Note also that the estimates for all pastures combined are the same as those for which all data were pooled (first section of Table 8.18). If exactly the same effort (points per unit area) is expended in each stratum, the two methods are equivalent. However, the current method (1) allows separate estimates by stratum, (2) is still valid if effort differs by stratum, and

(3) is preferable to the fully stratified analysis if sample sizes are too small to estimate effective detection radii reliably by stratum, although it assumes that detectability does not vary across strata.

The fourth section of Table 8.18 shows another method of analysing these data. In this case, data were stratified by cluster size (Quinn 1979, 1985). Detections were divided into two categories: single birds and at least two birds. The results suggest that clusters are more detectable than single birds, although the overall estimates of density differ very little from those obtained above.

The final section of Table 8.18 shows adjusted mean cluster size, estimated by regressing logarithm of cluster size on probability of detection, and from this regression, estimating mean cluster size when probability of detection is one. For each cluster, its probability of detection was estimated by substituting its detection distance into the fitted detection function from the analyses of the first section of Table 8.18. The correlation between log cluster size and detection probability was not significant for 1980 ($r = -0.022$, df $= 96$, $p > 0.1$), and estimation was barely affected. For 1981, the correlation was significant ($r = -0.193$, df $= 137$, $p < 0.05$), and estimated cluster size was reduced (1.49 birds per cluster, compared with 1.73 birds per cluster for the detected clusters).

Table 8.19 Analyses of savannah sparrow point transect data, Arapaho National Wildlife Refuge. Standard errors are given in parentheses

Year	Pasture	Effective detection radius $\hat{\rho}$	Encounter rate n/k	Estimated density	Log-based 95% confidence interval		Goodness of fit test p-value
1980	1	33.4	0.48	1.36	(0.89,	2.06)	0.02
		(2.6)	(0.07)	(0.29)			
	2	27.1	0.95	4.12	(3.14,	5.40)	0.17
		(1.3)	(0.10)	(0.57)			
	3	31.2	0.72	2.37	(1.73,	3.26)	0.86
		(2.0)	(0.07)	(0.39)			
	All			2.62	(2.18,	3.17)	
				(0.25)			
1981	0	27.0	0.31	1.36	(0.80,	2.29)	0.87
		(2.8)	(0.05)	(0.37)			
	1	47.0	0.32	0.46	(0.24,	0.89)	0.12
		(6.8)	(0.06)	(0.16)			
	2	31.6	0.51	1.63	(1.02,	2.58)	0.96
		(2.8)	(0.08)	(0.39)			
	3	34.6	0.48	1.27	(0.77,	2.11)	0.66
		(3.7)	(0.07)	(0.33)			
	All			1.18	(0.90,	1.54)	
				(0.16)			

The analyses for savannah and song sparrows presented no special difficulties, and the estimates are given in Tables 8.19 and 8.20, respectively. For red-winged blackbirds (Table 8.21), it was again necessary to analyse the detections as clusters, although average cluster size was small, so bias arising from possible greater detectability of groups of two or more birds would be small. Data were insufficient to stratify, either by pasture or by cluster size.

The final analyses from this example are of American robin (Table 8.22). Again, sample sizes were too small to stratify, but the data presented no additional problems. A single cosine adjustment to the

Table 8.20 Analyses of song sparrow point transect data, Arapaho National Wildlife Refuge. Standard errors are given in parentheses

Year	Pasture	Effective detection radius $\hat{\rho}$	Encounter rate n/k	Estimated density	Log-based 95% confidence interval	Goodness of fit test p-value
1980	1	27.0	0.38	1.66	(1.12, 2.45)	0.79
		(1.9)	(0.05)	(0.33)		
	2	23.7	0.40	2.29	(1.51, 3.47)	0.56
		(1.8)	(0.06)	(0.49)		
	3	23.8	0.41	2.34	(1.59, 3.42)	0.04
		(1.6)	(0.06)	(0.46)		
	All			2.10	(1.66, 2.65)	
				(0.25)		
1981	0	31.5	0.43	1.38	(0.83, 2.30)	0.23
		(3.4)	(0.06)	(0.36)		
	1	32.1	0.47	1.45	(0.94, 2.25)	0.99
		(2.8)	(0.07)	(0.33)		
	2	24.6	0.39	2.05	(1.26, 3.32)	0.12
		(2.5)	(0.06)	(0.51)		
	3	23.9	0.29	1.62	(0.98, 2.67)	0.16
		(2.2)	(0.05)	(0.42)		
	All			1.63	(1.27, 2.08)	
				(0.21)		

Table 8.21 Analyses of red-winged blackbird point transect data, Arapaho National Wildlife Refuge. Standard errors are given in parentheses

Year	Pasture	Effective detection radius $\hat{\rho}$	Encounter rate n/k	Mean cluster size $\bar{s}$	Estimated density	Log-based 95% confidence interval	Goodness of fit test p-value
1980	All	27.9	0.18	1.29	0.93	(0.63, 1.37)	0.38
		(1.6)	(0.03)	(0.09)	(0.19)		
1981	All	32.9	0.17	1.20	0.61	(0.39, 0.97)	0.39
		(3.2)	(0.02)	(0.07)	(0.14)		

half-normal fit was selected for both the 1980 and the 1981 data. To show the effects of allowing for estimation of the number of adjustments required on the confidence interval for density, both years' data were reanalysed selecting the bootstrap option for estimating the variance of $\hat{h}(0)$. The resulting confidence intervals were (0.15, 0.58) and (0.18, 0.64) respectively, wider than the intervals of Table 8.22, as might be expected.

The density estimates of Tables 8.17–8.22 are consistently higher than those of Knopf *et al.* (1988). They used the Fourier series model on squared distances, as recommended by Burnham *et al.* (1980). We no longer recommend this approach, as it can lead to underestimation of density (Buckland 1987a), so the differences between their estimates and those given here might be anticipated.

Table 8.22 Analyses of American robin point transect data, Arapaho National Wildlife Refuge. Standard errors are given in parentheses

Year	Pasture	Effective detection radius $\hat{\rho}$	Encounter rate n/k	Estimated density	Log-based 95% confidence interval	Goodness of fit test p-value
1980	All	30.2 (2.8)	0.09 (0.02)	0.30 (0.08)	(0.18, 0.50)	0.22
1981	All	36.1 (3.2)	0.14 (0.02)	0.34 (0.08)	(0.22, 0.52)	0.87

8.9 Assessing the effects of habitat on density

The design of line and point transect surveys was discussed in Chapter 7. Suppose estimates of object density are required by habitat. The study area should first be stratified by habitat type. Surveys may then be designed within each stratum as described in Chapter 7. A belt of width w, where w corresponds to the distance within which say 85–90% of detections are expected to lie, might be defined just within the border of each stratum. If line or point transects are constrained so that they do not lie within the belt, then differences in density between habitats will be easier to detect. If such provision is not made, density will be underestimated in habitat types holding high densities, and overestimated in habitats with low densities. Often, comparisons of density between uniform blocks of habitat and habitat edge are of interest. Point transects are more suited to such comparisons than line transects. As before, habitat edges should be determined by stratifying the area by habitat categories. Points should then be positioned randomly, or systematically (say every 200 m) with a random starting point, along each edge; this may be achieved by envisaging all the edges placed end-to-end and allocating the points along the entire length, before mapping these points back to the actual edges. Detections at these edge points should be recorded according to which side of the edge (i.e. which habitat type) they are in, to allow edge versus centre comparisons within the same habitat type. Thus there would be two analyses of edge points, each taking the fraction of the circle surveyed to be one half. Alternatively, data from each side can be pooled to obtain an estimate of 'edge' density to compare with densities in either or both habitat types, found as described above.

Note that density at a distance w from the edge may be appreciably different from that at the edge itself. This will not invalidate the above analysis, unless the trend in density is very great, although the estimated detection function will be biased. Provided the assumption $g(0) = 1$ holds, the method will give a valid estimate of density at the edge. For similar reasons, points from which density away from the edge is estimated could be taken to be at least a distance w from the edge, rather than say at least $2w$, although a larger value might be preferred for other reasons; for example, a value equal to the maximum likely territory diameter could be chosen.

In reality, habitat may be too patchy and heterogeneous to divide a study area into a small number of strata. In this case, density might be better considered as a function of habitat characteristics. One approach to this would be to include habitat information as covariates, as described in Chapter 3, so that the surface representing object density is modelled. We use as an example a different approach. The following is summarized from Bibby and Buckland (1987).

We consider here binomial count data collected during 1983 for a study into bird populations of recently restocked conifer plantations throughout north Wales. In total, 326 points were covered, divided among 62 forestry plots that had been restocked between 1972 and 1981. Further details are given by Bibby et al. (1985).

Each detected bird was recorded as to whether it was within or beyond 30 m of the point. The half-normal binomial model of Section 6.2.1 was applied to these data, together with a linear model, for which analytic results are also available, and a single parameter hazard-rate model, with power parameter b set equal to 3.3, fitted by numerical methods (Buckland 1987a). Table 8.23 (reproduced from Buckland 1987a) shows that the linear model consistently yields higher estimates of densities than does the half-normal model, which in turn yields higher estimates than the hazard-rate model in most cases. Standard errors of these estimates are similar for all three models. Note that all three models give very similar relative densities between species. For example, the ratio of willow warbler density to wren density is estimated as 2.15, 2.13 and 2.19 under the half-normal, linear and hazard-rate models, respectively. Indeed, all three models give exactly the same ranking of species by density. This suggests that binomial counts may be effective for estimating relative density, but yield potentially biased estimates of absolute density. Total counts, which fail to take account of variability in detectability between species or between habitats, give a markedly

Table 8.23 Analyses of binomial count data on songbirds from Welsh restocked conifer plantations under three models. Standard errors are given below estimates. n_1 is the number of birds detected within $c_1 = 30$ m of the point, and n_2 is the number beyond 30 m. Scientific names are given in Appendix A

Species	n_1	n_2	Linear model			Half-normal model			Hazard-rate model		
			$\hat{D}$	$\hat{r}_{1/2}$	$\hat{\rho}$	$\hat{D}$	$\hat{r}_{1/2}$	$\hat{\rho}$	$\hat{D}$	$\hat{r}_{1/2}$	$\hat{\rho}$
Willow warbler	421	504	6.09	32.1	38.5	6.65	31.9	36.8	5.23	32.2	43.3
			0.37	0.8	1.0	0.38	0.7	0.9	0.34	1.0	1.3
Wren	208	347	2.83	36.4	43.8	3.12	36.1	41.7	2.39	37.3	50.1
			0.22	1.3	1.5	0.23	1.2	1.4	0.18	1.4	1.9
Goldcrest	108	57	1.90	24.2	29.1	1.96	24.8	28.6	1.99	21.7	29.1
			0.26	1.2	1.5	0.25	1.1	1.2	0.38	1.9	2.5
Tree pipit	127	235	1.70	38.0	45.6	1.88	37.6	43.3	1.44	39.0	52.3
			0.18	1.7	2.0	0.19	1.6	1.9	0.15	1.8	2.5
Robin	78	89	1.14	31.5	37.8	1.24	31.4	36.2	0.98	31.5	42.4
			0.16	1.8	2.2	0.16	1.7	2.0	0.15	2.2	3.0
Chaffinch	73	141	0.97	38.7	46.4	1.07	38.2	44.1	0.82	39.7	53.3
			0.13	2.3	2.7	0.15	2.2	2.5	0.11	2.4	3.3
Garden warbler	58	87	0.80	34.9	42.0	0.88	34.6	40.0	0.68	35.6	47.8
			0.13	2.3	2.8	0.13	2.2	2.6	0.11	2.6	3.5
Siskin	36	74	0.47	39.7	47.6	0.52	39.2	45.3	0.40	40.7	54.7
			0.10	3.3	4.0	0.11	3.2	3.7	0.08	3.5	4.7
Whitethroat	33	48	0.46	34.5	41.5	0.51	34.2	39.5	0.39	35.1	47.2
			0.10	3.0	3.7	0.10	2.9	3.3	0.08	3.5	4.6
Coal tit	29	38	0.41	33.2	39.8	0.45	33.0	38.1	0.35	33.6	45.1
			0.10	3.1	3.7	0.10	2.9	3.4	0.09	3.7	4.9
Dunnock	28	32	0.41	31.5	37.8	0.45	31.4	36.3	0.35	31.6	42.4
			0.10	3.0	3.6	0.10	2.8	3.3	0.09	3.7	4.9
Song thrush	27	79	0.34	46.1	55.3	0.38	45.5	52.5	0.30	47.4	63.6
			0.08	4.4	5.3	0.08	4.4	5.1	0.06	4.6	6.1
Long-tailed tit	18	12	0.30	26.1	31.3	0.31	26.5	30.5	0.29	24.4	32.8
			0.10	3.2	3.8	0.10	2.8	3.3	0.12	4.6	6.2
Blackbird	15	40	0.19	44.3	53.2	0.21	43.7	50.4	0.16	45.5	61.1
			0.06	5.7	6.9	0.06	5.6	6.5	0.04	5.9	7.9
Blackcap	10	6	0.17	25.2	30.3	0.18	25.7	29.7	0.17	23.2	31.1
			0.07	4.1	5.0	0.07	3.7	4.2	0.10	6.2	8.3
Redpoll	12	20	0.16	36.4	43.8	0.18	36.1	41.6	0.14	37.3	50.0
			0.07	5.3	6.4	0.08	5.1	5.9	0.06	5.8	7.8
Chiffchaff	9	27	0.11	46.6	55.9	0.12	46.0	53.1	0.10	47.9	64.3
			0.04	7.8	9.4	0.05	7.7	8.9	0.03	8.0	10.8
Mistle thrush	6	41	0.07	67.6	81.2	0.08	67.1	77.5	0.06	70.9	95.2
			0.03	13.8	16.6	0.04	14.0	16.2	0.03	16.5	22.1

different ordering of species. Counts of birds within 30 m of the point give a better indication of relative densities, although the ordering of goldcrest (*Regulus regulus*) and tree pipit (*Anthus trivialis*), and of blackcap (*Sylvia atricapilla*) and redpoll (*Carduelis flammea*), is reversed relative to the density estimates.

Table 8.23 shows that estimates of $r_{1/2}$ agree remarkably well under the three models, considering their widely differing shapes, indicating that the estimates provide a useful guide to the relative detectability of species. Variation within a species in values of $\hat{\rho}$ suggests that estimation of ρ may be less robust than that of $r_{1/2}$.

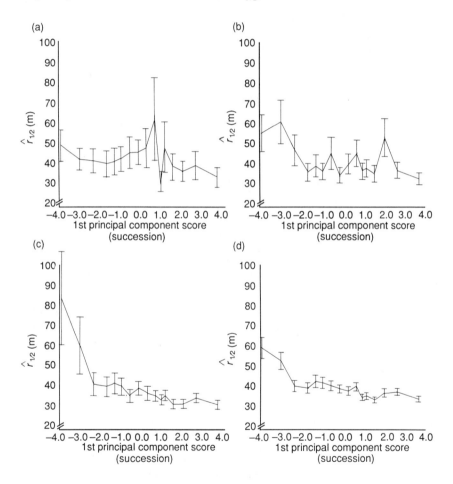

Fig. 8.13. Variation in detectability of (a) tree pipit, (b) wren, (c) willow warbler and (d) all species (pooled data) with habitat succession in conifer plantations aged between 2 and 11 years.

Various aspects of the habitat within a 30 m radius of each point were recorded, and a principal components analysis carried out. The first component was identified as succession. The plantations had been restocked between two and eleven years previously, so the environment ranged from open to very dense. Birds of each species were recorded according to whether they were within or beyond 30 m of the point. The binomial half-normal model for the detection function was assumed, and $r_{1/2}$, the distance at which probability of detection is one half, was used as a measure of detectability. Three species, the tree pipit, wren (*Troglodytes troglodytes*) and willow warbler (*Phylloscopus trochilus*)

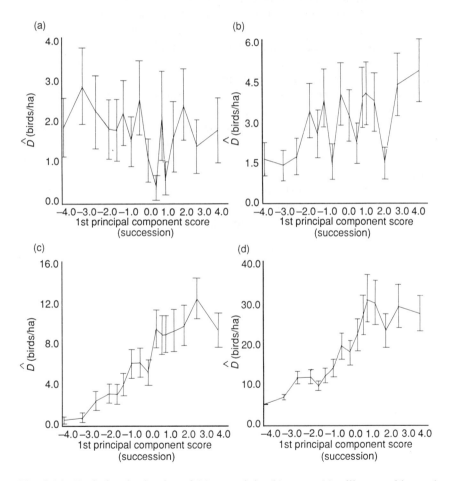

Fig. 8.14. Variation in density of (a) tree pipit, (b) wren, (c) willow warbler and (d) all species (pooled data) with habitat succession in conifer plantations aged between 2 and 11 years.

413

were present in sufficient numbers at each stage of development to examine their change in detectability and density with succession in habitat. Figure 8.13 shows the estimated change in detectability with succession for these three species, and for all species combined. Both the wren and the willow warbler appear to be more detectable in the very early stages of succession. The pattern for the tree pipit is less clear. Analyses of the combined data set show a similar pattern to those for wren and willow warbler.

To measure trends in bird density with succession, it is therefore necessary to adjust for greater detectability in more open habitats. In Fig. 8.14 estimated change in density with succession is shown for the same three species and for all species combined. Both the wren and the willow warbler show a trend to higher densities in the older plantations. The plot for all species combined shows roughly a fivefold increase in density for eleven year old restocks relative to two year old. If unadjusted counts of birds are used as measures of relative abundance, this increase is estimated to be just 1.4-fold, indicating the importance of adjusting counts for detectability.

Principal components analysis was used in the above because it proved effective at reducing the dimensionality of the habitat variables. The second component represented a trend from a more diverse habitat, with herbaceous plants and regenerating broadleaf trees, through to pure coniferous stand with little undergrowth. If the only aspects of interest were variation in detectability and density with succession, the analysis could have been simplified by replacing the first principal component by stand age.

Appendix A
List of common and scientific names cited

Common name	Scientific name
Abert's towhee	*Pipilo aberti*
American robin	*Turdus migratorius*
Ash-throated flycatcher	*Myiarchus cinerascens*
Baltic rush	*Juncus balticus*
Bewick's wren	*Thryomanes bewickii*
Blackbird	*Turdus merula*
Blackcap	*Sylvia atricapilla*
Black-tailed gnatcatcher	*Polioptila melanura*
Blue-winged teal	*Anas discors*
Bobolink	*Dolichonyx oryzivorus*
Brown-headed cowbird	*Molothrus ater*
Brown towhee	*Pipilo fuscus*
Bullrush	*Scirpus validus*
Bushtit	*Psaltriparus minimus*
Cactus wren	*Campylorhynchus brunneicapillus*
California grey whale	*Eschrichtius robustus*
Cassin's finch	*Carpodacus cassinii*
Cattail	*Typha latifolia*
Chaffinch	*Fringilla coelebs*
Chiffchaff	*Phylloscopus collybita*
Cinnamon teal	*Anas cyanoptera*
Clark's nutcracker	*Nucifraga columbiana*
Coal tit	*Parus ater*
Crissal thrasher	*Toxostoma crissale*
Dall's porpoise	*Phocoenoides dalli*
Dark-eyed junco	*Junco hyemalis*
Darkling beetle	*Eleodes* spp.
Deer	*Odocoileus* spp.
Dolphins	Delphinidae
Dunnock	*Prunella modularis*
Dusky flycatcher	*Empidonax oberholseri*
Eastern grey kangaroo	*Macropus giganteus*

415

Common name	Scientific name
Field mouse	*Peromyscus* spp.
Fin whale	*Balaenoptera physalus*
Fruit bat	*Chiroptera* spp.
Gadwall	*Anas strepera*
Garden warbler	*Sylvia borin*
Gila woodpecker	*Melanerpes uropygialis*
Goldcrest	*Regulus regulus*
Greasewood	*Sarcobatus vermiculatus*
Green-winged teal	*Anas carolinensis*
Grouse	Tetraoninae
Hares	*Lepus* spp.
Hermit thrush	*Catharus guttatus*
House wren	*Troglodytes aedon*
Jackrabbit	*Lepus* spp.
Kangaroo	Macropodidae
Ladder-backed woodpecker	*Picoides scalaris*
Lake trout	*Salvelinus namaycush*
Lion	Felidae
Long-billed curlew	*Numenius americanus*
Long-tailed tit	*Aegithalos caudatus*
Lucy's warbler	*Vermivora luciae*
Mallard	*Anas platyrhynchos*
Minke whale	*Balaenoptera acutorostrata*
Mistle thrush	*Turdus viscivorus*
Mountain chickadee	*Parus gambeli*
Northern bobwhite quail	*Colinus virginianus*
Northern oriole	*Icterus galbula*
Northern pintail	*Anas acuta*
Northern shoveler	*Anas clypeata*
Omao	*Phaeornis obscurus*
Pacific white-sided dolphin	*Lagenorhynchus obliquedens*
Pheasant	Phasianidae
Pine siskin	*Carduelis pinus*
Porpoise	Phocoenidae
Pronghorn	*Antilocapra americana*
Quail	Odontophorinae
Rabbitbrush	*Chrysothamnus* spp.
Rabbits	Leporidae
Red crab	*Grapsus grapsus*
Redhead	*Aythya americana*
Redpoll	*Carduelis flammea*
Red-winged blackbird	*Agelaius phoeniceus*
Risso's dolphin	*Grampus griseus*
Robin	*Erithacus rubecula*
Rockfish	*Sebastes* spp.
Ruby-crowned kinglet	*Regulus calendula*
Rufous-sided towhee	*Pipilo erythrophthalmus*
Sagebrush	*Artemisia* spp.
Saltgrass	*Distichlis stricta*

Common name	Scientific name
Savannah sparrow	*Passerculus sandwichensis*
Scrub jay	*Aphelocoma coerulescens*
Seal	Otariidae/Phocidae
Sedge	*Carex* spp.
Siskin	*Carduelis spinus*
Song sparrow	*Melospiza melodia*
Song thrush	*Turdus philomelos*
Spikerush	*Eleocharis macrosachya*
Spotted dolphin	*Stenella attenuata*
Tree pipit	*Anthus trivialis*
Tuna	*Thunnus* spp.
Verdin	*Auriparus flaviceps*
Whale	Balaenopteridae
White-crowned sparrow	*Zonotrichia leucophrys*
Whitethroat	*Sylvia communis*
Willow warbler	*Phylloscopus trochilus*
Wolf spider	*Atrax* spp.
Wren	*Troglodytes troglodytes*
Wrentit	*Chamaea fasciata*
Yellowfin tuna	*Thunnus albacares*
Yellow-rumped warbler	*Dendroica coronata*
Yellow warbler	*Dendroica petechia*

Appendix B
Notation and abbreviations, and their definitions

The following list is not exhaustive; notation is only included here if it is used through much of the text. Some of the notation listed below is occasionally used for another purpose; in such cases, the temporary definition is stated in the text. Standard mathematical and statistical symbols such as ∞, $\sum$ and $\hat{}$ are not listed.

μ effective strip width $= 1/f(0) = \int_0^w g(x)\,dx$; the half-width of the strip extending either side of a transect centreline such that as many objects are detected outside the strip as remain undetected within it

ν effective area $= 2\pi/h(0) = 2\pi\int_0^w r g(r)\,dr$ (point transect sampling); the area such that as many objects are detected outside it as remain undetected inside it

$\pi(s)$ probability distribution of cluster sizes in area A
$\pi^*(s)$ probability distribution of sizes of detected clusters; this differs from $\pi(s)$ when sampling of clusters is size-biased

ρ effective radius $= \sqrt{(\nu/\pi)}$; the radius of the circle around each point such that as many objects are detected beyond ρ as remain undetected within ρ

σ a scale parameter, used primarily in the half-normal and hazard-rate detection functions

θ sighting angle (subscript i, if present, denotes the ith detection)

a area within distance w of surveyed lines or points; the surveyed area
A size of study area, containing N objects; a sample of size a of this area is surveyed (subscript v, if present, denotes the vth stratum)
AIC Akaike's Information Criterion, used for model selection

b dispersion parameter, also called variance inflation factor
B number of bootstrap resamples

418

c the sampling fraction, usually equal to one, but equal to 0.5 if just one side of the line is recorded (line transect sampling), or $\phi/2\pi$ if just an arc of ϕ radians is counted (point transect sampling and, especially, cue counting)
c_i cutpoint i, separating interval i from interval $i + 1$, grouped distance data
cov sampling covariance
cv coefficient of variation = (standard error)/(estimate). When expressed numerically, usually converted to a percentage by multiplying by 100

D density of objects in study area = N/A (subscript v, if present, denotes the v th stratum)

$E(s)$ the mean size of the N_s clusters in the study area

$f(y)$ the probability density function of perpendicular distances (line transects) or detection distances (point transects)
$f(y, s)$ the joint probability density function of distances y and cluster sizes s
$f(y|s)$ the conditional probability density function of distances y given cluster size s
$f(0)$ the value of the probability density function of perpendicular distances, evaluated at zero distance (line transect sampling)

$g(y)$ the detection function; the probability that an object at distance y from the line or point is detected. If $g_0 < 1$, $g(y)$ is the conditional probability, scaled such that $g(0) = 1$
$g(y, s)$ the bivariate detection function; the probability that a cluster of size s and at distance y from the line or point is detected
$g(y|s)$ the conditional detection function; the probability that a cluster at distance y from the line or point is detected, given that it is of size s; functional expression is equivalent to $g(y, s)$
g_0 the probability that an object that is on the line or point ($y = 0$) is detected

$h(0)$ the slope of the probability density function of detection distances, evaluated at distance zero (point transect sampling) = $f'(0) = 2\pi/v = 1/\int_0^w rg(r)\,dr$

k number of replicate lines or points (subscript v, if present, denotes the v th stratum)

l_i the length of line i in a line transect survey, $i = 1, \ldots, k$
L the total line length in a line transect survey = $\sum_{i=1}^{k} l_i$ (subscript v, if present, denotes the v th stratum)
$\mathcal{L}$ the likelihood function for data arising from distance sampling

n sample size; number of objects detected (subscript v, if present, denotes the v th stratum)
N population size; total number of objects in the study area of size A (subscript v, if present, denotes the v th stratum)
N_s when objects occur in clusters, the total number of clusters in the study area

P_a the probability that an object in the surveyed area a is detected
pdf probability density function, for example $f(y)$

r the detection or radial distance; the distance of an object from the observer at the time the object is detected (subscript i, if present, denotes the ith detection)
$r_{1/2}$ the distance from a point at which probability of detection is one half

s the size of a cluster of objects (subscript i, if present, denotes the ith detection)
sd standard deviation
se standard error

V number of strata
var sampling variance

w the truncation point; distances exceeding w either are not recorded or are truncated before analysis

x the perpendicular distance; the distance of a detected object from the transect centreline (subscript i, if present, denotes the ith detection)

y the perpendicular distance x of a detected object from the centreline (line transect sampling) or the detection distance r of an object from the point (point transect sampling) (subscript i, if present, denotes the ith detection)

z distance parallel to the centreline of an object from the observer at the moment of detection (subscript i, if present, denotes the ith detection)

Bibliography

*Publications referenced in the text are indicated by an asterisk

*Akaike, H. (1973) Information theory and an extension of the maximum likelihood principle, in *International Symposium on Information Theory*, 2nd edn (eds B. N. Petran and F. Csàaki), Akadèemiai Kiadi, Budapest, Hungary, pp. 267–81.

*Akaike, H. (1985) Prediction and entropy, in *A Celebration of Statistics* (eds A. C. Atkinson and S. E. Fienberg), Springer-Verlag, Berlin, pp. 1–24.

*Alho, J. M. (1990) Logistic regression in capture–recapture models. *Biometrics*, **46**, 623–35.

*Alldredge, J. R. and Gates, C. E. (1985) Line transect estimators for left-truncated distributions. *Biometrics*, **41**, 273–80.

Andersen, D. E., Rongstad, O. J. and Mytton, W. R. (1985) Line transect analysis of raptor abundance along roads. *Wildlife Society Bulletin*, **13**, 533–9.

*Anderson, B. W. and Ohmart, R. D. (1981) Comparisons of avian census results using variable distance transect and variable circular plot techniques, in *Estimating Numbers of Terrestrial Birds. Studies in Avian Biology No. 6* (eds C. J. Ralph and J. M. Scott), Cooper Ornithological Society, pp. 186–92.

*Anderson, D. R., Burnham, K. P. and Crain, B. R. (1978) A log-linear model approach to estimation of population size using the line-transect sampling method. *Ecology*, **59**, 190–3.

*Anderson, D. R., Burnham, K. P. and Crain, B. R. (1979) Line transect estimation of population size: the exponential case with grouped data. *Communications in Statistics – Theory and Methods*, **A8**, 487–507.

*Anderson, D. R., Burnham, K. P. and Crain, B. R. (1980) Some comments on Anderson and Pospahala's correction of bias in line transect sampling. *Biometrical Journal*, **22**, 513–24.

*Anderson, D. R., Burnham, K. P. and Crain, B. R. (1985a) Estimating population size and density using line transect sampling. *Biometrical Journal*, **27**, 723–31.

*Anderson, D. R., Burnham, K. P. and Crain, B. R. (1985b) Some mathematical models for line transect sampling. *Biometrical Journal*, **7**, 741–52.

*Anderson, D. R., Burnham, K. P., White, G. C. and Otis, D. L. (1983) Density estimation of small-mammal populations using a trapping web and distance sampling methods. *Ecology*, **64**, 674–80.

*Anderson, D. R., Laake, J. L., Crain, B. R. and Burnham, K. P. (1979) Guidelines for line transect sampling of biological populations. *Journal of Wildlife Management*, **43**, 70–8.

*Anderson, D. R. and Pospahala, R. S. (1970) Correction of bias in belt transects of immotile objects. *Journal of Wildlife Management*, **34**, 141–6.

421

*Anganuzzi, A. A. and Buckland, S. T. (1989) Reducing bias in estimated trends from dolphin abundance indices derived from tuna vessel data. *Report of the International Whaling Commission*, **39**, 323–34.

*Anganuzzi, A. A., Buckland, S. T. and Cattanach, K. L. (1991) Relative abundance of dolphins associated with tuna in the eastern tropical Pacific, estimated from tuna vessel sightings data for 1988 and 1989. *Report of the International Whaling Commission*, **41**, 497–506.

Baggett, S. C. (1983) A comparison of two avian census methods: strip transect and fixed circular plot. MS Thesis, Stephen F. Austin State University, Nacogdoches, TX, USA. 68pp.

Balph, M. H., Stoddart, S. L. and Balph D. H. (1977) A simple technique for analyzing bird transect counts. *Auk*, **94**, 606–7.

Barlow, J. (1988) Harbor porpoise, *Phocoena phocoena*, abundance estimation for California, Oregon and Washington: I. Ship surveys. *Fishery Bulletin*, **86**, 417–31.

Barlow, J., Oliver, C. W., Jackson, T. D. and Taylor, B. L. (1988) Harbor porpoise, *Phocoena phocoena*, abundance estimation for California, Oregon, and Washington: II. Aerial surveys. *Fishery Bulletin*, **86**, 433–44.

*Barlow, R. E., Bartholomew, D. J., Bremner, J. M. and Brunk, H. D. (1972) *Statistical Inference under Order Restrictions*. Wiley, New York.

*Barndorff-Nielson, O. E. (1986) Inference on the full or partial parameters based on the standardized log likelihood ratio. *Biometrika*, **73**, 307–22.

Barnes, A., Hill, G. J. E. and Wilson, G. R. (1986) Correcting for incomplete sighting in aerial surveys of kangaroos. *Australian Wildlife Research*, **13**, 339–48.

Bart, J. and Herrick, J. (1984) Diurnal timing of bird surveys. *Auk*, **101**, 384–7.

*Bart, J. and Schoultz, J. D. (1984) Reliability of singing bird surveys: changes in observer efficiency with avian density. *Auk*, **101**, 307–18.

*Batcheler, C. L. (1975) Development of a distance method for deer census from pellet groups. *Journal of Wildlife Management*, **39**, 641–52.

*Beasom, S. L., Hood, J. C. and Cain, J. R. (1981) The effect of strip width on helicopter censusing of deer. *Journal of Range Management*, **34**, 36–7.

Bell, H. L. and Ferrier, S. (1985) The reliability of estimates of density from transect counts. *Corella*, **9**, 3–13.

*Bergstedt, R. A. and Anderson, D. R. (1990) Evaluation of line transect sampling based on remotely sensed data from underwater video. *Transactions of the American Fisheries Society*, **119**, 86–91.

*Best, L. B. (1981) Seasonal changes in detection of individual bird species, in *Estimating Numbers of Terrestrial Birds. Studies in Avian Biology No. 6* (eds C. J. Ralph and J. M. Scott), Cooper Ornithological Society, pp. 252–61.

*Best, P. B. and Butterworth, D. S. (1980) Report of the Southern Hemisphere Minke whale assessment cruise, 1978/79. *Report of the International Whaling Commission*, **30**, 257–83.

*Bibby, C. J. and Buckland, S. T. (1987) Bias of bird census results due to detectability varying with habitat. *Acta Ecologica*, **8**, 103–12.

*Bibby, C. J., Phillips, B. N. and Seddon, A. J. E. (1985) Birds of restocked conifer plantations in Wales. *Journal of Applied Ecology*, **22**, 619–33.

*Bollinger, E. K., Gavin, T. A. and McIntyre, D. C. (1988) Comparison of transects and circular-plots for estimating bobolink densities. *Journal of Wildlife Management*, **52**, 777–86.

Bonnell, M. L. and Ford, R. G. (1987) California sea lion distribution: a statistical analysis of aerial transect data. *Journal of Wildlife Management*, **51**, 13–20.

Bottenberg, H., Litsinger, J. A. and Kenmore, P. E. (1992) A line transect survey method for rice tungro virus. *Proceedings, Third International Conference on Plant Protection in the Tropics, Volume V*, Kuala Lumpur, Malaysia, March 1990.

*Breiwick, J. M., Rugh, D. J., Withrow, D. E., Dahlheim, M. E. and Buckland, S. T. (unpublished) Preliminary population estimate of gray whales during the 1987/88 southward migration. *Paper SC/40/PS12*, presented to the Scientific Committee of the International Whaling Commission, May 1988.

Brennan, L. A. and Block, W. M. (1986) Line transect estimates of mountain quail density. *Journal of Wildlife Management*, **50**, 373–7.

Briggs, K. T., Tyler, W. B. and Lewis, D. B. (1985) Aerial surveys for seabirds: methodological experiments. *Journal of Wildlife Management*, **49**, 412–7.

Brockelman, W. Y. (1980) The use of the line transect sampling method for forest primates, in *Tropical Ecology and Development* (ed. J. I. Furtado), The International Society of Tropical Ecology, Kuala Lumpur, Malaysia, pp. 367–71.

Broome, L. S., Bishop, K. D. and Anderson, D. R. (1984) Population density and habitat use by *Megapodius freyinet eremita* in West New Britain. *Australian Wildlife Research*, **11**, 161–71.

*Brunk, H. D. (1978) Univariate density estimation by orthogonal series. *Biometrika*, **65**, 521–8.

*Buckland, S. T. (1980) A modified analysis of the Jolly–Seber capture–recapture model. *Biometrics*, **36**, 419–35.

*Buckland, S. T. (1982) A note on the Fourier series model for analysing line transect data. *Biometrics*, **38**, 469–77.

*Buckland, S. T. (1984) Monte Carlo confidence intervals. *Biometrics*, **40**, 811–7.

*Buckland, S. T. (1985) Perpendicular distance models for line transect sampling. *Biometrics*, **41**, 177–95.

*Buckland, S. T. (1987a) On the variable circular plot method of estimating animal density. *Biometrics*, **43**, 363–84.

*Buckland, S. T. (1987b) An assessment of the performance of line transect models for fitting IWC/IDCR cruise data, 1978/79 to 1984/85. *Report of the International Whaling Commission*, **37**, 277–9.

*Buckland, S. T. (1987c) Estimation of Minke whale numbers from the 1984/85 IWC/IDCR Antarctic sightings data. *Report of the International Whaling Commission*, **37**, 263–8.

*Buckland, S. T. (1992a) Fitting density functions using polynomials. *Applied Statistics*, **41**, 63–76.

*Buckland, S. T. (1992b) Maximum likelihood fitting of Hermite and simple polynomial densities. *Applied Statistics*, **41**, 241–66.

Buckland, S. T. (1992c) Effects of heterogeneity on estimation of probability of detection on the trackline. *Report of the International Whaling Commission*, **42**.

*Buckland, S. T. and Anganuzzi, A. A. (1988a) Comparison of smearing methods in the analysis of Minke sightings data from IWC/IDCR Antarctic cruises. *Report of the International Whaling Commission*, **38**, 257–63.

*Buckland, S. T. and Anganuzzi, A. A. (1988b) Estimated trends in abundance of dolphins associated with tuna in the eastern tropical Pacific. *Report of the International Whaling Commission*, **38**, 411–37.

*Buckland, S. T. and Breiwick, J. M. (in press) Estimated trends in abundance of California gray whales from shore counts, 1967/68 to 1987/88. *Report of the International Whaling Commission (Special Issue)*.

*Buckland, S. T., Breiwick, J. M., Cattanach, K. L. and Laake, J. L. (in press) Estimated population size of the California gray whale. *Marine Mammal Science*.

*Buckland, S. T., Cattanach, K. L. and Anganuzzi, A. A. (1992a) Estimating trends in abundance of dolphins associated with tuna in the eastern tropical Pacific Ocean, using sightings data collected on commercial tuna vessels. *Fishery Bulletin*, **90**, 1–12.

*Buckland, S. T., Cattanach, K. L. and Gunnlaugsson, Th. (1992b) Fin whale abundance in the North Atlantic, estimated from Icelandic and Faroese NASS-87 and NASS-89 data. *Report of the International Whaling Commission*, **42**, 645–51.

*Buckland, S. T. and Garthwaite, P. H. (1990) Estimating confidence intervals by the Robbins–Monro search process. *Applied Statistics*, **39**, 413–24.

*Buckland, S. T. and Turnock, B. J. (1992) A robust line transect method. *Biometrics*, **48**, 901–9.

*Burdick, D. L. (1979) On estimation of the number of porpoise schools. *Publication 79–2*, San Diego State University, San Diego, CA, USA.

*Burnham, K. P. (1979) A parametric generalization of the Hayne estimator for line transect sampling. *Biometrics*, **35**, 587–95.

Burnham, K. P. (1981) Summarizing remarks: environmental influences, in *Estimating Numbers of Terrestrial Birds. Studies in Avian Biology No. 6* (eds C. J. Ralph and J. M. Scott), Cooper Ornithological Society, pp. 324–5.

*Burnham, K. P. and Anderson, D. R. (1976) Mathematical models for nonparametric inferences from line transect data. *Biometrics*, **32**, 325–36.

*Burnham, K. P. and Anderson, D. R. (1984) The need for distance data in transect counts. *Journal of Wildlife Management*, **48**, 1248–54.

*Burnham, K. P. and Anderson, D. R. (1992) Data-based selection of an appropriate biological model: the key to modern data analysis, in *Wildlife 2001: Populations* (eds D. R. McCullough and R. H. Barrett), Elsevier Science Publishers, London, pp. 16–30.

*Burnham, K. P., Anderson, D. R. and Laake, J. L. (1979) Robust estimation from line transect data. *Journal of Wildlife Management*, **43**, 992–6.

*Burnham, K. P., Anderson, D. R. and Laake, J. L. (1980) Estimation of density from line transect sampling of biological populations. *Wildlife Monograph Number 72*.

*Burnham, K. P., Anderson, D. R. and Laake, J. L. (1981) Line transect estimation of bird population density using a Fourier series, in *Estimating Numbers of Terrestrial Birds. Studies in Avian Biology No. 6* (eds C. J. Ralph and J. M. Scott), Cooper Ornithological Society, pp. 466–82.

*Burnham, K. P., Anderson, D. R. and Laake, J. L. (1985) Efficiency and bias in strip and line transect sampling. *Journal of Wildlife Management*, **49**, 1012–8.

*Burnham, K. P., Anderson, D. R., White, G. C., Brownie, C. and Pollock, K. H. (1987) *Design and analysis methods for fish survival experiments based on release–recapture*. American Fisheries Society, Monograph No. 5.

*Butterworth, D. S. (1982a) A possible basis for choosing a functional form for the distribution of sightings with right angle distance: some preliminary ideas. *Report of the International Whaling Commission*, **32**, 555–8.

*Butterworth, D. S. (1982b) On the functional form used for $g(y)$ for Minke whale sightings, and bias in its estimation due to measurement inaccuracies. *Report of the International Whaling Commission*, **32**, 883–8.

BIBLIOGRAPHY

*Butterworth, D. S. (1986) A note on the analysis of the 1980/81 variable speed experiment. *Report of the International Whaling Commission*, **36**, 485–9.

*Butterworth, D. S. and Best, P. B. (1982) Report of the Southern Hemisphere Minke whale assessment cruise, 1980/81. *Report of the International Whaling Commission*, **32**, 835–74.

*Butterworth, D. S., Best, P. B. and Basson, M. (1982) Results of analysis of sighting experiments carried out during the 1980/81 Southern Hemisphere Minke whale assessment cruise. *Report of the International Whaling Commission*, **32**, 819–34.

*Butterworth, D. S., Best, P. B. and Hembree, D. (1984) Analysis of experiments carried out during the 1981/82 IWC/IDCR Antarctic Minke whale assessment cruise in Area II. *Report of the International Whaling Commission*, **34**, 365–92.

*Butterworth, D. S. and Borchers, D. L. (1988) Estimates of $g(0)$ for Minke schools from the results of the independent observer experiment on the 1985/86 and 1986/87 IWC/IDCR Antarctic assessment cruises. *Report of the International Whaling Commission*, **38**, 301–13.

*Byth, K. (1982) On robust distance-based intensity estimators. *Biometrics*, **38**, 127–35.

Byth, K. and Ripley, B. D. (1980) On sampling spatial patterns by distance methods. *Biometrics*, **36**, 279–84.

*Carroll, J. R. and Ruppert, D. (1988) *Transformation and Weighting in Regression*, Chapman and Hall, London.

Caughley, G. (1972) Improving the estimates from inaccurate censuses. *Journal of Wildlife Management*, **36**, 135–40.

Caughley, G. (1974) Bias in aerial survey. *Journal of Wildlife Management*, **38**, 921–33.

Caughley, G., Sinclair, R. and Scott-Kemmis, D. (1976) Experiments in aerial survey. *Journal of Wildlife Management*, **40**, 290–300.

*Chafota, J. (1988) Effect of measurement errors in estimating density from line transect sampling. MS Paper, Colorado State University, Ft Collins, 47pp.

*Chapman, D. G. (1951) Some properties of the hypergeometric distribution with applications to zoological censuses. *University of California Publications in Statistics*, **1**, 131–60.

*Chatfield, C. (1988) *Problem Solving: a Statistician's Guide*, Chapman and Hall, London.

*Chatfield, C. (1991) Avoiding statistical pitfalls. *Statistical Science*, **6**, 240–68.

*Clark, P. J. and Evans, F. C. (1954) Distance to nearest neighbour as a measure of spatial relationships in populations. *Ecology*, **35**, 23–30.

*Cochran, W. G. (1977) *Sampling Techniques*, 3rd edn, Wiley, New York.

*Cooke, J. G. (1985) Notes on the estimation of whale density from line transects. *Report of the International Whaling Commission*, **35**, 319–23.

Cooke, J. G. (1987) Estimation of the population of Minke whales in Antarctic Area IVW in 1984/85. *Report of the International Whaling Commission*, **37**, 273–6.

Cottam, G. and Curtis, J. T. (1956) The use of distance measures in phytosociological sampling. *Ecology*, **37**, 451–60.

*Coulson, G. M. and Raines, J. A. (1985) Methods for small-scale surveys of grey kangaroo populations. *Australian Wildlife Research*, **12**, 119–25.

*Cox, D. R. (1969) Some sampling problems in technology, in *New Developments in Survey Sampling* (eds N. L. Johnson and H. Smith, Jr), Wiley-Interscience, New York, USA, pp. 506–27.

*Cox, D. R. and Snell, E. J. (1989) *Analysis of Binary Data*, 2nd edn, Chapman and Hall, London.

*Cox, T. F. (1976) The robust estimation of the density of a forest stand using a new conditioned distance method. *Biometrika*, **63**, 493–500.

Crain, B. R. (1974) Estimation of distributions using orthogonal expansions. *The Annals of Statistics*, **2**, 454–63.

*Crain, B. R., Burnham, K. P., Anderson, D. R. and Laake, J. L. (1978) *A Fourier Series Estimator of Population Density for Line Transect Sampling*, Utah State University Press, Logan, UT, USA.

*Crain, B. R., Burnham, K. P., Anderson, D. R. and Laake, J. L. (1979) Nonparametric estimation of population density for line transect sampling using Fourier series. *Biometrical Journal*, **21**, 731–48.

*Davison, A. C., Hinkley, D. V. and Schechtman, E. (1986) Efficient bootstrap simulation. *Biometrika*, **73**, 555–66.

*Dawson, D. G. (1981) Counting birds for a relative measure (index) of density, in *Estimating Numbers of Terrestrial Birds. Studies in Avian Biology No. 6* (eds C. J. Ralph and J. M. Scott), Cooper Ornithological Society, pp. 12–6.

DeJong, M. J. and Emlen, J. T. (1985) The shape of the auditory detection function and its implications for songbird censusing. *Journal of Field Ornithology*, **56**, 213–23.

*Dempster, A. P., Laird, N. M. and Rubin, D. B. (1977) Maximum likelihood from incomplete data via the EM algorithm (with Discussion). *Journal of the Royal Statistical Society, Series B*, **39**, 1–39.

*DeSante, D. F. (1981) A field test of the variable circular-plot censusing technique in a California coastal scrub breeding bird community, in *Estimating Numbers of Terrestrial Birds. Studies in Avian Biology No. 6* (eds C. J. Ralph and J. M. Scott), Cooper Ornithological Society, pp. 177–85.

*DeSante, D. F. (1986) A field test of the variable circular-plot censusing method in a Sierran subalpine forest habitat. *The Condor*, **88**, 129–42.

DeVries, P. G. (1973) A general theory on line transect sampling with applications to logging residue inventory. *Mededelingen Landbouwhogeschool Wageningen*, **73**, 1–23.

DeVries, P. G. (1974) Multistage line intersect sampling. *Forest Science*, **20**, 129–33.

DeVries, P. G. (1979a) Line intersect sampling – statistical theory, applications, and suggestions for extended use in ecological inventory, in *Sampling Biological Populations* (eds R. M. Cormack, G. P. Patil and D. S. Robson), International Co-operative Publishing House, Fairland, MD, USA, pp. 1–70.

DeVries, P. G. (1979b) A generalization of the Hayne-type estimator as an application of line intercept sampling. *Biometrics*, **35**, 743–8.

DeYoung, C. A., Guthery, F. S., Beasom, S. L., Coughlin, S. P. and Heffelfinger, J. R. (1989) Improving estimates of white-tailed deer abundance from helicopter surveys. *Wildlife Society Bulletin*, **17**, 275–9.

*Dice, L. R. (1938) Some census methods for mammals. *Journal of Wildlife Management*, **2**, 119–30.

Diggle, P. J. (1975) Robust density estimation using distance methods. *Biometrika*, **62**, 39–48.

Diggle, P. J. (1977) A note on robust density estimation for spatial point patterns. *Biometrika*, **64**, 91–5.

*Diggle, P. J. (1983) *Statistical Analysis of Spatial Point Patterns*, Academic Press, London.

Dodd, C. K. (1990) Line transect estimation of Red Hills salamander burrow density using a Fourier series. *Copeia*, **2**, 555–7.

Dohl, T. P., Bonnell, M. L. and Ford, R. G. (1986) Distribution and abundance of common dolphin, *Delphinus delphis*, in the Southern California Bight: a quantitative assessment based upon aerial transect data. *Fishery Bulletin*, **84**, 333–43.

*Doi, T. (1971) Further development of sighting theory on whales. *Bulletin of Tokai Regional Fisheries Research Laboratory*, **68**, 1–22.

*Doi, T. (1974) Further development of whale sighting theory, in *The Whale Problem: a Status Report* (ed. W. E. Schevill), Harvard University Press, Cambridge, MA, USA, pp. 359–68.

*Doi, T., Kasamatsu, F. and Nakano, T. (1982) A simulation study on sighting survey of Minke whales in the Antarctic. *Report of the International Whaling Commission*, **32**, 919–28.

*Doi, T., Kasamatsu, F. and Nakano, T. (1983) Further simulation studies on sighting by introducing both concentration of sighting effort by angle and aggregations of Minke whales in the Antarctic. *Report of the International Whaling Commission*, **33**, 403–12.

*Drummer, T. D. (1985). *Size-bias in line transect sampling*. PhD Thesis, University of Wyoming, Laramie, WY, USA. Available from University Microfilms International, 300 N. Zeeb Road, Ann Arbor, MI 48106.

*Drummer, T. D. (1990) Estimation of proportions and ratios from line transect data. *Communications in Statistics – Theory and Methods*, **19**, 3069–91.

*Drummer, T. D. (1991) SIZETRAN: analysis of size-biased line transect data. *Wildlife Society Bulletin*, **19**, 117–8.

*Drummer, T. D., Degange, A. R., Pank, L. L. and McDonald, L. L. (1990) Adjusting for group size influence in line transect sampling. *Journal of Wildlife Management*, **54**, 511–4.

*Drummer, T. D. and McDonald, L. L. (1987) Size bias in line transect sampling. *Biometrics*, **43**, 13–21.

Duffy, D. C. and Schneider, D. C. (1984) A comparison of two transect methods of counting birds at sea. *Cormorant*, **12**, 95–8.

*Eberhardt, L. L. (1967) Some developments in 'distance sampling'. *Biometrics*, **23**, 207–16.

*Eberhardt, L. L. (1968) A preliminary appraisal of line transect. *Journal of Wildlife Management*, **32**, 82–8.

*Eberhardt, L. L. (1978a) Transect methods for population studies. *Journal of Wildlife Management*, **42**, 1–31.

*Eberhardt, L. L. (1978b) Appraising variability in population studies. *Journal of Wildlife Management*, **42**, 207–38.

*Eberhardt, L. L. (1979) Line-transects based on right-angle distances. *Journal of Wildlife Management*, **43**, 768–74.

Eberhardt, L. L., Chapman, D. G. and Gilbert, J. R. (1979) A review of marine mammal census methods. *Wildlife Monographs Number 63*.

*Edwards, D. K., Dorsey, G. L. and Crawford, J. A. (1981) A comparison of three avian census methods, in *Estimating Numbers of Terrestrial Birds. Studies in Avian Biology No. 6* (eds C. J. Ralph and J. M. Scott), Cooper Ornithological Society, pp. 170–6.

*Efron, B. (1979) Bootstrap methods: another look at the jackknife. *Annals of Statistics*, **7**, 1–16.

*Efron, B. (1981) Nonparametric standard errors and confidence intervals (with discussion). *Canadian Journal of Statistics*, **9**, 139–72.

*Emlen, J. T. (1971) Population densities of birds derived from transect counts. *Auk*, **88**, 323–42.

*Emlen, J. T. (1977) Estimating breeding season bird densities from transect counts. *Auk*, **94**, 455–68.

Emlen, J. T. and DeJong, M. J. (1981) The application of song detection threshold distance to census operations, in *Estimating Numbers of Terrestrial Birds. Studies in Avian Biology No. 6* (eds C. J. Ralph and J. M. Scott), Cooper Ornithological Society, pp. 346–52.

Erwin, R. M. (1982) Observer variability in estimating numbers: an experiment. *Journal of Field Ornithology*, **53**, 159–67.

Estes, J. A. and Gilbert, J. R. (1978) Evaluation of an aerial survey of Pacific walruses. *Journal of the Fisheries Research Board of Canada*, **35**, 1130–40.

Fowler, G. S. (1985) *An evaluation of line transect methods for estimation of large mammal populations in heterogeneous habitats*, MS Thesis, University of California, Berkeley, CA, USA, 74pp.

*Franzreb, K. E. (1976) Comparison of variable strip transect and spot-map methods for censusing avian populations in a mixed-coniferous forest. *Condor*, **78**, 260–2.

*Franzreb, K. E. (1981) Determination of avian densities using the variable-strip and fixed-width transect surveying methods, in *Estimating the Numbers of Terrestrial Birds. Studies in Avian Biology No. 6* (eds C. J. Ralph and J. M. Scott), Cooper Ornithological Society, pp. 139–45.

Fuller, R. J. and Langslow, D. R. (1984) Estimating numbers of birds by point counts: how long should counts last? *Bird Study*, **31**, 195–202.

*Garthwaite, P. H. and Buckland, S. T. (1992) Generating Monte Carlo confidence intervals by the Robbins–Monro search process. *Applied Statistics*, **41**, 159–71.

Gaston, A. J., Collins, B. L. and Diamond, A. W. (1987) The 'snapshot' count for estimating densities of flying seabirds during boat transects: a cautionary comment. *Auk*, **104**, 336–8.

*Gates, C. E. (1969) Simulation study of estimators for the line transect sampling method. *Biometrics*, **25**, 317–28.

*Gates, C. E. (1979) Line transect and related issues, in *Sampling Biological Populations* (eds R. M. Cormack, G. P. Patil and D. S. Robson), International Co-operative Publishing House, Fairland, MD, USA, pp. 71–154.

*Gates, C. E. (1980) LINETRAN, a general computer program for analyzing line transect data. *Journal of Wildlife Management*, **44**, 658–61.

*Gates, C. E., Evans, W., Gober, D. R. *et al.* (1985) Line transect estimation of animal densities from large data sets, in *Game Harvest Management* (eds S. L. Beasom and S. F. Roberson), Caesar Kleberg Wildlife Research Institute, Texas A&I University, Kingsville, TX, USA, pp. 37–50.

*Gates, C. E., Marshall, W. H. and Olson, D. P. (1968) Line transect method of estimating grouse population densities. *Biometrics*, **24**, 135–45.

*Gates, C. E. and Smith, P. W. (1980) An implementation of the Burnham–Anderson distribution free method of estimating wildlife densities from line transect data. *Biometrics*, **36**, 155–60.

Geimsdell, J. J. R. and Westley, S. B. (eds) (1979) *Low-Level Aerial Survey Techniques*, International Livestock Centre for Africa, Addis Ababa, Ethiopia, Africa.

*Genstat 5 Committee (1987) *Genstat 5 Reference Manual*, Clarendon Press, Oxford.

*Gilbert, D. W., Anderson, D. R., Ringelman, J. K. and Szymczak, M. R. (in prep.) Response of nesting ducks to habitat management on the Monte Vista National Wildlife Refuge, Colorado. To be submitted to *Wildlife Monographs*.

Gogan, P. J., Thompson, S. C., Pierce, W. and Barrett, R. H. (1986) Line-transect censuses of fallow and black-tailed deer on the Point Reyes Peninsula. *California Game and Fish*, **72**, 47–61.

Good, I. J. and Gaskins, R. A. (1980) Density estimation and bump-hunting by penalized likelihood method exemplified by scattering and meteorite data. *Journal of the American Statistical Association*, **75**, 42–73.

Granholm, S. L. (1983) Bias in density estimates due to movement of birds. *Condor*, **85**, 243–8.

*Gray, H. L. and Schucany, W. R. (1972) *The Generalized Jackknife Statistic*, Marcel Dekker, New York, 308pp.

*Green, G. A., Brueggeman, J. J., Bowlby, C. E., Grotefendt, R. A., Bonnell, M. L. and Balcomb, K. T. (1992) Cetacean distribution and abundance off Oregon and Washington, 1989–1990. Chapter I, in *Oregon and Washington Marine Mammal and Seabird Surveys* (ed. J. J. Brueggeman), final report prepared by Ebasco Environmental, Bellevue, WA, and Ecological Consulting, Inc., Portland, OR, for the Minerals Management Service, Pacific OCS Region. OCS Study MMS 91–0072.

Gross, J. E., Stoddart, L. C. and Wagner, F. H. (1974) Demographic analysis of a northern Utah jackrabbit population. *Wildlife Monographs Number 40*.

*Gunnlaugsson, Th. and Sigurjónsson, J. (1990) NASS-87: estimation of whale abundance based on observations made on board Icelandic and Faroese survey vessels. *Report of the International Whaling Commission*, **40**, 571–80.

*Guthery, F. S. (1988) Line transect sampling of bobwhite density on rangeland: evaluation and recommendations. *Wildlife Society Bulletin*, **16**, 193–203.

Hamel, P. B. (1984) Comparison of variable circular-plot and spot-map censusing methods in temperate forest. *Ornis Scandinavica*, **15**, 266–74.

Hamel, P. B. (1990) Response to Tomiatojc and Verner. *Auk*, **107**, 451–3.

*Hammond, P. S. (1984) An investigation into the effects of different techniques of smearing the IWC/IDCR Minke whale sighting data and the use of different models to estimate density of schools. *Report of the International Whaling Commission*, **34**, 301–7.

*Hammond, P. S. and Laake, J. L. (1983) Trends in estimates of abundance of dolphins (*Stenella* spp. and *Delphinus delphis*) involved in the purse-seine fishery for tunas in the eastern Pacific Ocean. *Report of the International Whaling Commission*, **33**, 565–88.

Hammond, P. S. and Laake, J. L. (1984) Estimates of sperm whale density in the eastern tropical Pacific, 1974–1982. *Report of the International Whaling Commission*, **34**, 255–8.

Hanowski, J. M., Niemi, G. J. and Blake, J. G. (1990) Statistical perspectives and experimental design when counting birds on line transects. *The Condor*, **92**, 326–35.

Harden, R. H., Muir, R. J. and Milledge, D. R. (1986) An evaluation of the strip transect method for censusing bird communities in forests. *Australian Wildlife Research*, **13**, 203–11.

Harris, R. B. (1986) Reliability of trend lines obtained from variable counts. *Journal of Wildlife Management*, **50**, 165–71.

Haukioja, E. (1968) Reliability of the line survey method in bird census with reference to reed bunting and sedge warbler. *Ornis Fennica*, **45**, 105–13.

*Hayes, R. J. (1977) *A critical review of line transect methods.* MSc thesis, University of Edinburgh.

*Hayes, R. J. and Buckland, S. T. (1983) Radial-distance models for the line-transect method. *Biometrics*, **39**, 29–42.

*Hayne, D. W. (1949) An examination of the strip census method for estimating animal populations. *Journal of Wildlife Management*, **13**, 145–57.

Healey, W. M. and Welsh, C. J. E. (1992) Evaluating line transects to monitor gray squirrel populations. *Wildlife Society Bulletin*, **20**, 83–90.

Heitjan, D. F. (1989) Inference from grouped continuous data: a review. *Statistical Science*, **4**, 164–83.

Hemingway, P. (1971) Field trials of the line transect method of sampling large populations of herbivores, in *The Scientific Management of Animal and Plant Communities for Conservation* (eds E. Duffey and A. S. Watts), Blackwell Scientific Publications, Oxford, England, pp. 405–11.

*Hiby, A. R. (1982) Using average number of whales in view to estimate population density. *Report of the International Whaling Commission*, **32**, 562–5.

*Hiby, A. R. (1985) An approach to estimating population densities of great whales from sighting surveys. *IMA Journal of Mathematics Applied in Medicine and Biology*, **2**, 201–20.

*Hiby, A. R. (1986) Results of a hazard rate model relevant to experiments on the 1984/85 IDCR Minke whale assessment cruise. *Report of the International Whaling Commission*, **36**, 497–8.

*Hiby, A. R. and Hammond, P. S. (1989) Survey techniques for estimating current abundance and monitoring trends in abundance of cetaceans, in *The Comprehensive Assessment of Whale Stocks: the early years* (ed. G. P. Donovan), International Whaling Commission, Cambridge, pp. 47–80.

*Hiby, A. R. and Lovell, P. A. (unpublished) A suggested function for the detection probabilities of surfacings. *Paper SC/42/NHMi33*, presented to the Scientific Committee of the International Whaling Commission, June 1990.

*Hiby, A. R., Martin, A. R. and Fairfield, F. (1984) IDCR cruise/aerial survey in the north Atlantic 1982: aerial survey. *Report of the International Whaling Commission*, **34**, 633–44.

*Hiby, A. R. and Ward, A. J. (1986a) Simulation trials of a cue-counting technique for censusing whale populations. *Report of the International Whaling Commission*, **36**, 471–2.

*Hiby, A. R. and Ward, A. J. (1986b) Analysis of cue-counting and blow rate estimation experiments carried out during the 1984/85 IDCR Minke whale assessment cruise. *Report of the International Whaling Commission*, **36**, 473–6.

*Hiby, A. R. and Ward, A. J. (unpublished) An update of the cue-counting method of abundance estimation. *Paper SC/41/SHMi20*, submitted to the Scientific Committee of the International Whaling Commission, 1989.

*Hiby, A. R., Ward, A. J. and Lovell, P. (1989) Analysis of the 1987 north Atlantic sightings survey: aerial survey results. *Report of the International Whaling Commission*, **39**, 447–55.

Hilden, O. and Järvinen, A. (1989) Efficiency of the line-transect method in mountain birch forest. *Annales Zoologici Fennici*, **26**, 185–90.

*Hogg, R. V. and Craig, A. T. (1970) *Introduction to Mathematical Statistics*, 3rd edn, Macmillan, New York, USA.

*Holgate, P. (1964) The efficiency of nearest neighbour estimators. *Biometrics*, **20**, 647–9.

Holt, R. S. and Cologne, J. (1987) Factors affecting line transect estimates of dolphin school density. *Journal of Wildlife Management*, **51**, 836–43.

*Holt, R. S. and Powers, J. E. (1982) *Abundance Estimation of Dolphin Stocks Involved in the Eastern Tropical Pacific Yellowfin Tuna Fishery Determined from Aerial and Ship Surveys to 1979*, United States Department of Commerce, NOAA Technical Memorandum NOAA-TM-NMFS-SWFC-23, 95pp.

*Hone, J. (1986) Accuracy of the multiple regression method for estimating density in transect counts. *Australian Wildlife Research*, **13**, 121–6.

*Hone, J. (1988) A test of the accuracy of line and strip transect estimators in aerial survey. *Australian Wildlife Research*, **15**, 493–7.

*Huggins, R. M. (1989) On the statistical analysis of capture experiments. *Biometrika*, **76**, 133–40.

*Huggins, R. M. (1991) Some practical aspects of a conditional likelihood approach to capture experiments. *Biometrics*, **47**, 725–32.

Hutto, R. L., Pletschet, S. M. and Hendricks, P. (1986) A fixed radius point count method for nonbreeding and breeding season use. *Auk*, **103**, 593–602.

Järvinen, O. (1976) Estimating relative densities of breeding birds by the line transect method. II. Comparison between two methods. *Ornis Scandinavica*, **7**, 43–8.

Järvinen, O. (1978) Estimating relative densities of land birds by point counts. *Annales Zoologici Fennici*, **15**, 290–3.

*Järvinen, O. (1978) Species-specific efficiency in line transects. *Ornis Scandinavica*, **9**, 164–7.

*Järvinen, O. and Väisänen, R. A. (1975) Estimating relative densities of breeding birds by the line transect method. *Oikos*, **26**, 316–22.

Järvinen, O. and Väisänen, R. A. (1976) Between-year component of diversity in communities of breeding land birds. *Oikos*, **27**, 34–9.

Järvinen, O. and Väisänen, R. A. (1976) Estimating relative densities of breeding birds by the line transect method. IV. Geographical constancy of the proportion of main belt observations. *Ornis Fennica*, **53**, 87–90.

Järvinen, O. and Väisänen, R. A. (1976) Finnish line transect censuses. *Ornis Fennica*, **53**, 115–8.

Järvinen, O. and Väisänen, R. A. (1983) Confidence limits for estimates of population density in line transects. *Ornis Scandinavica*, **14**, 129–34.

Järvinen, O. and Väisänen, R. A. (1983) Correction coefficients for line transect censuses of breeding birds. *Ornis Fennica*, **60**, 97–104.

Järvinen, O., Väisänen, R. A. and Haila, Y. (1976) Estimating relative densities of breeding birds by the line transect method. III. Temporal constancy of the proportion of main belt observations. *Ornis Fennica*, **53**, 40–5.

Jett, D. A. and Nichols, J. D. (1987) A field comparison of nested grid and trapping web density estimators. *Journal of Mammalogy*, **68**, 888–92.

*Johnson, B. K. and Lindzey, F. G. (unpublished) *Guidelines for Estimating Pronghorn Numbers Using Line Transects*. Unpublished report, Wyoming Cooperative Fish and Wildlife Research Unit, University of Wyoming, Laramie. 15pp. plus appendices.

Johnson, B. K., Lindzey, F. G. and Guenzel, R. J. (1991) Use of aerial line transect surveys to estimate pronghorn populations in Wyoming. *Wildlife Society Bulletin*, **19**, 315–21.

*Johnson, E. G. and Routledge, R. D. (1985) The line transect method: a nonparametric estimator based on shape restrictions. *Biometrics*, **41**, 669–79.

Johnson, F. A., Pollock, K. H. and Montalbano, F., III. (1989) Visibility bias in aerial surveys of mottled ducks. *Wildlife Society Bulletin*, **17**, 222–7.

*Johnson, N. L. and Kotz, S. (1969) *Discrete Distributions*, Houghton Mifflin Company, Boston.

Kaiser, L. (1983) Unbiased estimation in line-intercept sampling. *Biometrics*, **39**, 965–76.

*Kelker, G. H. (1945) *Measurement and interpretation of forces that determine populations of managed deer*, PhD Dissertation, University of Michigan, Ann Arbor, MI, USA.

*Kishino, H., Kasamatsu, F. and Toda, T. (1988) On the double line transect method. *Report of the International Whaling Commission*, **38**, 273–9.

Kishino, H., Kato, H., Kasamatsu, F. and Fujise, Y. (1991) Detection of heterogeneity and estimation of population characteristics from the field survey data: 1987/88 Japanese feasibility study of the Southern Hemisphere minke whales. *Annals of the Institute of Statistical Mathematics*, **43**, 435–53.

*Knopf, F. L. (1986) Changing landscapes and the cosmopolitism of the eastern Colorado avifauna. *Wildlife Society Bulletin*, **14**, 132–42.

*Knopf, F. L., Sedgwick, J. A. and Cannon, R. W. (1988) Guild structure of a riparian avifauna relative to seasonal cattle grazing. *Journal of Wildlife Management*, **52**, 280–90.

*Koopman, B. O. (1956) The theory of search II. Target detection. *Operations Research*, **4**, 503–31.

*Koopman, B. O. (1980) *Search and Screening: General Principles with Historical Applications*, Pergamon Press, New York, USA, 369pp.

Koster, S. H. (1985) *An evaluation of line transect census methods in West African wooded savanna*, PhD Dissertation, Michigan State University, Ann Arbor, MI, USA, 207pp.

Kovner, J. L. and Patil, S. A. (1974) Properties of estimators of wildlife population density for the line transect method. *Biometrics*, **30**, 225–30.

*Laake, J. L. (1978) *Line transect estimators robust to animal movement*, MS Thesis, Utah State University, Logan, UT, USA, 55pp.

Laake, J. L. (1981) Abundance estimation of dolphins in the eastern Pacific with line transect sampling – a comparison of the techniques and suggestions for future research, in *Report of the Workshop on Tuna–Dolphin Interactions* (ed. P. S. Hammond), Inter-American Tropical Tuna Commission Special Report Number 4, pp. 56–95.

*Laake, J. L., Buckland, S. T., Anderson, D. R. and Burnham, K. P. (1993) *DISTANCE User's Guide*. Colorado Cooperative Fish and Wildlife Research Unit, Colorado State University, Fort Collins, CO 80523, USA.

*Laake, J. L., Burnham, K. P. and Anderson, D. R. (1979) *User's Manual for Program TRANSECT*, Utah State University Press, Logan, UT, USA.

Leatherwood, S., Gilbert, J. R. and Chapman, D. G. (1978) An evaluation of some techniques for aerial censuses of bottlenosed dolphins. *Journal of Wildlife Management*, **42**, 239–50.

Leatherwood, S. and Show, I. T., Jr (1982) Effects of varying altitude on aerial surveys of bottlenose dolphins. *Report of the International Whaling Commission*, **32**, 569–75.

Leatherwood, S., Show, I. T., Jr, Reeves, R. R. and Wright, M. B. (1982) Proposed modification of transect models to estimate population size from aircraft with obstructed downward visibility. *Report of the International Whaling Commission*, **32**, 577–9.

*Lebreton, J. D., Burnham, K. P., Clobert, J. and Anderson, D. R. (1992) Modeling survival and testing biological hypotheses using marked animals: case studies and recent advances. *Ecological Monographs*, **62**, 67–118.

*Lehmann, E. L. (1959) *Testing Statistical Hypotheses*, Wiley, New York.

*Leopold, A. (1933) *Game Management*, Charles Schribner's Sons, New York.

LeResche, R. E. and Rausch, R. A. (1974) Accuracy and precision of aerial moose censusing. *Journal of Wildlife Management*, **38**, 175–82.

Lucas, H. A. and Seber, G. A. F. (1977) Estimating coverage and particle density using line intercept method. *Biometrika*, **64**, 618–22.

Mackintosh, N. A. and Brown, S. G. (1956) Estimates of the southern population of the larger baleen whales. *The Norwegian Gazette*, **45**, 469.

Marsh, C. W. and Wilson, W. L. (1981) *A Survey of Primates in Peninsular Malaysian Forests*, Final Report for the Malaysian Primates Research Programme, Universiti Kebangsaan, Malaysia.

Matz, A. W. (1978) Maximum likelihood parameter estimation for the quartic exponential distribution. *Technometrics*, **20**, 475–84.

*Mayfield, H. F. (1981) Problems in estimating population size through counts of singing males, in *Estimating Numbers of Terrestrial Birds. Studies in Avian Biology No. 6* (eds C. J. Ralph and J. M. Scott), Cooper Ornithological Society, pp. 220–4.

*McCullagh, P. and Nelder, J. A. (1989) *Generalized Linear Models*, 2nd edn, Chapman and Hall, London.

McDonald, L. L. (1980) Line-intercept sampling for attributes other than coverage and density. *Journal of Wildlife Management*, **44**, 530–3.

McIntyre, G. A. (1953) Estimation of plant density using line transects. *Journal of Ecology*, **41**, 319–30.

Mikol, S. A. (1980) *Field Guidelines for Using Transects to Sample Nongame Bird Populations*, US Fish and Wildlife Service, Report FWS/OBS-80/58.

*Miller, R. G. (1974) The Jackknife – a review. *Biometrika*, **61**, 1–15.

*Milliken, G. A. and Johnson, D. E. (1984) *Analysis of Messy Data*, Lifetime Learning Publications, Belmont, California.

*Morgan, B. J. T. and Freeman, S. N. (1989) A model with first-year variation for ring-recovery data. *Biometrics*, **45**, 1087–101.

*Morgan, D. G. (1986) *Estimating Vertebrate Population Densities by Line Transect Methods*, Occasional Papers Number 11, Melbourne College, Melbourne, Australia.

Morrison, M. L., Mannan, R. W. and Dorsey, G. L. (1981) Effects of number of circular plots on estimates of avian density and species richness, in *Estimating Numbers of Terrestrial Birds. Studies in Avian Biology No. 6* (eds C. J. Ralph and J. M. Scott), Cooper Ornithological Society, pp. 405–8.

Morrison, M. L. and Marcot, B. G. (1984) Expanded use of the variable circular-plot census method. *Wilson Bulletin*, **96**, 313–5.

433

Myrberget, S. (1976) Field tests of line transect census methods for grouse. *Norwegian Journal of Zoology*, **24**, 307–17.

National Research Council (1981) *Techniques for the Study of Primate Population Ecology*, National Academy Press, Washington, DC, USA.

Nichols, J. D., Tomlinson, R. E. and Waggerman, G. (1986) Estimating nest detection probabilities for white-winged dove nest transects in Tamaulipas, Mexico. *Auk*, **103**, 825–8.

North, P. M. (1977) A novel clustering method for estimating numbers of bird territories. *Journal of the Royal Statistical Society, Series C*, **26**, 149–55.

North, P. M. (1979) A novel clustering method for estimating numbers of bird territories: an addendum. *Journal of the Royal Statistical Society, Series C*, **28**, 300–1.

*O'Meara, T. E. (1981) A field test of two density estimators for transect data, in *Estimating the Numbers of Terrestrial Birds. Studies in Avian Biology No. 6* (eds C. J. Ralph and J. M. Scott), Cooper Ornithological Society, pp. 193–6.

*Otis, D. L., Burnham, K. P., White, G. C. and Anderson, D. R. (1978) Statistical inference from capture data on closed animal populations. *Wildlife Monographs*, **62**, 1–135.

Otten, A. and deVries, P. G. (1984) On line-transect estimates for population density, based on elliptic flushing curves. *Biometrics*, **40**, 1145–50.

*Otto, M. C. and Pollock, K. H. (1990) Size bias in line transect sampling: a field test. *Biometrics*, **46**, 239–45.

*Overton, W. S. and Davis, D. E. (1969) Estimating the number of animals in wildlife populations, in *Wildlife Management Techniques* (ed. R. H. Giles, Jr), The Wildlife Society, Washington, DC, pp. 405–55.

Parker, K. R. (1979) Density estimation by variable area transect. *Journal of Wildlife Management*, **43**, 484–92.

*Parmenter, R. R., MacMahon, J. A. and Anderson, D. R. (1989) Animal density estimation using a trapping web design: field validation experiments. *Ecology*, **70**, 169–79.

*Patil, G. P. and Ord, J. K. (1976) On size-biased sampling and related form-invariant weighted distributions. *Sankhyā B*, **38**, 48–61.

*Patil, G. P. and Rao, C. R. (1978) Weighted distribution and size-biased sampling with applications to wildlife populations and human families. *Biometrics*, **34**, 179–89.

*Patil, G. P., Taillie, C. and Wigley, R. L. (1979a) Transect sampling methods and their application to deep-sea red crab, in *Environmental Biomonitoring, Assessment, Prediction, and Management – Certain Case Studies and Related Quantitative Issues* (eds J. Cairns, Jr, G. P. Patil and W. E. Waters), International Co-operative Publishing House, Fairland, MD, USA, pp. 51–75.

*Patil, S. A., Burnham, K. P. and Kovner, J. L. (1979b) Nonparametric estimation of plant density by the distance method. *Biometrics*, **35**, 597–604.

*Patil, S. A., Kovner, J. L. and Burnham, K. P. (1982) Optimum nonparametric estimation of population density based on ordered distances. *Biometrics*, **38**, 243–8.

Persson, O. (1971) The robustness of estimating density by distance measurements, in *Statistical Ecology*, Vol. 2 (eds G. P. Patil, E. C. Pielou and W. E. Waters), Pennsylvania State University Press, University Park, PA, USA, pp. 175–90.

Peterhofs, E. and Priednieks, J. (1989) Problems in applying the line-transect method without repeated counts when the breeding season is long. *Annales Zoologici Fennici*, **26**, 181–4.

*Petersen, C. G. J. (1896) The yearly immigration of young plaice into the Limfjord from the German Sea. *Report of the Danish Biological Station*, **6**, 1–48.

Pinkowski, B. (1990) Performance of Fourier series in line transect simulations. *Simulation*, **54**, 211–2.

*Polacheck, T. and Smith, T. D. (unpublished) Simulation results on the effects of dive time and movements on line transect estimates. *Paper SC/42/NHMi31*, presented to the Scientific Committee of the International Whaling Commission, June 1990.

Pollard, J. H. (1971) On distance estimators of density in randomly distributed forests. *Biometrics*, **27**, 991–1002.

*Pollock, K. H. (1978) A family of density estimators for line-transect sampling. *Biometrics*, **34**, 475–8.

Pollock, K. H. and Kendall, W. L. (1987) Visibility bias in aerial surveys: a review of estimation procedures. *Journal of Wildlife Management*, **51**, 502–9.

Quang, P. X. (1990a) Confidence intervals for densities in line transect sampling. *Biometrics*, **46**, 459–72.

*Quang, P. X. (1990b) Nonparametric estimators for variable circular plot surveys. *Unpublished Report*, 18pp.

*Quang, P. X. (1991) A nonparametric approach to size-biased line transect sampling. *Biometrics*, **47**, 269–79.

*Quang, P. X. and Lanctot, R. B. (1991) A line transect model for aerial surveys. *Biometrics*, **47**, 1089–102.

*Quinn, T. J., II (1977) *The effects of aggregation on line transect estimators of population abundance with application to marine mammal populations*, MS thesis, University of Washington, WA, USA, 116pp.

*Quinn, T. J., II (1979) The effects of school structure on line transect estimators of abundance, in *Contemporary Quantitative Ecology and Related Ecometrics* (eds G. P. Patil and M. L. Rosenzweig), International Co-operative Publishing House, Fairland, MD, USA, pp. 473–91.

*Quinn, T. J., II (1985) Line transect estimators for school populations. *Fisheries Research*, **3**, 183–99.

*Quinn, T. J., II and Gallucci, V. F. (1980) Parametric models for line transect estimators of abundance. *Ecology*, **61**, 293–302.

*Ralph, C. J. (1981) An investigation of the effect of seasonal activity levels on avian censusing, in *Estimating Numbers of Terrestrial Birds. Studies in Avian Biology No. 6* (eds C. J. Ralph and J. M. Scott), Cooper Ornithological Society, pp. 265–70.

*Ralph, C. J. and Scott, J. M. (eds) (1981) *Estimating Numbers of Terrestrial Birds. Studies in Avian Biology No. 6*, Cooper Ornithological Society.

*Ramsey, F. L. (1979) Parametric models for line transect surveys. *Biometrika*, **66**, 505–12.

*Ramsey, F. L. (1981) Introductory remarks: data analysis, in *Estimating Numbers of Terrestrial Birds. Studies in Avian Biology No. 6* (eds C. J. Ralph and J. M. Scott), Cooper Ornithological Society, p. 454.

*Ramsey, F. L., Gates, C. E., Patil, G. P. and Taillie, C. (1988) On transect sampling to assess wildlife populations and marine resources, in *Handbook of*

Statistics 6: Sampling (eds P. R. Krishnaiah and C. R. Rao), Elsevier Publishers, Amsterdam, pp. 515–32.

*Ramsey, F. L. and Scott, J. M. (1978) Use of circular plot surveys in estimating the density of a population with Poisson scattering. *Technical Report 60*, Department of Statistics, Oregon State University, Corvallis, OR, USA.

*Ramsey, F. L. and Scott, J. M. (1979) Estimating population densities from variable circular plot surveys, in *Sampling Biological Populations* (eds R. M. Cormack, G. P. Patil and D. S. Robson), International Co-operative Publishing House, Fairland, MD, USA, pp. 155–81.

*Ramsey, F. L. and Scott, J. M. (1981a) Tests of hearing ability, in *Estimating Numbers of Terrestrial Birds. Studies in Avian Biology No. 6* (eds C. J. Ralph and J. M. Scott), Cooper Ornithological Society, pp. 341–5.

*Ramsey, F. L. and Scott, J. M. (1981b) Analysis of bird survey data using a modification of Emlen's methods, in *Estimating Numbers of Terrestrial Birds. Studies in Avian Biology No. 6* (eds C. J. Ralph and J. M. Scott), Cooper Ornithological Society, pp. 483–7.

*Ramsey, F. L., Scott, J. M. and Clark, R. T. (1979) Statistical problems arising from surveys of rare and endangered forest birds, in *Proceedings of the 42nd Session of the International Statistical Institute*, pp. 471–83.

*Ramsey, F. L., Wildman, V. and Engbring, J. (1987) Covariate adjustments to effective area in variable-area wildlife surveys. *Biometrics*, **43**, 1–11.

*Rao, C. R. (1973) *Linear Statistical Inference and its Applications*, 2nd edn. Wiley, New York.

*Rao, P. V. (1984) Density estimation based on line transect samples. *Statistics and Probability Letters*, **2**, 1–57.

*Rao, P. V. and Portier, K. M. (1985) A model for line transect sampling clustered populations. *Statistics and Probability Letters*, **3**, 89–93.

Rao, P. V., Portier, K. M. and Ondrasik, J. A. (1981) Density estimation using line transect sampling, in *Estimating Numbers of Terrestrial Birds. Studies in Avian Biology No. 6* (eds C. J. Ralph and J. M. Scott), Cooper Ornithological Society, pp. 441–4.

Raphael, M. G. (1987) Estimating relative abundance of forest birds: simple versus adjusted counts. *Wilson Bulletin*, **99**, 125–31.

Raphael, M. G., Rosenberg, K. V. and Marcot, B. G. (1988) Large-scale changes in bird populations of Douglas-fir forests, Northwestern California, in *Bird Conservation* (ed. J. A. Jackson), University of Wisconsin Press, Madison, WI, USA, pp. 63–83.

Ratti, J. T., Smith, L. M., Hupp, J. W. and Laake, J. L. (1983) Line transect estimates of density and the winter mortality of Gray Partridge. *Journal of Wildlife Management*, **47**, 1088–96.

*Redmond, R. L., Bicak, T. K. and Jenni, D. A. (1981) An evaluation of breeding season census techniques for long-billed curlews (*Numenius americanus*), in *Estimating the Numbers of Terrestrial Birds. Studies in Avian Biology No. 6* (eds C. J. Ralph and J. M. Scott), Cooper Ornithological Society, pp. 197–201.

*Reilly, S. B., Rice, D. W. and Wolman, A. A. (1980) Preliminary population estimate for the California gray whale based upon Monterey shore censuses, 1967/68 to 1978/79. *Report of the International Whaling Commission*, **30**, 359–68.

*Reilly, S. B., Rice, D. W. and Wolman, A. A. (1983) Population assessment of the gray whale, *Eschrichtius robustus*, from California shore censuses, 1967–80. *Fishery Bulletin*, **81**, 267–81.

*Reynolds, R. T., Scott, J. M. and Nussbaum, R. A. (1980) A variable circular-plot method for estimating bird numbers. *Condor*, **82**, 309–13.

*Richards, D. G. (1981) Environmental acoustics and censuses of singing birds, in *Estimating Numbers of Terrestrial Birds. Studies in Avian Biology No. 6* (eds C. J. Ralph and J. M. Scott), Cooper Ornithological Society, pp. 297–300.

Rinne, J. (1985) Density estimates and their errors in line transect counts of breeding birds. *Ornis Fennica*, **62**, 1–8.

*Robbins, C. S. (1981) Effect of time of day on bird activity, in *Estimating Numbers of Terrestrial Birds. Studies in Avian Biology No. 6* (eds C. J. Ralph and J. M. Scott), Cooper Ornithological Society, pp. 275–86.

*Robinette, W. L., Jones, D. A., Gashwiler, J. S. and Aldous, C. M. (1956) Further analysis of methods for censusing winter-lost deer. *Journal of Wildlife Management*, **20**, 75–8.

*Robinette, W. L., Loveless, C. M. and Jones, D. A. (1974) Field tests of strip census methods. *Journal of Wildlife Management*, **38**, 81–96.

*Roeder, K., Dennis, B. and Garton, E. O. (1987) Estimating density from variable circular plot censuses. *Journal of Wildlife Management*, **51**, 224–30.

Rotella, J. J. and Ratti, J. T. (1986) Test of a critical density index assumption: a case study with gray partridge. *Journal of Wildlife Management*, **50**, 532–9.

*Routledge, R. D. and Fyfe, D. A. (1992) Computing confidence limits for line transect estimates based on shape restrictions. *Journal of Wildlife Management*, **56**, 402–7.

Royama, T. (1960) The theory and practice of line transects in animal ecology by means of visual and auditory recognition. *Yamashina Institute of Ornithology and Zoology*, **2(14)**, 1–17.

Russell, V. S. (ed.) (1979) *Aerial Surveys of Fauna Populations*, Australian National Parks and Wildlife Service Special Publication 1, Australian Government Publishing Service, Canberra, Australia.

*Sakamoto, Y., Ishiguro, M. and Kitawaga, G. (1986) *Akaike Information Criterion Statistics*, KTK Scientific Publishers, Tokyo, Japan.

*Satterthwaite, F. E. (1946) An approximate distribution of estimates of variance components. *Biometrics Bulletin*, **2**, 110–4.

*Schweder, T. (1974) Transformations of point processes: applications to animal sighting and catch problems, with special emphasis on whales. Unpublished dissertation, University of California, Berkeley, CA, USA.

*Schweder, T. (1977) Point process models for line transect experiments, in *Recent Developments in Statistics* (eds J. R. Barba, F. Brodeau, G. Romier and B. Van Cutsem), North-Holland Publishing Company, New York, USA, pp. 221–42.

*Schweder, T. (1990) Independent observer experiments to estimate the detection function in line transect surveys of whales. *Report of the International Whaling Commission*, **40**, 349–55.

*Schweder, T., Øien, N. and Høst, G. (1991) Estimates of the detection probability for shipboard surveys of northeastern Atlantic Minke whales, based on a parallel ship experiment. *Report of the International Whaling Commission*, **41**, 417–32.

*Scott, J. M., Jacobi, J. D. and Ramsey, F. L. (1981) Avian surveys of large geographical areas: a systematic approach. *Wildlife Society Bulletin*, **9**, 190–200.

*Scott, J. M. and Ramsey, F. L. (1981) Length of count period as a possible source of bias in estimating bird densities, in *Estimating Numbers of Terrestrial*

Birds. Studies in Avian Biology No. 6 (eds C. J. Ralph and J. M. Scott), Cooper Ornithological Society, pp. 409–13.

*Scott, J. M., Ramsey, F. L. and Kepler, C. B. (1981) Distance estimation as a variable in estimating bird numbers, in *Estimating Numbers of Terrestrial Birds. Studies in Avian Biology No. 6* (eds C. J. Ralph and J. M. Scott), Cooper Ornithological Society, pp. 334–40.

*Seber, G. A. F. (1973) *The Estimation of Animal Abundance*, Hafner, New York.

*Seber, G. A. F. (1979) Transects of random length, in *Sampling Biological Populations* (eds R. M. Cormack, G. P. Patil and D. S. Robson), International Co-operative Publishing House, Fairland, MD, USA, pp. 183–92.

*Seber, G. A. F. (1982) *The Estimation of Animal Abundance and Related Parameters*, Macmillan, New York.

*Seber, G. A. F. (1986) A review of estimating animal abundance. *Biometrics*, **42**, 267–92.

Seber, G. A. F. (1992) A review of estimating animal abundance II. *International Statistical Review*, **60**, 129–66.

*Sedgwick, J. A. and Knopf, F. L. (1987) Breeding bird response to cattle grazing of a cottonwood bottomland. *Journal of Wildlife Management*, **51**, 230–7.

*Sen, A. R., Smith, G. E. J. and Butler, G. (1978) On a basic assumption in the line transect method. *Biometrical Journal*, **20**, 363–9.

Sen, A. R., Tourigny, J. and Smith, G. E. J. (1974) On the line transect sampling method. *Biometrics*, **30**, 329–41.

*Shupe, T. E., Guthery, F. S. and Beasom, S. L. (1987) Use of helicopters to survey northern bobwhite populations on rangeland. *Wildlife Society Bulletin*, **15**, 458–62.

*Silverman, B. W. (1982) Algorithm AS 176: kernel density estimation using the fast Fourier transform. *Applied Statistics*, **31**, 93–9.

*Silverman, B. W. (1986) *Density Estimation for Statistics and Data Analysis*, Chapman and Hall, London.

Skellam, J. G. (1958) The mathematical foundations underlying the use of line transects in animal ecology. *Biometrics*, **14**, 385–400.

*Skirvin, A. A. (1981) Effect of time of day and time of season on the number of observations and density estimates of breeding birds, in *Estimating Numbers of Terrestrial Birds. Studies in Avian Biology No. 6* (eds C. J. Ralph and J. M. Scott), Cooper Ornithological Society, pp. 271–4.

*Smith, G. E. (1979) Some aspects of line transect sampling when the target population moves. *Biometrics*, **35**, 323–9.

Smith, G. W. and Nydegger, N. C. (1985) A spot-light, line transect method for surveying jack rabbits. *Journal of Wildlife Management*, **49**, 699–702.

Smith, G. W., Nydegger, N. C. and Yensen, D. L. (1984) Passerine bird densities in shrubsteppe vegetation. *Journal of Field Ornithology*, **55**, 261–4.

Smith, T. D. (1981) Line-transect techniques for estimating density of porpoise schools. *Journal of Wildlife Management*, **45**, 650–7.

*Stoyan, D. (1982) A remark on the line transect method. *Biometrical Journal*, **24**, 191–5.

*Stuart, A. and Ord, J. K. (1987) *Kendall's Advanced Theory of Statistics, Volume 1*, Griffin, London.

*Swartz, S. L., Jones, M. L., Goodyear, J., Withrow, D. E. and Miller, R. V. (1987) Radio-telemetric studies of gray whale migration along the California coast: a preliminary comparison of day and night migration rates. *Report of the International Whaling Commission*, **37**, 295–9.

Szaro, R. C. and Jakle, M. D. (1982) Comparison of variable circular-plot and spot-map methods in desert riparian and scrub habitats. *Wilson Bulletin*, **94**, 546–50.

*Tasker, M. L., Hope Jones, P., Dixon, T. and Blake, B. F. (1984) Counting seabirds at sea from ships: a review of methods employed and a suggestion for a standardized approach. *Auk*, **101**, 567–77.

*Thompson, S. K. (1990) Adaptive cluster sampling. *Journal of the American Statistical Association*, **85**, 1050–9.

*Thompson, S. K. and Ramsey, F. L. (1987) Detectability functions in observing spatial point processes. *Biometrics*, **43**, 355–62.

Tilghman, N. G. and Rusch, D. H. (1981) Comparison of line-transect methods for estimating breeding bird densities in deciduous woodlots, in *Estimating Numbers of Terrestrial Birds. Studies in Avian Biology No. 6* (eds C. J. Ralph and J. M. Scott), Cooper Ornithological Society, pp. 202–8.

Tomiatojc, L. and Verner, J. (1990) Do point counts and spot mapping produce equivalent estimates of bird densities? *Auk*, **107**, 447–50.

*Turnock, B. J. and Quinn, T. J. II. (1991) The effect of responsive movement on abundance estimation using line transect sampling. *Biometrics*, **47**, 701–15.

Upton, G. J. G. and Fingleton, B. (1985) *Spatial Data Analysis by Example, Volume 1*, Wiley, New York.

*Velleman, P. F. and Hoaglin, D. C. (1981) *Applications, Basics and Computing of Exploratory Data Analysis*. Duxbury Press, Boston, MA, USA.

*Verner, J. (1985) Assessment of counting techniques, in *Current Ornithology, Volume 2* (ed. R. F. Johnston), Plenum Press, New York, USA, pp. 247–302.

Verner, J. (1988) Optimizing the duration of point counts for monitoring trends in bird populations. *U.S. Forest Service Research Note PSW-395*.

*Verner, J. and Ritter, L. V. (1985) A comparison of transects and point counts in oak–pine woodlands of California. *Condor*, **87**, 47–68.

Verner, J. and Ritter, L. V. (1986) Hourly variation in morning point counts of birds. *Auk*, **103**, 117–24.

Verner, J. and Ritter, L. V. (1988) A comparison of transect and spot mapping in oak–pine woodlands of California. *Condor*, **90**, 401–19.

*Ward, A. J. and Hiby, A. R. (1987) Analysis of cue-counting and blow rate estimation experiments carried out during the 1985/86 IDCR Minke whale assessment cruise. *Report of the International Whaling Commission*, **37**, 259–62.

*Warren, W. G. and Batcheler, C. L. (1979) The density of spatial patterns: robust estimation through distance methods, in *Spatial and Temporal Analysis in Ecology* (eds R. M. Cormack and J. K. Ord), International Co-operative Publishing House, Fairland, pp. 240–70.

*Weir, B. S. (1990) *Genetic data analysis*, Sinauer Associates, Inc., Sunderland, MA, USA.

Wegman, E. J. (1972a) Nonparametric probability density estimation: I. A summary of available methods. *Technometrics*, **14**, 533–46.

Wegman, E. J. (1972b) Nonparametric probability density estimation: II. A comparison of density estimation methods. *Journal of Statistical Computer Simulation*, **1**, 225–45.

Wetmore, S. P., Keller, R. A. and Smith G. E. P. (1985) Effects of logging on bird populations in British Columbia as determined by a modified point-count method. *Canadian Field Naturalist*, **99**, 224–33.

*White, G. C., Anderson, D. R., Burnham, K. P. and Otis, D. L. (1982) Capture–recapture and removal methods for sampling closed populations. Los Alamos National Laboratory, Los Alamos, NM, USA. *Rep. LA-8787-NERP*.

*White, G. C., Bartmann, R. M., Carpenter, L. H. and Garrott, R. A. (1989) Evaluation of aerial line transects for estimating mule deer densities. *Journal of Wildlife Management*, **53**, 625–35.

Whitesides, G. H., Oates, J. F., Green, S. M. and Kluberdanz, R. P. (1988) Estimating primate densities from transects in a west African rain forest: a comparison of techniques. *Journal of Animal Ecology*, **57**, 345–67.

*Wiens, J. A. and Nussbaum, R. A. (1975) Model estimation of energy flow in northwestern coniferous forest bird communities. *Ecology*, **56**, 547–61.

Wigley, R. L., Theroux, R. B. and Murray, H. E. (1975) Deep-sea red crab, *Gervon quinquedens*, survey off northeastern United States. *Marine Fisheries Review 37*.

Wildman, V. J. (1983) *A new estimator of effective area surveyed in wildlife studies*, PhD Dissertation, Department of Statistics, Oregon State University, Corvallis, OR, USA.

*Wildman, V. J. and Ramsey, F. L. (1985) Estimating effective area surveyed with the cumulative distribution function. Department of Statistics, Oregon State University, Corvallis, OR, USA, *Technical Report No. 106*, 37pp.

*Wilson, K. R. and Anderson, D. R. (1985a) Evaluation of two density estimators of small mammal population size. *Journal of Mammalogy*, **66**, 13–21.

*Wilson, K. R. and Anderson, D. R. (1985b) Evaluation of a density estimator based on a trapping web and distance sampling theory. *Ecology*, **66**, 1185–94.

Wywialowski, A. P. and Stoddart, L. C. (1988) Estimation of jack rabbit density: methodology makes a difference. *Journal of Wildlife Management*, **52**, 57–9.

*Yapp, W. B. (1956) The theory of line transects. *Bird Study*, **3**, 93–104.

*Zahl, S. (1989) Line transect sampling with unknown probability of detection along the transect. *Biometrics*, **45**, 453–70.

*Zahl, S. (unpublished) Analysis of the 1984/85 Antarctic variable speed experiment. *Paper SC/41/SHMi8*, presented to the Scientific Committee of the International Whaling Commission, May 1989.

Index